Science and Society

Professor A. Rupert Hall

A. Rupert Hall

Science and Society

Historical Essays on the Relations
of Science, Technology and Medicine

Routledge
Taylor & Francis Group

LONDON AND NEW YORK

First published 1994 by Variorum, Ashgate Publishing

Published 2017 by Routledge
2 Park Square, Milton Park, Abingdon, Oxon OX14 4RN
52 Vanderbilt Avenue, New York, NY 10017

Routledge is an imprint of the Taylor & Francis Group, an informa business

British Library CIP data
 Hall, A. Rupert
 Science and Society: Historical Essays on the
 Relations of Science, Technology and Medicine.
 (Variorum Collected Studies Series; CS 434)
 I. Title II. Series
 306.4

U.S. Library of Congress CIP Data
 Hall, A. Rupert (Alfred Rupert), 1920–
 Science and Society: Historical Essays on the Relations of
 Science, Technology and Medicine/A. Rupert Hall.
 p. cm. -- (Collected Studies Series: CS434)
 ISBN 0-86078-400-2
 1. Science-Social aspects-History. 2. Technology-Social aspects-
 History. 3. Life sciences-Social aspects-History. 4. War-Social
 aspects-History. 5. Royal Society (Great Britain)-History. I. Title.
 II. Series: Collected Studies Series: CS434.
 Q175.5.H37 1994
 608-dc20   93-46266
                                                      CIP

ISBN 13: 978-0-86078-400-5 (hbk)

COLLECTED STUDIES SERIES CS434

CONTENTS

SCIENCE AND WARFARE

SCIENCE, MEDICINE AND THE ROYAL SOCIETY

This volume contains x + 324 pages

INTRODUCTION

The fifteen lectures and papers republished in this collection reflect facets of a career stretching over nearly half a century: as an undergraduate before the war I was fascinated by Albert Neuburger's *Technical Arts of the Ancients* (1930), and after it an *Observer* review by G. B. S. led me to Douglas Guthrie's *History of Medicine* (1945). Early on I bought and read some now half-forgotten books like Beckmann's *History of Inventions* (1814) and no student attending the lectures of Professor Postan and John Saltmarsh of King's could be loftily oblivious of the material substratum below economic history. My thesis (1949) brought me into contact with some special aspects of the history of technology, while my first teaching of the history of science impelled me to examine some well-known episodes in the history of medicine; who, with an interest in the seventeenth century, could pass by William Harvey? And who could forbear to look into Joseph Needham's *History of Embryology* (1934)?

In 1951 an invitation from Charles Singer and E. J. Holmyard to join their editorial team preparing a massive co-operative *History of Technology*, generously financed by I. C. I., brought me seven years of fruitful and agreeable engagement with this topic. At that time, with the exception of a limited curatorial staff in a handful of museums, professional historians of technology were virtually non-existent, though there were many expert authorities upon particular issues whose profession might be archaeology or soil mechanics, architecture or economic history. Among these, here and there, one like A. P. Usher aimed at developing general views about the nature of long-continuing processes such as mechanical invention, while another like Bertrand Gille strove to create a more accurate material foundation for the much-studied political and ecclesiastical history of the Middle Ages. Needham was already at work upon his even more general comparative investigation of technology in the East and West which has so admirably complemented our own volumes, focused on the Mediterranean and the Atlantic. I make this point because it is too often assumed that the history of technology in a broad sense was conceived only in the 1960s and later.

I should not have written the essays in the first section of this book, several of then by invitation, if I had not been involved in the great *History of Technology*, now extended into the twentieth century by T. I. Williams. The problems arising from the great variety of interactions between pure science, applied science and craftsmanship — to avoid, if possible, the perplexing ambiguities associated with the multivalued modern-language derivatives from the Greek word *technè* — have always been of great interest to me. In recent decades some scholars though approaching these problems with antithetical presuppositions and looking to opposite objectives, have combined in striving, as a matter of principle, to separate natural science from engineering and industry. While agreeing with them that the conflation of *knowing* and *doing* may be perilous, I have long held that these problems of interaction or independence are not invariant but rather are historically determined: they were not the same in 1750 as in 1850, or in 1850 as they were in 1950. It is a mistake to attempt to discuss them as it were philosophically, apart from a defined historical context.

I also record here a debt to Professors Melvin Kranzberg and Lynn White, who further stimulated my interest in the history of technology during my years in the United States, and later.

At Imperial College, London, with the cheerful and learned collaboration of my wife and of Norman Smith — who has done much to enlarge my horizons — we had an opportunity to develop a systematic course of teaching in the history of technology. The history of medicine, on the other hand, has remained rather a personal than a professional concern. I am grateful to those scholars in that field (among them Allen Debus, Noel Poynter and Mirko Grmek) who encouraged my ventures. Again, but for an unexpected niche created for me by Dr Horace W. Magoun at the University of California's Los Angeles Medical Center, I would never have been able to devote even a short time to some problems in medical history, and generally improve my knowledge. This admittedly brief experience was of great service to me later during a long relationship with the Wellcome Institute for the History of Medicine, brought about by the confidence of the then Director of the Wellcome Trust, Dr P. O. Williams. The team of historical scholars at the Wellcome Institute and its Library staff greatly enriched my knowledge of their special field.

Though the range of tropics considered in these lectures and papers is highly disparate, they are to some extent integrated by a consideration never far from my own mind, that if we exclude the history of the present century — with which I am not concerned — topical boundaries were far less significant in the past than they now seem to be. This was true not only in the sense that until little more than a century ago one man could excel in many fields (as did both Newton and Leibniz — one need not multiply example of polymathy) but in the sense that technical linguistic and conceptual barriers between different aspects of the investigation or exploitation of nature were not serious. So elastic and loosely defined a concept as that of attractive force between corpuscles of matter could readily be transferred from mechanics to chemistry, from chemistry to physiology.

Similarly with the conjunction of the seemingly antithetical abilities of knowing and doing. It would be utterly anachronistic to wonder that Agricola was a physician, Spinoza a spectacle-maker and Lavoisier a tax-farmer. Satirists like Shadwell and Swift might ridicule the notion of theoretical swimming or fixing solar energy in a cucumber, but at a more serious level the idea of the unity of all things, pure and coarse, ugly and beautiful, lofty and base, within the divine cerebrations went unquestioned, and no natural theologian thought it impious to enlighten the divine process of creation by an analogy drawn with watchmaking. To the age in which aesthetic creation was firmly linked with geometry — and verses were numbers — the practical skills of William Herschel and Edward Jenner did not seem strange. We may flatter ourselves that in our time distinctions in the social order have diminished but our society has nevertheless raised intellectual and professional distinctions by which our ancestors were not troubled. If these lectures and papers display a variety now unusual, they conform to the greater internal unity of the period with which they are concerned.

A. RUPERT HALL

Tackley, Oxon
1993

PUBLISHER'S NOTE

The articles in this volume, as in all others in the Collected Studies Series, have not been given a new, continuous pagination. In order to avoid confusion, and to facilitate their use where these same studies have been referred to elsewhere, the original pagination has been maintained wherever possible.

Each article has been given a Roman number in order of appearance, as listed in the Contents. This number is repeated on each page and quoted in the index entries.

ACKNOWLEDGEMENTS

I am grateful to the following persons, publishers, societies and journals for permission to publish papers included in this volume: the University of Wisconsin Press (for Paper 1); *Technology and Culture* (II); *History of Technology* (IV, V); *History and Philosophy of the Life Sciences* (VI); the Royal Society for the Encouragement of Arts, Manufactures and Commerce (VII); The Regents of the University of California (VIII, X, XIV); the Newcomen Society for the Study of the History of Engineering and Technology (XI); the Athlone Press (XII); the Trustee, the Wellcome Trust (XIII, XV). I also thank my wife for permitting the republication here of a short joint paper (XV).

I

THE SCHOLAR AND THE CRAFTSMAN

IN THE SCIENTIFIC REVOLUTION

Never has there been such a time as that during the later sixteenth and the seventeenth centuries for the great diversity of men in the forefront of scientific achievement.[1] A proportion of those who contributed to the swelling literature of science were in a broad sense professionals: indeed, a sizable proportion, since many minor figures enlarge this group. Among these professionals were university teachers, professors of mathematics, anatomy, and medicine; teachers of these subjects, especially applied mathematics, outside the universities; and their various practitioners —physicians, surveyors, mariners, engineers and so on; and lastly the instrument-makers, opticians, apothecaries, surgeons, and other tradesmen, though their great period in science is to be found rather in the eighteenth century than in the seventeenth.[2] These men, widely divergent as they were in social origins and intellectual attainments, at least occupied positions in a recognizable scientific hierarchy. Some had won them through academic study, others through private education and research, others again by apprenticeship and pursuit of an occupation closely related to scientific inquiry. All were trained men in some way, whether in mathematics, physic and dissection, or the exercise of a manual craft. Now it is surprising enough, whether we make comparison with the scientific world of recent times, or with that of the later Middle Ages, to find such disparity in the professional group, that is, to find that the definition of scientific professionalism must be so loosely drawn; yet it is still more astonishing that many minor figures in the history of seventeenth-century science, and not a few notable ones, constitute an even more heterogeneous collection. Among these true "amateurs" of science (the distinction has really little meaning), some, it is true, had been exposed to scientific influences of a kind in college or university; yet the creation of a permanent interest thus, in an ordinary passing student, must have been as rare then as the acquisition of a taste for Latin verse is now. A few also, no doubt,

Reprinted from *Critical Problems in the History of Science*, ed. M. Clagett, by permission of the publisher. © 1959 University of Wisconsin Press, Madison, Wisconsin.

were quietly encouraged by discerning fathers or by private patrons. The rest remain as "sports"; diffusionist and environmental principles hardly suffice to explain their appearance on the scene. One thinks of such men as William Petty, son of a clothier, Otto von Guericke, Mayor of Magdeburg, John Flamsteed, an independent gentleman of modest means, or, most extraordinary of all, Antony van Leeuwenhoek, an unschooled borough official.

Thus one can never predict the social circumstances or personal history of a seventeenth-century scientist. Given the taste, the ability, and freedom from the immediate necessities of the struggle for subsistence, any man who could read and write might become such. Latin was no longer essential, nor mathematics, nor wide knowledge of books, nor a professorial chair. Publication in journals, even membership in scientific societies, was open to all; no man's work needed the stamp of academic approval. This was the free age between the medieval M.A. and the modern Ph.D. In the virtual absence of systematic scientific training, when far more was learned from books than from lectures, the wholly self-educated man was hardly at a disadvantage as compared with his more fortunate colleague who had attended the seats of learning, except perhaps in such special fields as theoretical astronomy or human anatomy. There were no important barriers blocking entry into the newer areas of exploration, such as chemistry, microscopy, qualitative astronomy, where all types of ability, manual and intellectual, were almost equally required. Obviously it was statistically more probable that a scientist would spring from the gentry class (if I may use this disputed term) than any other, and that he would be a university man rather than not. But the considerations determining the probability were sociological rather than scientific; if the texture of science was almost infinitely receptive of first-rate ability of any kind, the texture of society was such that it was more likely to emerge from some quarters than from others.

It is needful to traverse this familiar ground in order to set in perspective the dichotomy to which I shall turn—that of craftsman and scholar. It is a quadruple dichotomy—social, intellectual, teleological, and educational. It marks off, broadly, men of one class in society from another— those who earn their bread from scientific trades of one kind or another from those who do not. It distinguishes likewise those achievements in science which are in the main practical or operational from those which are cerebral or conceptual. Thirdly, it draws attention to the different

The Scholar and the Craftsman

objects of those who seek mainly practical success through science, and those who seek mainly understanding. And finally, if we consider only the group whom I have previously called professional, we may discern on the one hand the "scholars" who have been introduced to science by university or similar studies, and on the other the "craftsmen" who have learnt something of practical science in a trade. But we must be cautious in detecting polar opposites where there is in reality a spectrum. The scientific movement of the seventeenth century was infinitely varied, its successes demanded an infinite range of different qualities, and it is against this background of wide inclusion that we must set any attempt at analysis in particular terms.

By far the most closely-knit, homogeneous, and intellectually influential of the groups I have described was that of the university men, including both those who remained as teachers and those who departed to other walks of life. Some of the harshest critics of the contemporary "schools," like Bacon, Descartes, or Webster, were nevertheless their products. The opponents of the Aristotelian "forms and qualities" had been firmly grounded in that doctrine; many future scientists found stimulus in the universally required mathematical studies. To exemplify this point, one may consider the earliest membership of the Royal Society in 1663. Of the 115 names listed,[3] I find that 65 had definitely attended a university, while only 16 were certainly non-academic. The remaining 34 are doubtful, but at any rate the university men had the majority. It is still more telling to single out the names which have a definite association-value on inspection; I rate 38 on this test, of whom 32 are "U" and only 6 "non-U." Whether or not we term such men "scholars" is largely a rather unimportant question of definition: at any rate they had in common a knowledge of Latin, some training in mathematics, and an introduction at least to logic and natural philosophy; quite a proportion would also have had such experience of the biological and medical sciences as was available at the time.

It appears then that the medieval association of scientific activity with the universities was weakened, but not disrupted, in the seventeenth century, though the association certainly became less strong as the century advanced. It was weakened not only by the importance in science of men who were not academically trained at all, but by the shift in the locus of scientific activity from the universities, where it had remained securely fixed throughout the Middle Ages,[4] to new institutions like Gresham College, to the scientific societies meeting in capital cities, and to the circles basking in the patronage of a Montmor or a Medici.[5] If a majority of creative scientists had been at the university, they were so no longer in their mature age. Moreover, while in the medieval university there had been little disparity between the instruction given to the student, and the advanced researches of the master, this was no longer the case in the seventeenth century. In the schools of the fourteenth century the master who remained to teach pushed forward his knowledge, in the main, within the framework of ideas, and through study of authorities, with which he had become familiar at a more elementary level. The seventeenth-century university, on the other hand, almost ignored observational and experimental science. The unprecedented advances in scientific technique occurring in physics, astronomy, botany and zoology, and chemistry were not made widely available to students: there was a fairly good grounding only in mathematics and human medicine. The potential investigator had to learn the techniques he required from practice, by the aid of books, and through personal contact with an experienced scientist, often only obtainable elsewhere. Perhaps even more serious was the absence from university courses of the leading principles of the scientific revolution and of the ideas of the new natural philosophy. In the last quarter of the seventeenth century Cartesian science was indeed expounded in some of the colleges of France, and less widely elsewhere,[6] but dissemination of the thought of Galileo, of Bacon, and of the exponents of the mechanical philosophy owed little to university courses. Occasional examples of a university teacher having a decided influence upon a circle of pupils—as was the case with John Wilkins at Wadham College, Oxford, and Isaac Barrow at Trinity, Cambridge—hardly vitiate the general conclusion that the activities of various societies, books, and journals were far more potent vehicles of proselytization, which is supported by many personal biographies. However stimulating the exceptional teacher, formal courses were commonly conservative and pedestrian: it is curious to note that the two greatest scientists of the age who were also professors, Galileo and Newton, seem to have been singularly unremarkable in their public instruction. If the universities could produce scholars, they were ill-adapted to turning out scientists; the scientist had to train himself. Many who accomplished this transition regarded it, indeed, as a revulsion from the ordinary conception of scholarship. The learning they genuinely prized, in their own scientific disciplines, they had hardly won for themselves. It would surely be absurd to argue that Newton was less a self-made

The Scholar and the Craftsman

scientist than Huyghens, or Malpighi than Leeuwenhoek, because the former had attended a university and the latter not.

It lies outside my brief to discuss the fossilization of the universities, which, from what I can learn, the Renaissance did little to diminish so far as science was concerned, nor the rise of the new science as a rejection of academic dogma. Recent investigations would, I believe, tend to make one hesitant in concluding that the innovations and criticisms in the academic sciences—astronomy, physics, anatomy—which we call the scientific revolution, were the product solely, or even chiefly, of forces and changes operating outside the universities. Rather it would seem that, in relation to these subjects, it was a case of internal strife, one party of academic innovators trying to wrest the field from a more numerous one of academic conservatives. Certainly this was the case with Vesalius and his fellow-anatomists, with Copernicus, with Galileo. It was the academic and professional world that was passionately divided on the question of the inviolability of the Galenic, Aristotelian, or Ptolemaic doctrines; these quarrels of learned men had as little to do with capitalism as with the protestant ethic. Only towards the middle of the seventeenth century were they extended through the wider range of the educated class.

In the long run—that is to say within a century or so in each instance —the innovators won. In the short run they were defeated; academic conservatism prevented the recognition and implementation of the victories of the revolution in each science until long after they were universally applauded by thoughtful men outside. Whereas in the thirteenth century the schools had swung over to the Greeks and Muslims, despite their paganism and their often unorthodox philosophy, whereas in the fourteenth century the development of mechanics, of astronomy theoretical and practical, of anatomical and other medical studies, had been centered upon them, in the later sixteenth and seventeenth centuries teaching failed to adapt itself to the pace with which philosophy and science were moving. In the mid-sixteenth century the universities could still have formed the spear-head of this astonishing intellectual advance; in Galileo's life-time the opportunity was lost, and despite the invaluable efforts of individual teachers, as institutions the universities figured only in the army of occupation, a fantastic position not reversed until the nineteenth century. The innovators really failed, at the critical period, to capture the universities and bring them over to their side as centers of teaching and research in the new scientific manner. There were, for instance, many schemes in the seventeenth century for organizing scientific research, and for the provision of observatories, museums, laboratories and so on: yet no one, I think, thought of basing such new institutions on a university. That would have seemed, during the last century, a natural course to follow; and it would presumably have seemed equally natural in the Middle Ages.[7]

Hence it happened that the academic type, the scholar, book-learned in Aristotle or Galen, the Simplicius, the professor who could see the holes in the septum of the heart but was blind to the spots on the face of the sun, became the butt of the scientific revolutionaries.

> Oxford and Cambridge are our laughter,
> Their learning is but pedantry,

as the ballad has it. The passage in the *Discourse on Method* may be recalled, in which Descartes reviews critically the content of education and learning as ordinarily understood:

> Of philosophy I will say nothing, except that when I saw it had been cultivated for many ages by the most distinguished men, and that yet there is not a single matter within its sphere which is not still in dispute, and nothing therefore which is above doubt, I did not presume to anticipate that my success would be greater in it than that of others; and further, when I considered the number of conflicting opinions touching a single matter that may be upheld by learned men, while there can be but one true, I reckoned as well-nigh false all that was only probable.

After observing that the other sciences derived their principles from philosophy, which was itself infirm, so that "neither the honour nor the gain held out by them was sufficient to determine one to their cultivation," Descartes abandoned the study of letters "and resolved no longer to seek any other science than the knowledge of myself, or of the great book of the world." With this one may compare Bacon's "surprise, that among so many illustrious colleges in Europe, all the foundations are engrossed by the professions, none being left for the free cultivation of the arts and sciences." This restriction, he declares, "has not only dwarfed the growth of the sciences, but been prejudicial to states and governments themselves." The candid appraisal of the first chapter of the *Advancement of Learning* could have been applied to many academic institutions more than two centuries after it was penned.

The Scholar and the Craftsman

Admittedly the period when Bacon and Descartes formed such adverse opinions was one early in the scientific revolution; but there is little evidence to show that academic reform progressed rapidly thereafter, and it would not be difficult to quote parallel judgments from a later time. It was not the case, of course, that learned conservatives could see no merit in the study of science. This was no science *versus* humanities wrangle, for the conservatives were themselves teachers of science, of Renaissance science in fact. Their science was Aristotelian and formal; it denounced both Copernicanism and the mechanical philosophy, and distrusted the new instruments and experiments. An analogous situation existed in medicine, where the modernists who were experimenting with chemical preparations and new drugs such as a guaiacum and Jesuits' bark, who followed Harvey and attempted the transfusion of blood, were opposed by the entrenched faculties of so-called Galenists, enemies of every innovation. Nevertheless, the effect was much the same: The "new philosophy" and science were forced to take root outside the academic garden where they should have found most fertile soil.

The effect of the development of new scientific ideas and methods in diminishing the role of the universities as creative centers reinforced rather than initiated the decline of their intellectual prestige, which had begun with the Renaissance. Then, too, new movements in learning and scholarship were at least as much associated with the activities of private scholars, as with those of university teachers. Private patrons had been as energetic in encouraging neo-classical modes of writing and sculpture, as they were to be in promoting science in the seventeenth century. Already in the Renaissance academic learning was reproached for its inelegance in the classical tongues, its imported Arabisms, its lingering attachment to imperfect texts, its barren philosophy. No one was more scathing of academic pedantry than Erasmus, not to say Paracelsus. "Then grew the learning of the schoolmen to be utterly despised as barbarous," says Bacon, so that when he himself attacked the fine philosophic web of scholasticism—too many words spun out of too little matter—he was but repeating an old canard. This revulsion of the Renaissance scholar from the "barbarousness" of still-medieval universities was, as is well known, a linguistic and textual one in the main; it did not touch so much the content of thought as its expression, nor did it, in particular, greatly disturb the pre-eminent position of the ancient masters of science. This aspect of the Renaissance can most clearly be seen in the history of medicine during the first half of the sixteenth century. Some of the lost ground the universities recovered; they began to teach Greek and Ciceronian Latin; more attention was paid to history and literature and less to disputative philosophy. But they could not recover their medieval pre-eminence as cultural centers—particularly perhaps in northern Europe—and the scientists of the seventeenth century had only, in a sense, to follow the path which Renaissance humanists had trodden, in rejecting it.

The object of the preceding remarks is to justify my conception of the scientific scholar of the sixteenth and seventeenth centuries, as a man learned not merely in recent scientific activities and methods, but in the thought of the past. It seems superfluous to argue that the majority of the scientists of the time were of this type, neither technicians nor ignorant empiricists. Certainly the learning of Galileo, or Mersenne, or Huyghens, or Newton, was not quite like learning in the medieval or Renaissance conception; they may have been as deficient in the subtleties of Thomist philosophy as in the niceties of Greek syntax; but to deny that they were learned scholars in their field and in their outlook, would be to deny that any scientist is entitled to be called learned.

I have tried also to trace in outline the way in which, at this time, scientific learning diverged from other branches of scholarship, without wholly severing its affiliations with academic institutions. One might also ask the question: how far was the new scientific spirit of the seventeenth century brought into being by activities of a purely scholarly kind—for example, through the evolution of certain principles of logic during the Middle Ages, or through the activities of the persistent students of Greek science in the Renaissance?

The latter especially furnished the core of an interpretation of the scientific revolution which held favor until recent times. To put it crudely, the scientific revolution was seen, according to this view, as the terminal stage of a scientific renaissance beginning about the mid-fifteenth century, and characterized chiefly by its full exploration of classical scientific texts, which was aided particularly by the invention of printing; the scientific renaissance was itself regarded as a classical reaction against the gothic barbarity of the Middle Ages.[8] This interpretation is in effect an extension of Bacon's, to which I referred earlier; an extension which Bacon himself was unable to make because he did not know that the revolution he sought was going on around him. Clearly, if such a view is accepted, it attaches a very great importance indeed to the activity of

The Scholar and the Craftsman

the scholar-scientists of the Renaissance, who besides polishing and extending the works of the most authoritative ancient authors, shed a full light on others, such as Lucretius, Celsus, and Archimedes, whose writings had not previously been widely studied.

The merits of this hypothesis of the origin of the scientific revolution are as obvious as its defects. It draws attention to the weight of the contribution of sheer scholarship, and of the amazing Hellenophile instinct of the Renaissance, to the change in science which occurred between 1550 and 1700. No one would deny the connection between the mechanical, corpuscular philosophy of the seventeenth century, and *De natura rerum;* nor the significance for anatomy of the intensive study of Galen; nor would he dispute that the virtual rediscovery of Archimedes transformed geometry, and ultimately algebra. Equally, however, it is clear that this is far from being the whole story: the instances I have quoted are not universally typical ones. The history of mechanics before Galileo, which has been so elaborately worked out in the present half-century, proves the point. Medieval science was not abruptly cut short by a classical revival called a renaissance: it had much—how much must be the subject of continuing research—to contribute to the formation of modern science. Very important threads in the scientific revolution are not really traceable to antiquity at all, at least not through the channels of scholarship; here the chemical sciences furnish examples. Above all, the renaissance-scholarship interpretation fails to account for the *change* in science. If anything is fairly certain, it is that the intention of the Renaissance was the imitation of antiquity, and there is evidence that this ideal extended to the scholar-scientists. Yet the pursuit of this ideal seems to have endured least long in science, of all the learned subjects; it had ceased to have force long before the end of the sixteenth century. There never was a true Palladian age in science, and the limitations that had bound the Greeks themselves were relatively soon transcended in Europe. Why this was so is really the whole point at issue, and the Renaissance-scholarship interpretation does not squarely face it.

Nevertheless, if that view is not completely adequate, it must serve as an element in any more complete interpretation. The different view of the importance of scholarly activities, this time in the Middle Ages, that I mentioned previously, has won ground much more recently. It is one that the non-medievalist has some difficulty in evaluating, and it would be inappropriate for him to criticize it. I had better state my conception of its tenets at the risk of oversimplification: that medieval philosophers evolved a theory for the investigation of natural phenomena which was essentially that applied with success in the scientific revolution. It is not claimed for those who elaborated this theory that they were themselves as eminent in experiment, or observation, or the use of mathematics as their successors; its applications—other than to optical phenomena and the discussion of impetus—seem to have been few and sporadic. It was a scholar's method of science, vindicated by some successes, which only awaited general application to transform the whole exploration of nature, and this the method found in the late sixteenth and seventeenth centuries. Again, then, great importance is attached to the role of the scholar; the scientific revolution, it might be said, is the direct consequence of a philosophic revolution.

At the same time, it is evident that there is a measure of incompatibility between these two alternative appraisals of the supposed contribution of scholars to the genesis of modern science. One lays emphasis on content, the other on method; if the medieval ideas on method are pre-eminently important, then the Renaissance revival of classical science is irrelevant, and vice versa. One view, if allowed to fill the whole picture, tends to obscure the other. We are not forced to an exclusive choice, however, and I think it may be granted that a compromise which allows room for both views is possible. It would seem to be the case that while one theory best accounts for certain aspects of the development of science in the sixteenth and seventeenth centuries, the other best explains other aspects. Nor should it be forgotten that changes of emphasis within the scope of the classical tradition may be attributable, in part at least, to changed ideas of method derived from the Middle Ages. I mentioned earlier the rediscovery of Lucretius in 1417, and the connection of this with the mechanical philosophy of two and a half centuries later; it may well be that new ideas on the form and structure of a scientific theory had much to do with the preference for the atomistic tradition in Greek thought over the Aristotelian thus evinced. The fuller acquaintance that textual scholarship conferred with those relics of Greek science which best exemplified the newer medieval notions of scientific procedure might have built up a greater pressure for the further application of those notions. For example, if Galileo, unknown to himself, inherited a method of scientific enquiry from medieval philosophers, he thought of himself (on occasion) as practicing a method used with success by Archimedes.

The Scholar and the Craftsman

The medievalist view, if I may so term it, raises in a peculiarly acute form the question which seems central to my problem. Is the effective and creative impulse, which urged men to abandon not merely the philosophy and doctrines of medieval science but even their Greek foundations, to be found in the dissatisfaction of learned men with established modes of inquiry, and the theories and practices to which they gave rise? In short, was the scientific revolution in the main the product of a sense of intellectual frustration and sterility? If we think this was the case, and that it was the same philosophers, scholars and intellectuals who suffered this frustration, who found a way of breaking through it to a more rewarding kind of inquiry and a more satisfying mode of scientific explanation, then our historical seeking is at an end. We might of course go on to enquire where this frustration originated and what brought it into being, and we might also ask what factors enabled the learned men of science to break through it, but at least we should have established their crucial role in the actual break-through, and all else would be ancillary.

That such frustration was experienced hardly requires demonstration. It is expressed by Vesalius, when he laments—with whatever element of exaggeration and ingratitude—the wasted effort into which too uncritical a confidence in the exactitude of Galen had led him; by Copernicus, when he speaks of the disagreement of mathematicians, and their ineptitude: "Rem quoque praecipuam, hoc est mundi formam, ac partium eius certam symmetriam non potuerunt invenire, vel ex illis colligere"; and surely conspicuously enough by Galileo. The latter's is the attitude of one who has broken out of the dead circle of ancient thought, and who can, from reliance on his own new knowledge, pity as well as condemn those still bound by the chains of authority:

> Oh, the inexpressible baseness of abject minds! To make themselves slaves willingly; to accept decrees as inviolable; to place themselves under obligation and to call themselves persuaded and convinced by arguments that are so "powerful" and "clearly conclusive" that they themselves cannot tell the purpose for which they were written, or what conclusion they serve to prove! . . . Now what is this but to make an oracle out of a log of wood, and run to it for answers; to fear it, revere it, and adore it?[9]

Now what the medievalists contend for is, I take it, that such an attitude to authority was already nascent in the Middle Ages, and that it was not merely negative but creative. I quote Dr. Crombie's very plain statement, from the first page of his book on *Grosseteste and Experimental Science* (Oxford, 1953):

> Modern science owes most of its success to the use of these inductive and experimental procedures, constituting what is often called "the experimental method." The thesis of this book is that the modern, systematic understanding of at least the qualitative aspects of this method was created by the philosophers of the West in the thirteenth century. It was they who transformed the Greek geometrical method into the experimental science of the modern world.

Why was it necessary to devise new inductive and experimental procedures at all at this point? Dr. Crombie finds the answer to this question in the problem presented to Western natural philosophers by the scientific texts recently made available to them: "How is it possible to reach with the greatest possible certainty true premises for demonstrated knowledge of the world of experience, as for example the conclusions of Euclid's theorems are demonstrated?"

This view places the genesis of the scientific revolution at a very high level of intellectual achievement, which is still maintained if we transfer our attention to a somewhat different field from Crombie's, namely the history of theories of mechanics from the Middle Ages down to the time of Galileo. Here again we may note, not merely striking dissatisfaction with the Aristotelian explanation of continued motion founded on the total separation of the moving force from the moved inanimate body, as well as with certain other features of mechanics of supposedly Aristotelian formulation, but definite and partially successful steps towards more satisfactory concepts.[10] When we come to the critical point, with Galileo himself, we contemplate an intellectual struggle of the most sublime kind, which Professor Koyré has analyzed for us. If the ultimate victory here is not the result of prolonged and arduous cerebration, then it is difficult to see what successes could be attributable to thought and reason in science. Just as the medieval criticism of Aristotle had come from scholars, so also it was in the minds of scholars that the battle between old and new in science had to be fought. I should find it difficult to cite an exponent of the "new philosophy" who did not visualize its fate in those terms.

There is a point here, however, that deserves fuller consideration, and allows the craftsman to enter on the scene. For while we recognize science as a scholarly activity, and the reform of science as an act of

The Scholar and the Craftsman

learned men, it may plausibly be asked whether the impulse to reform was spontaneously generated among the learned. Was it perhaps stimulated elsewhere? Some support for this suspicion might seem to spring from the emphasis that has been laid on empiricism, not merely in the scientific revolution itself, but among its philosophical precursors. Thus, to quote Dr. Crombie again: "The outstanding scientific event of the twelfth and thirteenth centuries was the confrontation of the empiricism long present in the West in the practical arts, with the conception of rational explanation contained in scientific texts recently translated from Greek and Arabic." It is unnecessary to dwell on the well-known interest of at least a few learned men, during the Middle Ages, in such fruits of empirical invention as the magnetic compass, the grinding of lenses, and above all, the important advances in the chemical and metallurgical arts.[11] Similarly, everyone is familiar with the arguments of the Baconian school: that true command—and therefore real if unwitting knowledge—of natural processes had been won by the arts rather than by sciences, and that the scholar would often become more learned if he would consent to apprentice himself to the craftsman. All this might suggest that the increasingly spectacular achievements of empirical technology arrested the attention of scholarly scientists, enforcing some doubt of the rectitude of their own procedures, and still more, leading them to accept as an ideal of science itself that subjection of the natural environment to human purposes which had formerly seemed to belong only to the arts and crafts.

There are two issues here. One is the fact of technological progress, which some philosophical critics contrasted with the stagnation of science.[12] The other is the reaction of learned men to the state of technology, and this is more properly our concern. Technological progress was not simply a feature of the Middle Ages and Renaissance: it occurred in the ancient empires, in the Greek world, under the Roman dominion, and even in the so-called "Dark Ages." It would be difficult to think of a long period of complete technical stagnation in European history, though individual arts suffered temporary periods of decline. Some craftsmen at some places seem always to have been making their way forward by trial and error. In short, a philosopher of antiquity had as great an opportunity of appreciating the inventiveness of craftsmen as his successors of the sixteenth and seventeenth centuries, and of drawing the same lessons as were drawn then. Indeed, ancient writers were aware of the importance of the crafts in creating the means of civilized existence, and praised works of ingenuity and dexterity; where they differed from the moderns was in their preservation of the distinction between *understanding* and *doing*. They did not conclude that the progressive success of the crafts set up any model of empiricism for philosophy to emulate. They would not have written, as Francis Bacon did, in the opening lines of the *Novum Organum:* "Man, as the minister and interpreter of nature, does and understands as much as his observations on the order of nature, either with regard to things or the mind, permit him, and neither knows nor is capable of more. The unassisted hand and the understanding left to itself possess but little power. . . . Knowledge and human power are synonymous."

It is the philosopher who has modified his attitude, not the craftsman, and the change is essentially subjective. The success of craft empiricism was nothing new in late medieval and early modern times, and if the philosopher became conscious of its significance for science it was not because such success was more dramatic now than in the past. It was always there to be seen, by those who had eyes to see it, and the change was in the eye of the beholder. It is absurd, for instance, to suppose that the introduction of gunpowder and cannon into warfare was in any serious sense the cause of a revival of interest in dynamics, and especially in the theory of the motion of projectiles, during the sixteenth and early seventeenth century. The ancient torsion artillery provided equally dramatic machines in its day, not to mention the crossbow, mangonel and trebuchet of the Middle Ages. The simplest methods of hurling projectiles —the human arm, the sling, the bow—pose problems of motion no less emphatically than more complex or powerful devices, and as everyone knows, appeal to practical experience of this primitive kind was the basis for the development of the concept of impetus. The earliest "scientific" writers on explosive artillery, such as Tartaglia, did no more than transfer this concept to the operation of a different device.

Such an example reminds us that it may be naive to assume that even major technological advances suggested, contemporaneously, such questions worthy of scientific enquiry as would, indeed, immediately spring to our own minds. The scientific examination of the three useful forms of iron—cast-iron, wrought iron, and steel—did not begin until the early eighteenth century; the geometrical theory of gear-wheels was initiated about fifty years earlier; the serious study of the chemistry of the ceramics industry was undertaken a little later. I choose deliberately examples

The Scholar and the Craftsman

of practical science each associated with notable developments in late-medieval craftsmanship: the introduction, respectively, of the effective blast-furnace; of the gear-train in the windmill, water-mill, mechanical clock, and other devices; and of fine, brightly pigmented, tin-glazed earthenware.[13] The time-lag in each instance between the establishment of a new craft-skill, and the effective appearance of scientific interest in it, is of the order of 250 years, and in each of these examples it appears *after* the scientific revolution was well under way. If there is some truth in the view that interest in crafts promoted a change in scientific procedures, it is also true that, at a later date, the very success of the new scientific knowledge and methods opened up the possibility of examining craft procedures systematically, which had not existed before.

It would be a *non sequitur* to argue that, because an important measure of technological progress occurred in the Middle Ages (as *we* are aware), medieval scholars recognized the fact and appreciated its significance. Clearly in many instances they did not—that is why the history of medieval technology is so difficult to reconstruct. Our literary records of the Middle Ages were in large part compiled by scholars; the paucity in them of technological documentation—concerning not merely the use of tools like the carpenter's brace and lathe, but major industries such as paper-making and iron-working—is very conspicuous. The historian of medieval technology is notably better served by the artist than by the scribe. This could hardly have happened, had more than a very few scholars been impressed by the empiricism which brought in the mindmill, the magnetic compass, the mechanical clock, and so on.

In any case, I hesitate to conclude that the behavior of an empirical scientist—that is, I take it, one who observes and experiments, both to discover new information and to confirm his statements and ideas—is derivable by virtually direct imitation from the trial-and-error, haphazard, and fortuitous progress of the crafts. This seems to me to be the defect of the view that sees the new scientist of the seventeenth century as a sort of hybrid between the older natural philosopher and the craftsman. It is easy enough to say that the philosopher thought much and did little, while the craftsman did much but had no ideas, and to see the scientist as one who both thinks and does. But is such a gross simplification really very helpful in describing or explaining a complex historical transition? Neither Copernicus, nor Vesalius, nor Descartes, to name only three, were more craftsmanlike than Ptolemy, Galen, or Aristotle. Surely scientific empiricism is itself a philosophical artefact, or at least the creation of learned men—here I believe Dr. Crombie has a very strong point—and it stands in about the same relation to craftsmanship as the theory of evolution does to the practices of pigeon-fanciers. It is a highly sophisticated way of finding out about the world in which we live; on the other hand, the notion that direct immersion in the lore of tradesmen was the essential baptism preceding scientific discovery was one of the sterile by-paths from which the scientists of the seventeenth century fortunately emerged after a short time. Modern studies combine in revealing that the empirical element in the scientific revolution, taking the word in its crudest, least philosophical and most craftsmanlike sense, has been greatly exaggerated; correspondingly we are learning to attach more and more significance to its conceptual and intellectual aspects.

This is not to deny that the processes of artisans constituted an important part of the natural environment. If, by an internal displacement, the attention of the natural philosopher was more closely directed to this, and less to his own consciousness and limited academic horizon, he could learn much of what the world is like. As the Middle Ages verged on the Renaissance, an increasingly rich technological experience offered ample problems for enquiry, and besides, much knowledge of facts and techniques. This, apart from their direct technological importance, was the significance for science of the great works of craft-description and invention by Cellini, Agricola, Biringuccio, Palissy, Ercker, Ramelli and others that appeared in the sixteenth century, for while their own scientific content was slight, these authors provided materials and methods for the use of others more philosophically equipped than themselves. Science indeed owes much to technology: but we must remember that the debt was itself created by natural philosophers and other men of learning.

There is no straightforward answer to any question about the whole nature of the scientific revolution. Here it may again be useful to recall the deep distinction between the academic sciences (astronomy, anatomy, mechanics, medicine) and the nonacademic (experimental physics, chemistry, botany and zoology, metallurgy)—the latter group being so described because it had no regular place in university studies. Comparing paradigm cases from the two groups, say, astronomy and chemistry, we note that the former was already highly organized, with an elaborate theoretical structure, in the Middle Ages; it used relatively sophisticated techniques, both instrumental and mathematical; searching criticism of

The Scholar and the Craftsman

one of its fundamental axioms, that is, the stability of the Earth, occurred in the fourteenth century (and indeed long before) while dissatisfaction with its existing condition was vocal and definite before the end of the fifteenth. A fundamental change in ideas came early—in 1543—and was followed, not preceded, by great activity in the acquisition of new factual material, which in turn prompted fresh essays in theory. All this was the work of learned men, and there was little possibility of craft-influence; even if the pivotal invention of the telescope were a craft invention, its scientific potentialities were perceived by scholars.[14] Chemistry reveals a very different historical pattern, in which almost everything said of astronomy is negated. There was no organized chemical science before a comparatively late stage in the scientific revolution; there was no coherent theory of chemical change and reaction; there was no clearly definable classical and medieval tradition to challenge; the conception of chemistry as a branch of natural philosophy was late in establishing itself, and involved a lengthy fact-gathering stage that preceded the formulation of general theories; and in all these developments the influence of craft-empiricism was strong. It can hardly be doubted that the range of chemical phenomena known to craftsmen about 1550 was much greater than that known to scholars, and that, as Professor C. S. Smith has pointed out, craftsmen had developed both qualitative and quantitative techniques of vital necessity to the growth of chemistry as an exact science.[15]

Sometimes, when one turns from considering the history of such a science as mechanics or astronomy to that of, say, chemistry or a biological subject, it seems as though the transition is from one discipline to another completely alien to the first. Nor is it enough simply to admit that some sciences developed more slowly than others; the situations are really different, so that Lavoisier's work in chemistry cannot be made strictly analogous, point by point, to that of Newton in celestial mechanics or optics. Hence all generalizations concerning the scientific revolution require qualification when the attempt is made to apply them to a particular science.

Perhaps I may illustrate this in the following way. The contributions of craftsmanship to the development of scientific knowledge in the sixteenth and seventeenth centuries seem to be analyzable under five heads:

1 the presentation of striking problems worthy of rational and systematic enquiry;

2   the accumulation of technological information susceptible to scientific study;

3   the exemplification of techniques and apparatus adaptable from the purposes of manufacture to those of scientific research;

4   the realization of the scientific need for instruments and apparatus;

5   the development of topics not embraced in the organization of science proper.

The incidence of these contributions is highly variable among the individual sciences. None are strongly relevant in anatomy, medicine, or indeed any biological science, except that 4 would apply to microscopy. All the sciences demonstrate an increasing dependence on the instrument-maker's craft. Again, 4 is relevant to astronomy, while mechanics draws very slightly upon 1 and 2. Chemistry, on the other hand, exemplifies all these possible contributions, and most forms of applied science—other than mathematical sciences—owe much to the fifth contribution. All we can conclude, therefore, is an obvious truism: that those sciences in whose development empiricism played the greatest part are those in which elements derived from craftsmanship have the most effect. It does not follow, however, that the empirical sciences are those that best exhibit the profundity or the nature of the change in scientific thought and work, nor that the theoretical function of scholars is insignificant even in these sciences. Rather the converse would seem to be true, namely that some of those scientists, like Robert Boyle, who at first sight seem to be highly empirical in their scientific work and attitude, were in fact deeply engaged in the search for general theories and laws.[16] The academic and above all the mathematical sciences were not only those that advanced fastest, but they were already regarded as the models for the structure of other sciences, when these should have reached a sufficiently mature stage. In an ascending scale of sophistication, it was regarded as desirable to render all physical science of the same pattern as mechanics and astronomy, and to interpret all other phenomena in terms of the basic physical laws. The first great step towards the attainment of such an ambition was Newton's *Principia,* a work soon regarded by many as the ultimate manifestation of man's capacity for scientific knowledge. I believe it would be wrong to suppose that the scientists of the late seventeenth century, with such rich examples before them, were content to remain indefinitely at the level of empiricism or sublimated craftsmanship, though indeed in many branches of enquiry it was not yet possible to soar far above it. They were aware

The Scholar and the Craftsman

that the more abstruse and theoretical sciences, where the contributions of learned men had been greatest, were of a higher type than this.

Perhaps I may now summarize the position I have sought to delineate and justify in the following six propositions, in which it is assumed as an axiom that a science is distinguished by its coherent structure of theory and explanation from a mass of information about the way the world is, however carefully arranged.

1 The scientific revolution appears primarily as a revolution in theory and explanation, whether we view it in the most general fashion, considering the methods and philosophy of the new scientists, or whether we consider the critical points of evolution in any single science.

2 There is a tradition of logical (or, more broadly, philosophical) preoccupation with the problem of understanding natural phenomena of which the later stages, from the thirteenth to the seventeenth century, have at the lowest estimate some bearing on the attitudes to this problem of seventeenth century scientists.

3 Some of the most splendid successes of the scientific revolution sprang from its novel treatment of questions much discussed by medieval scholars.

4 These may be distinguished from the "contrary instances" of success (or an approximation to it) in handling types of natural phenomena previously ignored by philosophers, though familiar in technological experience.

5 While "scholars" showed increasing readiness to make use of the information acquired by craftsmen, and their special techniques for criticizing established ideas and exploring phenomena afresh, it is far less clear that craftsmen were apt or equipped to criticize the theories and procedures of science.

6 Though the early exploitation of observation and experiment as methods of scientific enquiry drew heavily on straightforward workshop practice, the initiative for this borrowing seems to be with scholars rather than craftsmen.

I dislike dichotomies: of two propositions, so often neither *a* nor *b* by itself can be wholly true. The roles of the scholar and the craftsman in the scientific revolution are complementary ones, and if the former holds the prime place in its story, the plot would lack many rich overtones had the latter also not played his part. The scholar's function was active, to transform science; the craftsman's was passive, to provide some of the raw material with which the transformation was to be effected. If science is not constructed from pure empiricism, neither can it be created by pure thought. I do not believe that the scientific revolution was enforced

I

by a necessity for technological progress, but equally in a more backward technological setting it could not have occurred at all. If the genesis of the scientific revolution is in the mind, with its need and capacity for explanation, as I believe, it is also true that the nascent movement would have proved nugatory, had it not occurred in a world which offered the means and incentive for its success.

References

1 Robert K. Merton, "Science, Technology and Society in Seventeenth-Century England," *Osiris*, IV (1938), 360–632. This is the major study of the sociology of science in a single country.
2 The most useful work on the instrument-making craft generally is Maurice Daumas, *Les instruments scientifiques aux XVIIe et XVIIIe siècles* (Paris, 1953). Cf. also the two chapters by Derek J. Price in Singer, Holmyard, Hall, and Williams, *A History of Technology* (Oxford, 1957), Volume III.
3 *The Record of the Royal Society of London* (3rd ed.; London, 1912), pp. 309–11.
4 The more "practical" departments of science, such as alchemy, metallurgy, and cartography, admittedly had little direct dependence on the universities; but it should be remembered that knowledge of fundamental texts in these as well as other topics was derived from them. The universities played an important role in the development of the mathematical and astronomical techniques required for practical ends.
5 On the origins of scientific societies: Martha Ornstein, *The Role of Scientific Societies in the Seventeenth Century* (Chicago, 1938); Harcourt Brown, *Scientific Organization in Seventeenth Century France* (Baltimore, 1934).
6 Paul Mouy, *Le Developpement de la physique cartésienne, 1646–1712* (Paris, 1934).
7 Universities had their anatomy theatres and libraries (often of medieval foundation) and, later, museums and laboratories; the latter were, however, private creations (as at Bologna and Oxford) and failed to become living and growing features of academic life.
8 For bibliographical details on the scientific activities of Renaissance scholars, cf. George Sarton, *The Appreciation of Ancient and Medieval Science during the Renaissance (1450–1600)* (Philadelphia, 1955).
9 Galileo Galilei, *Dialogues Concerning the Two Chief World Systems—Ptolemaic & Copernican*, tr. Stillman Drake (Berkeley and Los Angeles, 1953), p. 112.
10 Besides the extensive studies of the history of medieval mechanics by Pierre Duhem, Anneliese Maier and others, convenient short discussions (with further bibliographical details) are in A. C. Crombie, *Augustine to Galileo* (London, 1952); René Dugas, *Histoire de la mécanique* (Neuchâtel, 1950). On Galileo and impetus, cf. Alexandre Koyré, *Etudes Galiléennes* (Paris, 1939), *Actualités scientifiques et industrielles*, 852–54.
11 The extent to which the technological progress of Europe during the Middle

The Scholar and the Craftsman

Ages was due to transmission rather than indigenous invention is immaterial here, since such transmission seems to have occurred at the level of craftsmanship rather than scholarship. Cf. Thomas Francis Carter, *The Invention of Printing in China and its Spread Westward,* ed. L. Carrington Goodrich (New York, 1955); Joseph Needham, *Science and Civilization in China* (Cambridge, Eng., 1954—).

12 The idea that ancient technological secrets had been lost (in the same way as scientific knowledge and artistic skill) was, however, voiced during the Renaissance.

13 Here again it does not affect the argument that three or more of the inventions mentioned were made outside Europe; each became significant for European technology in the period *c.* 1300–1500.

14 If the evidence for the invention of the telescope in Italy *c.* 1590 is accepted (as seems reasonable), it would appear that this invention was connected with the experiments of Giovanbaptista Porta and other scholars shortly before this date. It is thus difficult to believe that the discovery was completely accidental.

15 *Lazarus Ercker's Treatise on Ores and Assaying,* tr. Anneliese Grünhaldt Sisco and Cyril Stanley Smith (Chicago, 1951), pp. xv–xix. Charles Singer *et al., A History of Technology,* III, 27–68.

16 On Boyle, I am much indebted to the work of Marie Boas, "Boyle as a Theoretical Scientist," *Isis,* XLI (1950); "The Establishment of the Mechanical Philosophy," *Osiris, X* (1952); "La Méthode scientifique de Robert Boyle," *Rev. d'Hist. des Sciences,* IX (1956). Cf. also her *Robert Boyle and Seventeenth Century Chemistry* (Cambridge, 1958).

II

Engineering and the Scientific Revolution

HARDLY MORE THAN a couple of generations ago sound engineering practice consisted for the most part of a combination of manual dexterity with a rather large number of empirical rules approved by long, evolving experience. Yet engineering had already been invaded by formulae or principles which were not merely empirically derived: thermodynamical calculations, the procedures for determining structural stresses, part at least of the theory of fluid-flow as well as the whole new development of electrical engineering were products of theoretical investigation which had been confirmed by experiments and in due course by experience. In such cases as these we may properly speak of engineering's being dependent upon science. It is true that the practical engineer of the nineteenth century derived his figures rather from tables and graphs than by going back to pure theory; what matters is the origin of the rules he applied in a theoretical analysis rather than in a codification of successful empirical practice. And it is significant that all the time the trend was away from customary rules to theoretically-derived ones; by and large the engineer did not distrust science: he welcomed it.

To all this the sixteenth and seventeenth centuries offer no parallel. Theoretically-derived rules were employed in no branch of engineering, and there is little sign that practical men felt any need for them. Custom was powerful; next to that the most important consideration was not scientific but aesthetic. Alteration in taste is really the only fact that need be invoked to account for changes in building practice at this time—though the greater use of metals for taking tensile stresses, for example, ensured that Baroque architecture was not quite like that of the ancients. Perhaps the nearest parallel to the nineteenth century situation—and it is outside engineering—was the fact that gunners were offered (from 1638 onwards) tables of elevations and ranges calculated on the assumption that projectiles follow a parabolic trajectory. But the assumption was false, since projectiles are subject to air-resistance, and the tables were quite inapplicable under seventeenth-century conditions. Before scientific ballistics could be useful it was necessary to develop both a more complex mathematical theory and artillery of predictable, long-range performance; neither was achieved before the nineteenth century.[1] Tables of proportions were used in engineering, naturally; by shipwrights, possibly by bridge-builders, certainly by architects. But such tables were governed by rule-of-thumb (or rule of three), or by aesthetic canons of proportionality; they had no analytical justification.

For all its swift advance seventeenth century science still lacked the depth of precise, quantitative information that alone is useful to engineers. Its achievements were either conceptual or at sublime heights of mathematical theory. In his *Principia* (1687) Newton could equate exactly the centrifugal force of each planet at any point in its orbit with the sun's gravitational pull at that point; yet he erred by a factor of two in calculating the height to which a given head of water would spout.[2] The great discoveries of mathematical physicists were not merely over the heads of practical engineers and craftsmen; they were useless to them. Christian Huygens's theoretical enquiry into the isochronism of the pendulum had no effect at all on the actual improvement of clocks, although his mechanical concept (the regulation of a train of clockwork by a pendulum) was of the first importance and the starting-point for a succession of practical improvements by working clockmakers.[3] The well-known studies of the engagement of gear-teeth by the Danish astronomer Ole Römer and others in the French Academy of Sciences fall into a different category of practical usefulness, for millwrights (if they ever heard of them) were quite uninterested in constructing more perfectly shaped gears, and even if they had wished to do so they lacked machinery for cutting the required epicycloidal profiles.

In this instance the millwrights were much more concerned with substance than with form. Wood was an unsuitable material to use for gears; on the other hand, metals such as iron were impossibly difficult to work except on the small scale, and were almost equally subject to fracture. Such a limitation of materials and of means for making the best use of available materials was a principal problem confronting the seventeenth century engineer, as all his predecessors. It cut down the extent of his building-spans; it reduced the carrying-capacity of his ships and forced him to rebuild them after as little as ten years' service; it excluded the use of high temperatures and pressures; it made accurate dimensions unattainable; it made his machinery either excessively cumbersome or weak and rickety. This limitation has been overcome subsequently chiefly through the use of concrete and metals, that is, by chemical knowledge. Now chemistry took swift strides through the seventeenth century both in theory and practice, but again not in such terms as to yield cheap steel or phosphor-bronze bearings. Even the problem of smelting a coarse iron with coal could not be solved—and it was a practical man, not a chemist, who took the first steps to making coal-smelting practicable, in 1709. Before the mid-nineteenth century there was no useful body of chemical theory from which useful consequences could be drawn to benefit metallurgy; hence the one great amelioration in material resources that came before then (the availability of cheap cast-iron) came as yet another triumph of craft empiricism. Sophisticated engineering was thus out of the question until materials superior to wood, brick, stone, and crude iron were at hand in quantity, so that the use of physical theories in engineering was in any case delayed by the rather slower progress of chemical theory.

At the same time the engineer himself was notably blind to the importance of mathematics, which was the foundation of the seventeenth century success in physics. The association of the engineer with the slide rule is quite recent. True, there was little enough for the practical man to compute: but what he did he did badly, clinging to antiquated rules-of-thumb, as Pepys found when he looked into the ways used to compute footage of timber for the Admiralty. No one ever knew how many gallons a cask would hold, or what the displacement of a ship would be. Until science took a hand, levelling and survey were carried out extremely crudely, and the reckoning of quantities on any job was full of arithmetical errors. Until after about 1650 the engineer, like every other technologist, was only rarely educated or mathematically literate. So long as multiplication and division remained almost intolerable burdens it was hopeless for engineers to attempt to borrow from science. For the ability to reason " If one windlass will raise 100 lbs., I shall need ten to raise half a ton," or some acquaintance with the laws of pulley and lever can hardly be reckoned as indebtedness to science at this stage of the game. The engineer, indeed, was hardly yet equipped to formulate his difficulties and his hopes for progress in a way that a mathematician or physicist could comprehend.

In consequence, the initiative towards mathematical accuracy, and often towards other improvements too, had to come from above. It was the philosopher Galileo who conceived the possibility of a mathematical theory of structures, showing in his *Mathematical Discourses* (1638) how strength could be calculated in proportion to size, not merely guessed at.[4] It was the statistician-economist William Petty who endeavored to introduce a radically new principle of ship-building, the catamaran hull.[5] It was the physicist Huygens who saw the possibility of making heat do work, while numerous mathematicians from Galileo's pupil Benedetto Castelli onwards investigated fluid dynamics, whether in the more specialized field of river-control or in the more general applications tackled by Huygens and Newton. Indeed, scientists seem to have been far more awake to the broad prospects of applying science to technology than were the technical experts. It was unfortunate that their most ambitious attempts in this direction failed. Hydraulic engineering was not put on a scientific basis. Huygens abandoned his experiments with the gunpowder engine so that, although there is a direct connection between his device and Newcomen's, the successful fruition of the heat-engine owes virtually everything but the concept of atmospheric pressure to practical engineers.[6] Petty utterly failed to win over the shipwrights with his experimental boats—and in any case his new design was not born of scientific principles. As for Galileo's theory of structures, it involved a fundamental misconception and would have led to failures if it had even been applied.[7] And in spite of further advances in the same theory from the end of the seventeenth century and later, a long time elapsed before calculation impinged on bridge and building design. It is significant that it is hard to hit upon anything that renders the mathematician Wren or the experimenter Hooke more scientific in their approach to design than were other contemporary architects.

In brief, the naive notion of some philosophers that the mathematical and experimental methods of science would revolutionize civil engineering and the construction of machinery within a few decades was falsified. The new-made world of which they dreamed was not to be attained so easily. Late seventeenth century engineering was much like that of the time of Agricola and the first machine-books.[8] Some recent feats were indeed beyond comparison with anything in the immediate past, notably the Languedoc Canal; but they involved no more than slogging at pick-and-shovel, masonry, and carpentry. Perhaps the greatest civil engineering improvement of the seventeenth century was the introduction of the wheel-barrow, which the Chinese had used for centuries! Wind- and water-power were more extensively exploited than before but with machines of established design and construction. Ships about doubled in size, but shipyards did not alter. Towns were still watered by placing a wheel in the stream of a river and connecting it to a set of backyard pumps, and so on. Indeed, it is difficult to imagine how, without new sources of energy or new materials, there could be any very rapid change in engineering—even if more complete and reliable scientific theorization had been accessible. As it was, technical invention could consist of little more than a re-shuffling of existing, familiar knowledge and practice; just as some engineering enterprises and some machines could become larger merely because greater wealth and continuity of purpose were at hand to push them through. Not that the age of early science was lacking in technical inventiveness—though often enough devoted to such chimaeras as perpetual motion and the flying-machine; it is found in the engineering of every craft, from improvements in the printing-press, through the stocking-frame and the clockmaker's gear-cutting engine to the use of shearing and rolling mills in the iron industry. The scope and resource of the machine-maker were continually increasing, preparing the way (one might say) for his next great crashing of barriers in the textile industry. But for the sixteenth and seventeenth centuries the pattern remained largely unmodified; there was no engineering revolution alongside the scientific revolution, barely indeed a minor disturbance.

Does this mean that engineering owed nothing to science beyond the concept of the pendulum clock and of the atmospheric/heat engine? Not quite, though some of the debt was potential rather than actual. In rather indirect ways the scientific notion of exactitude began to affect engineers. This is most evident in survey and levelling, where serious errors in measurement could have brought catastrophe to such schemes as the Languedoc Canal and the draining of the Fens.[9] Partly it was simply a question of training field-surveyors in basic geometrical procedures.[10] Not less important, though it came rather later, was the borrowing of the astronomer's tools, notably the telescopic sight with cross-hairs devised by Picard in 1668. To this must be added the bubble-level invented by the scholar Thevenot a few years earlier.[11] More years passed before the surveyor became much concerned to refine his reading of angles, but when he did he once more borrowed the astronomer's vernier and micrometer screw. In turn, through *his* dependance upon the skills of the instrument-maker and clockmaker, the astronomer impelled fresh exactitude into their miniature craft methods. Just as Boyle's airpump (1658),[12] constructed for him by Robert Hooke (who was later Curator of Experiments to the Royal Society and the progenitor of many other devices, including the universal joint), was the antecedent of all later machines depending on a piston accurately fitting in a true cylinder—and in turn derived from the backyard pump—so the clockmaker's gear-cutter and the instruments-maker's dividing-engine [13] were the ancestors of many later machine tools. Long before large-scale engineering could attempt such tasks these craftsmen were making fairly accurate gears, turning exact steel shafts, cutting good screws, measuring to an accuracy of one part in a quarter of a million or better, and so on, for their scientific employers and even for the open market. From their ranks some great engineers like Smeaton and Watt were recruited, and the total effect of their craftsmanship on productive mechanical engineering may well have been considerable. For besides their craftsmanship the leading men were scientifically educated (in England a few of them attained to the Fellowship of the Royal Society) and well able to look beyond the bounds of manual dexterity; indeed, they were the first of all craftsmen to perceive the fundamental truth that a machine can be more precise than any pair of hands.[14]

However, the fruition of this trend only occurred towards the end of the eighteenth century when a new race of engineers was already active on the scene, on whom instrument-makers like Jesse Ramsden or watchmakers like Thomas Mudge probably had little direct influence. It may be more proper to enquire whether, from a much earlier time, science did not influence engineering in a more subtle, pervasive fashion. Since it is fairly clear that engineers did not readily adopt mathematical analysis from science, might they not at least have borrowed the experimental method, or inhaled a still more nebulous "scientific spirit"? The difficulty is to know how this could possibly be estimated. Building a medieval cathedral was in some sense an experiment; it might stay up or fall down. Craftsmen have always experimented and talked of experiments: but not scientifically. It would be impossible to prove that there were more engineering experiments between, say, 1500 and 1700 than between 1200 and 1400, and meaningless if it could be done because the two periods differ so greatly in other respects than their science. Suppose one could show that bridge-builders were more daring in the seventeenth century than the fourteenth: what would this prove with regard to science? Is there any meaning in saying that Palladio's trusses or de l'Orme's composite beams owe anything at all to science? And so forth: everything becomes intangible, so that one might as well try to prove the opposite, that an empirical, engineering spirit begins in the sixteenth century to pervade science (some people seem to want to say both at the same time!)

To be more specific, most people would agree that the notebooks of Leonardo da Vinci undoubtedly demonstrate a true scientific outlook and the sound, original design ability of a true engineer. To some extent also it is obvious that these two interests or qualities are related in Leonardo; the Leonardo who was interested in dynamics (leaning very heavily on the medieval philosophers) overlapped with the Leonardo who was interested in ballistics, the construction of cannon, and fortification, but that his sketches and notes on machinery and other engineering topics were greatly marked by his concern for abstract scientific questions would be hard to demonstrate. Leonardo did what other inventors of new machines did; he re-assembled the familiar parts (tappets, screws, cams, gears, and so forth) to achieve new combinations of motions, and so new mechanical effects. As an engineering designer Leonardo was most nearly modern in his notes on stresses in structures, for he tried to formulate problems geometrically and so to solve them; but his lack of success meant that he could not have applied geometry to architecture, even had he had the opportunity. Leonardo, it is true, is an enigma; but what is obscure in him, who was engineer and scientist combined, and concerning whom ample evidence survives, must be far more opaque with respect to other engineers of whose scientific interest and attainments nothing is known, and whose personalities are almost wholly obscure.

Nebulous questions can only receive nebulous answers. We may be content, if we choose, with an intuition that practicing engineers and inventors absorbed something of the more scientific environment of the seventeenth century, even though this can hardly be proved; but we must at the same time allow for other economic and educational factors that pressed upon them, and not forget the effects of accumulating knowledge and experience in evaluating their attainments. But what of the other side of the question—what did engineering contribute to science? Again, the wider the terms in which such a question is posed the more vaporous and unanswerable it becomes. There is evidence that scientific propaganda in the seventeenth century stressed the potential benefits flowing to medicine and technology from scientific enquiry; there is far bulkier evidence that in their work scientists investigated only the questions that took their fancy, from mathematical physics to the physiology of reproduction; in other words, science was apt to be concerned at any moment not with urgent problems of engineering and agriculture but with matters that scientists considered significant and interesting. And there are distinct statements that the Fellows of the Royal Society and the scientific Academicians in Paris alike resisted attempts to tie them down to useful researches.[15] In any case, as an object for the improving attentions of science, engineering came a rather poor third after agriculture and medicine, in

340

England at any rate; in Louis XIV's France some attention was given to military engineering—and François Blondel was encouraged to write *L'Art de Jetter les Bombes*—as well as to the ornamental hydraulics at Versailles, but solely in response to orders from above.

It is true that, as Robert Boyle was fond of saying, scientists could learn from craftsmen, rather as Darwin was to learn from pigeon-fanciers. Experimental science had to master certain techniques: distillation, lens-grinding, glass-blowing, even turning and metal-working and the art of assay, though then as now the scientist often left this sort of thing to carefully supervised technicians. To that extent a minimum level of technological and engineering competence was necessary before serious experimental science could begin; a competence to be had far more readily in Florence, Paris, London, or Nuremberg about 1600 than it was in Aristotle's Athens. And engineers, like dyers and gaugers, suggested problems that demanded solution. So Galileo in the *Discourses* remarks on the mechanical wonders accomplished in the Arsenal at Venice, which leads him on to discuss the theory of structures; and later in the same book the observation that water in a suction-pump will rise no more than 30 feet is brought into a discussion of the vacuum. It was an observation that led through the barometer to Boyle's pneumatics and so back through the steam-pump to the mine again. But such few examples have been vastly overworked. In fact, the scientist and the engineer rarely coincided, for each preserved his own autonomy, the one of science, the other of craft. By the end of the seventeenth century, indeed, the scientist was already beginning to assume a position of conscious superiority to the " rude mechanicals "; he felt he knew enough to tell the engineer why his surveys were bad, why his estimates of water-delivery proved wrong, and why his vaults sometimes collapsed. But the corrections to customary practice—if such they were—that the scientist could suggest were as yet few, and were to remain few until the whole nature of technology had begun to change through the advent of new materials and a new source of energy.

REFERENCES

[1] See A. R. Hall, *Ballistics in the Seventeenth Century* (Cambridge, 1952).

[2] Of course it should be added that Newton was easily able to correct this error when it was pointed out to him, as he did in the second edition. It remains true, however, that some rather obvious physical effects of concern to the engineer were surprisingly resistant to theoretical analysis; see C. Truesdell, " A Program towards Rediscovering the Rational Mechanics of the Age of Reason," *Archive for History of Exact Sciences*, I (1960), pp. 3-36.

[3] Cf. *Horologium Oscillatorium*, 1673. Huygens at first believed that (theoretically correct) cycloidal motion was essential to give the pendulum time-

keeping accuracy; he had no part in the development of escapements (beginning with anchor, of William Clements, about 1671), though he did (like Hooke) also invent a spring-controlled balancewheel escapement. No scientist paid any serious attention to temperature compensation, which was dealt with by the clockmakers.

[4] English translation by Henry Crew and A. de Salvio, 1914 (and Dover reprint).

[5] The Marquess of Lansdowne, *The Double-Bottom Ship of Sir William Petty* (Oxford, Roxburghe Club, 1931).

[6] See pp. 381 ff.

[7] Because Galileo placed the neutral axis at the base of the beam; apart from this assumption his geometry is correct.

[8] Georgius Agricola, *De re metallica* (1556); Eng. trans. by Herbert Hoover and Lou Henry Hoover (1912; Dover reprint, 1950). Some further machines in Vannoccio Biringuccio, *Pirotechnia* (1540), Eng. trans. by Cyril Stanley Smith and Martha Teach Gnudi (1943; Basic Books, 1959) and other metallurgical authors (for bibliography see C. S. Smith in Singer *et al.*, *A History of Technology*, vol. III). The chief writers on machinery, Ramelli, Besson, Zonca, Veranzio, etc., are listed in William Barclay Parsons, *Engineers and Engineering in the Renaissance* (Baltimore, 1939), p. 118, in which also some features of their books are discussed.

[9] Cf. Parsons, *op. cit.*, pp. 438, 454.

[10] Cf. Maurice Daumas, *Les Instruments Scientifiques aux XVIIe et XVIIIe Siècles* (Paris, 1953); Edmond R. Kiely, *Surveying Instruments* (New York, 1947); E. G. R. Taylor, *The Mathematical Practitioners of Tudor and Stuart England* (Cambridge, 1954).

[11] The first surveying level with telescope, cross-wires, and bubble that I have seen was made by John Rowley and dated 1703; but there may be earlier ones extant. Huygens considered the use of such instruments for accurate survey many years before.

[12] Described in *New Experiments Physico-Mechanical touching the Spring of the Air and its Effects* (1660). In another book, *The Usefulness of Experimental Natural Philosophy* (Tome II, 1671), Boyle specifically mentions the instrument-making crafts as ones that had profited from scientific discoveries, giving as instances the telescope and microscope and the fact "that we daily see the shops of clockmakers and watchmakers more and more furnished with those useful instruments, pendulum clocks, as they now are called, which but a few years ago were brought into request by that ingenious gentleman, who discovered the new planet about Saturn" (*Works*, ed. T. Birch, London, 1772, III, p. 399).

[13] There is, however, almost a century's interval between these two machines, the development of the dividing-engine beginning c. 1750.

[14] Many scientists gave particular instructions or general directions to craftsmen about the construction of particular instruments; for one such "collaboration" see Henry W. Robinson and Walter Adams (eds.), *The Diary of Robert Hooke* (London, 1935), Index *s. v.* "Tompion."

[15] On this question in general see Robert K. Merton, "Science, Technology and Society in Seventeenth Century England," *Osiris*, IV, 1938—who inclines to the view that science was closely linked to technology—and G. N. Clark, *Science and Social Welfare in the Age of Newton* (Oxford, 1949), who suggests many reasons for not supposing either one to stand in a causal relation to the other. The contemporary attitude is well depicted in Boyle's *Usefulness* (above, n. 12).

III

What did the Industrial Revolution
in Britain owe to Science?

Discussion of the intellectual origins of the Industrial Revolution has, like medieval philosophy, provoked a division between nominalists and realists. One school of historians seeks to apply, in one formulation or another, the word *scientific* either to the process of technical change in the eighteenth century or at least to the mental habits of those who effected these changes; others, the realists, search without success for precise examples of a technical innovation's being derived consciously from pre-existent theoretical knowledge of a non-trivial character, or at least write of the relations between science and technology in the eighteenth century as being so subtle and complex that such terms as *application* and *derivation* become wildly inappropriate.

It may be difficult, initially, to convince oneself that the two positions are separated by more than words. The facts about technological development in the eighteenth century are not in dispute, nor are the biographies of the inventors, and in any case the main burden of analysing and accounting for the occurrence of the early Industrial Revolution in Britain falls upon economic historians.[1] No one today

---

[1] Although T. S. Ashton followed Mantoux in the supposition that an English tradition of empirical inductive science contributed in a notable way to the technological changes of the Industrial Revolution, recent economic historians have adopted a more sceptical position. Thus David Landes (*Cambridge Economic History of Europe*, VI, C.U.P., 1965, p. 293) writes: 'in spite of some efforts to tie the Industrial Revolution to the Scientific Revolution of the sixteenth and seventeenth centuries, the link would seem to have been an extremely diffuse one. . . . Indeed, if anything, the growth of scientific knowledge [i.e., in thermodynamics] owed much to the concerns and achievements of technology; there was far less flow of ideas or methods the other way.' And

III

would argue (in this case, at least) that brilliant technical improvisation compelled industry to reorganize itself; rather, everyone agrees that economic opportunity offered unprecedented incentives to hasten in Britain a secular process of technological development common to all Europe. M. Maurice Daumas has presented forceful arguments to support his contention that even if the historian may be permitted to speak of an Industrial Revolution in the eighteenth century, he should certainly not postulate a concomitant *technical* revolution. Indeed, the well-known tradition of early, but futile, attempts to smelt iron with coal, to make textiles with machines (William Lee's stocking-frame being the earliest successful example) or to raise water by fire – all going back to Elizabeth's reign – illustrate M. Daumas's view that invention by a Darby or a Newcomen is not a unique, simple historical event, but rather 'une opération complexe qui bénéficie d'une expérience parfois longue de plusieurs siècles, accumulée de génération en génération'.[2] In a different, but strictly analogous manner, Professor Nef has in a lifetime's work traced the technological antecedents of the industrial revolution to the fuel famine of the sixteenth century, showing thereby how the line that led to Etruria and Paisley became differentiated from that leading to Sèvres and Gobelins. If one were to push this historical view to its logical limits it would seem that there is nothing out of the ordinary for the historian of technology to account for in the Industrial Revolution, nothing but a smoothly continuing course of evolution. Few historians have been able, however, to contemplate the intro-duction of steam-power or the beginnings of a synthetic chemical industry quite so calmly, even though they may feel that mechanical

---

he rightly draws attention to the crucial importance of the *engineering* development of the steam-engine (p. 333). Similarly, S. G. Checkland (*The Rise of Industrial Society in England, 1815–1885*, Longman, 1964, pp. 73–4) maintains that: 'It was impossible to bring together at this stage the brilliant generalized mechanics of d'Alembert and Lagrange and the experience of the artisan. . . . This is not to deny that the mechanical inventors were scientific in the sense of deliberately construing and solving their problems, nor to deny that they drew on the observations and experiences of others. But it is to say that the engineering progress of the day was made by men with little philosophical knowledge, working with what contemporaries called "mechanical instinct".' Phyllis Deane (*The First Industrial Revolution*, C.U.P., 1965) makes no men-tion at all of science as a predisposing factor to invention.

[2] M. Daumas in *Revue d'Histoire des Sciences*, XVI, 1963, 291–302; cf. a second article by the same author, ibid., XXII, 1969. To what remote ancestry the steam-engine may be traced is evident from Joseph Needham in *Trans. Newcomen Soc.*, XXXV, 1962–3, 3–56 (=*Clerks and Craftsmen in China and the West*, C.U.P., 1970, pp. 136–202).

III

ingenuity in the textile industry or innovations in the manufacture of iron do exhibit this 'normal' character. Moreover, there is strong evidence of a contemporary awareness both in France and England that technical change had assumed a new character in which scientific self-consciousness formed an important element (though one can, equally, find contemporary assertions of the not necessarily antithetical point of view that this change was not effected by cabinet philosophers). Whether one prefers to speak of a 'revolution' in productive technology or, instead, of an accelerating rate of change – and if it is appreciated that all revolutions have their historical antecedents the difference between these manners of expression is not very great – many historians write with the conviction that there is a phenomenon here deserving particular explanation.

The temptation, and it is a perilous one as Charles Gillispie has pointed out already,[3] is to assume that the explanation can be couched in simple terms. If a historian postulates as an explanation of innovation, either in science or technology, a direct, self-evident translation of one into the other, he is liable to slide over the essential logical distinction between *episteme* and *techne*, evident already in the Greek writings on machines. In societies prior to our own it was the concern of the natural philosopher (and is still that of the theoretical scientist today) to analyse and rationalize phenomena; to show that, far from being capricious, they may be set within a universal and acknowledged system of ideas. The scientist publishes the results of his investigations openly for the information and criticism of his peers, unlike the technologist who is normally most anxious to keep novel processes as secret as possible, and to prevent their dissemination by patent protection; if James Watt, for example, thought like a scientist, he behaved as secretively as any traditional craftsman. Scientific theories have been demonstrated as successful by their breadth and capacity for fruitful extension, by their internal consistency, and by their applicability to a multitude of confirmatory instances. Success is largely a question of esteem by fellow-scientists; it does not spring from popular acclaim – often voiced in favour of moribund theories – nor from practical usefulness, nor even from the absence of contrary instances. The technologist, on the other hand, is not esteemed according to the votes of his peers, or according to the consistency of his reasoning, or according to the skill with which

[3] In 'The Natural History of Industry', *Isis*, 48, 1957, 398–407.

he derives his methods from general principles. He is required only to succeed in what he undertakes to do; his opportunity to attempt a project and the measure of his achievement in it may be very much a matter of popular acclaim. Hence, in considering the history of technical development, it is no more necessary to enquire after abstract rationality than, in similarly considering the history of science, to enquire after utility. It is temptingly easy to over-intellectualize technical invention: even in the twentieth century careful analysis indicates that science is only one source of such innovation, sometimes a necessary source, never a sufficient one.[4]

No one can need much convincing that the origin and progress of technology were the work of unlettered craftsmen. Despite the myths of ancient civilizations, the names of the obscure individuals who were the first inventors remain unknown to us. So are the authors of the carpenter's plane, the rotary grindstone, the water-wheel and the water-raising machines of classical antiquity (unless we attribute the screw to Archimedes); in medieval times the mechanical clock, the wind-mill and the many water-powered devices are similarly anonymous. In all the process of transmission of technology from East to West only Gutenberg's name survives, and he may therefore be reckoned the first of the known inventors of modern times. Even in the early Industrial Revolution period the biographies of some early inventors are almost lost. We know their successors better because the importance of their work was recognized. But it would be rash, after this long history of development, to underestimate the capacity for innovation in the craft tradition, to imagine that to call a man an 'unlettered mechanic' was a term of abuse before a changing society made it so, or to believe that the inventions of the Industrial Revolution were of so high an order of sophistication that they could never have been made without assistance from a superior, scientific plane. Even where new working principles may have been drawn from science, experience and mechanical skill were needed for their realization in practical form; and the number of instances of this occurrence seems to be small. Even historians who have particularly striven to elicit the scientific contribution to the Industrial Revolution can write: 'We do not wish to push this thesis too far – to under-estimate the contributions of unscientific though intelligent practical craftsmen – but the evidence

[4] John Jewkes, David Sawers, Richard Stillerman, *The Sources of Invention* (rev. ed., London, Macmillan, 1962).

appears to necessitate some modification of the traditional view of the Industrial Revolution.'[5]

It is, I believe, particularly misleading to translate the logical and objective distinctions between science and craftsmanship into a social distinction between craftsmen and scholars, as though the former were of no significance in the historical interrelations between science and technology, and all failure of the one to merge with the other was to be ascribed to an artificial social barrier. So far as the study of the history of technology from literary sources is concerned, the postulation of this barrier is, for early times at least, self-defeating: these sources were, by definition, compiled by scholars and gentlemen. But there is much good reason for doubting the importance of class-division in this context. We have Needham's evidence from China of the ascription of important inventions to high officials and scholars who were, nevertheless, technically informed also.[6] In the classical Greek and Roman tradition educated men familiar with philosophy and mathematics engaged practically in both building and mechanical engineering; Vitruvius, one of this group, seems to find it noteworthy that Ctesibios was the son of a barber. A few of the high officials charged with superintendence of Rome's water-supply (Frontinus among them) were technical experts. Again, the common illusion that medieval architecture was entirely the creation of illiterate masons is a fallacy; medieval architects understood and applied clear design principles, though these were not based on scientific knowledge. So did shipwrights.[7] The role of monastic houses in the exploitation of water-power and the development of time-keepers is well known. All such examples illustrate not only the unreality of a social barrier supposedly separating literacy and numeracy from manual skill, but the direction of skill by concepts or procedures that are not merely manual, and to which the trained eye and hand were irrelevant.

One may further argue on general grounds that it was not social distinction, causing ignorance of and contempt for the crafts among

---

[5] A. E. Musson and Eric Robinson, *Science and Technology in the Industrial Revolution* (Manchester U.P., 1969), p. 189.

[6] Joseph Needham, *Science and Civilization in China*, IV, Part 2 (Cambridge U.P., 1962), pp. 30–5.

[7] See F. C. Lane, *Venetian Ships and Shipbuilders of the Renaissance* (Baltimore, 1934, rev. ed. in French, 1965); R. L. Mainstone in *Architectural Review*, 1968, 303–10; and my survey article in *Society of Engineers Journal*, LX, 1969, 17–31.

educated men, that was the basic reason for the non-appearance of a fruitful interaction between science and technology in early societies. While elementary concepts may prove of some immediate value in mechanical engineering, most crafts depend on the empirical exploitation of properties of materials or chemical reactions that are quite complex; if shipbuilding had had to wait for rational fluid mechanics and brewing for biochemistry, men would have had neither ships nor beer. Technical progress in early science could occur only by fortuitous and empirical steps; in so far as the scholar attempted to immerse himself in craftsmanship he plunged himself more deeply into the craftsman's confusion. Military affairs have always been in the hands of men drawn from the upper strata of society without its being conspicuously obvious that military techniques have advanced faster than civilian ones. However far-reaching the tactical conceptions of a great engineer like Vauban, in their execution he was imprisoned by the craft skill of his age, not enabled to transcend it by his superior social status. Only Leonardo da Vinci – himself a product of the crafts – appears as an exception to this rule. Again, the temporary concern of the Royal Society for its histories of trades and that of the Académie Royale des Sciences for the *description des arts et métiers* was productive of little but materials for future historians. And, as it seems to me, necessarily so: one may solve intricate problems involving many variables by a lucky guess or by trial and error through the fruit of long experience, but one does not at once solve them by taking thought in the absence of a greatly extended theoretical structure, which must be created first.

For science must start at quite the other end, not by delving into the hopeless morass of agricultural or mineralogical detail but by investigating the apparently simplest phenomena and formulating the most general – and therefore practically useless – propositions. In antiquity and in recent centuries those branches of science in which such phenomena could be singled out, namely astronomy and mechanics, made the fastest progress; not because physiology or properties of matter were neglected for social reasons (in fact, the problems of these sciences, too, were constantly in debate) but because rationalization of the more simple must logically precede rationalization of the more intricate. On the other hand, where 'proto-chemists' and 'proto-geologists' deliberately prided themselves on a wide familiarity with craft secrets, the lore of still and mine, and cut themselves off from the general trends of natural philosophy, advance in these sciences was the more retarded. Whatever may be the case today, there was nothing to be gained in the past by maintaining that science and technology pursue the same ends, or that a *single* outlook, that of the 'scholar-craftsman', is sufficient to attain the ends of both.[8]

If it is neither profitable nor realistic, then, to pursue through the seventeenth and eighteenth centuries the gradual emergence of a single species of dual parentage, *Homo scientificus*, pursuing sometimes the purity of science, sometimes its applications to mundane affairs, one is left with a quite traditional conception that science was concerned with one class of problems and technology with another, and it remains to establish the relations between them. The commonest way of expressing a relation is to say that science is 'applied'; in their recent study of this question as it bears on the origins and course of the Industrial Revolution, A. E. Musson and Eric Robinson repeatedly refer to applied science as denoting a significant (though not the only significant) difference between technical innovation in that period and earlier innovation. Now the most obvious way of applying science would be to seek to practise the useful hints provided by scientists themselves. The seventeenth century is not without such hints: Descartes suggested that lenses should preferably be ground to a hyperbolic curvature; Galileo proposed the optimal shape for a loaded beam; Newton defined the solid of least resistance; Petty, if he may be classified as a scientist, advocated the catamaran construction for ships; Rømer and

---

[8] The opposite view has had its powerful advocates, notably among Marxist historians. Benjamin Farrington, starting from a definition of science as having its 'origin in techniques' with experience as its source and 'its aims practical, its only test that it works', has argued that the Milesians were successful because they were 'observers of nature whose eyes had been quickened, whose attention directed, and whose selection of phenomena to be observed had been conditioned, by familiarity with a certain range of techniques'. The 'positive content' of their science was 'drawn from the techniques of the age' (*Greek Science*, London, Penguin Books, 1944, I, pp. 14, 36–37). Similarly Edgar Zilsel, as well as Farrington himself, has emphasized the usefulness of the scholar-craftsman concept in accounting for the appearance of the 'modern' type of scientific investigator (see my 'Merton Revisited' in *History of Science*, 2, 1963, 1–15). Musson and Robinson (op. cit. note 5, pp. 11, 12) also incline towards the 'scholar-craftsman' hypothesis, suggesting that 'these distinctions between "pure" and "applied" science in the sixteenth, seventeenth and eighteenth centuries are to a large extent artificial'. All distinctions drawn by men are artificial, but they may well be conceptually important; and nothing could be more fatuous or more destructive of analytical enquiry than the view that 'after all, craftsmen and philosophers are all students of nature in their different ways'.

La Hire gave what they took to be an optimal profile for gear-teeth; Huygens and Newton worked out approximate solutions for the curve traced by a projectile in air. None of these and other suggestions from scientists for doing things better were adopted during the Industrial Revolution, though improved versions of these ideas have, of course, been applied in the nineteenth century or later.[9] The reasons are not far to seek: the novel designs were unnecessary or impracticable in the prevailing technical context. Evidently the scientists' notions of how to improve craft practice were not usually realistic; all Huygens's attempts to design a serviceable marine chronometer similarly failed, though his object was attained by working clockmakers in the following century. (The apparent exception to this general experience will be discussed later.)

It is true that the advocacy by writers on mathematics of more sophisticated and accurate methods of survey and navigation was rewarded with greater success, even though rudimentary methods were common through the Industrial Revolution. Their contribution to technology was minimal, however, and the advocacy was centuries, if not millennia, old. As Robert E. Schofield has put it, 'To demonstrate a relationship between aspects of an industrial revolution and science, we must show that those aspects influenced or were influenced by the development of contemporary theories in science.'[10] Trigonometry was no innovation of the eighteenth century. Surveyors and navigators also borrowed, through the scientific-instrument makers, practical improvements first devised for scientific purposes, such as telescopic sights; by the end of the eighteenth century navigation could be – but perhaps was not often – an 'applied science'. But this hardly goes near the root of the matter.

Turning now to the eighteenth century itself, one can again discover many scientific 'hints to craftsmen' concerning gunnery, optics, gear wheels, structures and so forth that were neglected for the time being, while it is correspondingly evident that an enormous proportion of scientific research in mechanics, astronomy, electricity, natural history, physiology and geology had no ascertainable bearing on technology

---

[9] The English mill-architect, Charles Bage, employed Galileo's theory of scaling in designing cast-iron beams for a mill at Shrewsbury (1796–97), and confirmed his results by full-scale breaking tests (A. W. Skempton in *Actes du VIIIe Congrès Inst. d'Hist. des Sciences*, Florence, 1958, III, pp. 1029–39).

[10] *Chymia*, 5, 1959, 186.

at all.[11] In fact the only obvious (and much discussed) instances of possible 'hints' from science that were gladly and effectively taken up by inventors and entrepreneurs both occurred at a time well after the inception of the Industrial Revolution in Britain, and perhaps played no crucial part in it: these 'hints' led to the introduction of chlorine bleaches, the manufacture of soda from salt, and the improved steam engine.

Chlorine bleaching was introduced in Britain in 1787. It is perhaps just worth noting that the name *chlorine* was conferred by Davy in 1810 when he established the elementary nature of this gas; the basic chemistry had been up to this point erroneous in that chlorine had been considered to be a compound, oxymuriatic acid. This basic investigation (by Scheele, Berthollet and Chaptal) was wholly continental, not British. The many confused and confusing attempts to make the new chemical agent effective in the absence of adequate scientific knowledge and manufacturing experience, involving the usual squabbles over patent rights, need not concern us here, save to note that commercial production of bleaching powder on a large scale was begun by Charles Tennant at Glasgow in 1799. Whatever the ethics of Tennant's commercial enterprise, it was certainly effective. By contrast, although from early 1787 James Watt (in association with his father-in-law James McGrigor) enjoyed the advantages of direct personal contact with Berthollet and other scientists in Paris as well as Joseph Black in Edinburgh, and possessed potentially valuable ideas about possible ways of manufacturing the bleach which he strove to keep as secret as possible, it does not seem that his acute interest led to commercial success. As usual in business, a great deal more was necessary for success than a short-lived lead in scientific knowledge. As early as 1790 Berthollet himself could claim that in this instance a scientific experiment had given birth to large manufactures but, whatever the propaganda value of this statement, it would be naïve to accept it as a complete summary of historical events. (In fact, in their first year of manufacture, Tennant's St Rollux works produced no more than 52 tons of bleaching powder.) A very great deal of what today would

[11] The discovery of electrostatic conduction generated, from 1753 onwards, a series of impractical proposals for telegraphs; this seems to be another example of frustrated scientific fertility in technical ideas. For a few years the electro-magnetic telegraph conception moved equally slowly, and one might argue that it was redeemed only by the needs of the railway.

III

138

be called 'development', much of it fruitless, had to be expended upon Berthollet's experiment before manufacture could be said to flourish. To my mind, this development would be more exactly classified as technical than as scientific research, for the problems to be solved by Watt and his multitudinous competitors were problems of economical, consistent and controllable manufacture to which science, though it indeed stimulated and inspired this technical research, was incapable of furnishing ready-made answers. Moreover, the solution of these problems contributed little or nothing to scientific knowledge whereas their technical profit was, in the end, considerable.

At a rather earlier period another technical investigation in chemical manufacture was begun in England, Scotland and France. Even before 1750 it was fairly generally known to chemists that soda could be prepared from other substances, chiefly common salt and Glauber's salt (that was itself made in the laboratory from common salt).[12] While an empirical relationship between these various materials was known, chemists were until after 1800 wholly ignorant of the element sodium common to them, and the nature of the carbonate ion ($CO_3^=$) in soda. Leblanc, author of the ultimately successful process for preparing soda synthetically, employed limestone because he knew that it was used as a flux in iron metallurgy. Thus he was guided to a useful reaction not by chemical science but by false analogy.[13] Similarly James Watt, working in collaboration with the Scottish chemist Joseph Black, attempted (like several continental chemists) to work an utterly impossible process for making soda from salt;[14] in fact the soda yielded by this process seems to have come, not from sodium chloride, but from the sodium sulphate (Glauber's salt) also present in common salt made by evaporating sea-water. Partial success, in other words, came from a happy accident, not from a chemical science inadequate for an understanding of the reactions involved. Black's part in the collaboration with Watt has been described as enigmatic. 'He does not seem to have grasped the chemistry of the process, though it was well within his abilities to do so. He probably decided at an early stage that the process was worthless, but was reluctant to say so to Watt, and

[12] The events have been reviewed recently by A. and N. Clow in *The Chemical Revolution* (Batchworth, 1952) and A. E. Musson and Eric Robinson (op. cit. note 5, pp. 251–337).
[13] See Charles Gillispie in *Isis*, 48, 1957, 152–70; 398–407.
[14] On this, see A. and N. Clow (op. cit. note 12) and Musson and Robinson (op. cit. note 5, pp. 352–71).

temporized.'[15] The obvious implication is that Watt was *not* able to perceive the futility of his experiments (in which he continued for many years), and the same might be said of several more distinguished chemists than he. In short, we may credit chemical science on this occasion with at most a partially correct observation, for the facts were never properly understood at the time, and no assistance at all towards the development of commercial manufacture.

Leblanc, whose process was at least feasible (though it yielded him no personal advantage), was probably the least chemically distinguished of the inventors of a dozen or so methods for preparing soda synthetically. Their history demonstrates, certainly, both collaboration between scientists and entrepreneurs and the activity of chemists like Guyton de Morveau as entrepreneurs. But their history shows too how empirical the first steps towards the synthetic alkali industry necessarily were; those engaged in this technical investigation were hardly better equipped intellectually than the early metal workers learning how to smelt copper from malachite. Their knowledge of the composition of the materials with which they worked was inadequate, and they had no means of predicting which reactions would be likely to succeed. As Charles Gillispie has remarked in his excellent study of the origins of the Leblanc process, by the time it was fully understood by chemists it was moribund.[16] And, as he has also remarked, both in France and in Britain this technical investigation was entirely unaffected by the 'chemical revolution' of Lavoisier which was the chief event in the history of chemical science when all this was going forward.

Finally, turning to Watt's invention of the separate condenser for the Newcomen steam engine, it is hardly necessary to do more than emphasize the fact that the myth of Watt's 'application' of an already formulated science of heat to the problem of the atmospheric engine's relative inefficiency has been exposed many times already.[17] The myth

---

[15] W. V. and K. R. Farrar in Musson and Robinson, p. 371. This passage is based on their interpretation of Watt's work.   [16] *Isis*, 48, 1957, 170.

[17] See Donald Fleming in *Isis*, 43, 1952, 3–5, and D. S. L. Cardwell, *Steam-power in the Eighteenth Century* (Sheed and Ward, London, 1963). It is perhaps unfortunate that, in his recent co-operative volume with Dr. Robinson (note 5), Mr. Musson (on pp. 79–81) in treating Watt as a scientific inventor does not firmly disavow this myth. While he writes that Watt 'appears to have rediscovered the principle of latent heat, independently of Black' he also writes of his consulting with Black in the course of his experiments on steam. Although this statement is formally correct, it might lead a reader to imagine that Black's superior scientific knowledge decisively stimulated Watt's invention.

was wholly created by Black himself and by John Robison who was in this affair a better friend to Black than to Watt.[18] It is abundantly evident that, before saying a word to Black, Watt was in possession of the salient facts: (a) a great deal of steam was wasted by the necessity for reheating the cylinder after condensation had produced a partial vacuum;[19] (b) a surprisingly small quantity of steam would suffice to raise water to its boiling-point; hence the need for a great quantity of injection-water to cool both the cylinder and the steam in it, and the consequent difficulty of obtaining a good vacuum so that the piston would descend.[20] Essentially, all that Black did was to assure Watt that his observation (b) was correct, and tell him of its agreement with his own unpublished theory of latent heat, of which until then Watt knew nothing. It would have been no logical impediment to Watt's invention of the separate condenser had he remained always unaware of the latent heat of steam, and a *general* conception of the latency of heat could be of no value at all in solving his problem.

It is easy to overlook the fact that while particular facts and *ad hoc* correlations may be no more than awkward or intriguing to a scientist, they may be quite sufficient to permit an engineer or technologist to take an important step; conversely generalization, essential to science, may be a feeble prop for the practitioner. The literate and sophisticated engineer, like Watt, may enjoy his use of scientific labels and his

[18] In his evidence on Watt's behalf against Hornblower and Maberley in 1796 Robison wrote (speaking of the work on the Glasgow model engine): 'Mr. Watt had learned from Dr. Black somewhat of his late discovery of the latent heat of Fluids and Steams . . . a subject of much conversation among the young Gentlemen at College. Mr. Watt was one of the most zealous partisans of this Theory. . . .' Watt's annotation on this passage reads: 'Dr. Robison is mistaken in this. I had not attended to Dr Black's experiment or theory on latent heat until I was led to it in the course of experiments upon the Engines when the fact proved a stumbling-block which the Dr. assisted me to get over. JW.' (A. E. Musson and Eric Robinson, *James Watt and the Steam Revolution*, Adams and Dart, 1969, pp. 26–7). See also Watt's letter of May 1814 in John Robison, *A System of Mechanical Philosophy* (Edinburgh, 1822), II, pp. iii–ix. This important autobiographical statement of some length is not reprinted by Musson and Robinson in their volume on Watt just cited, though Watt's contemporary experimental notes there printed (pp. 39–40) fully support Watt's much later recollections.

[19] Watt had already taken practical steps experimentally to reduce this heat-loss, which Cardwell has shown to be more significant in full-scale engines than the second factor, the latent heat of steam.

[20] Watt also knew that this effect was made the more serious by the lowering of the boiling point of water within the cylinder as the pressure within fell below the atmospheric.

mastery of scientific rationalization while always actually working with particular facts and ideas. *Any* scientific theory may envelop his actual immersion in particulars, right or wrong. R. E. Schofield has noted Wedgwood's distinction between two white clays (pipe-clay and kaolin) by their supposed difference in phlogiston content as an example of Wedgwood's use of chemical theory. But, in practice, Wedgwood did not – as quoted by Schofield – measure phlogiston in order to distinguish the clays; he distinguished them by firing specimens and observing that one was consistently darker than the other. His phlogiston-language was no more than a way of rationalizing what was physically observable; Wedgwood could just as well have said that the kaolin contained more *ying* than the pipe-clay. Being familiar with chemists' terminology, he used it, but it had no technological (or logical) significance.[21]

If, then, the application of science consisted in exploiting direct suggestions concerning utility coming from scientists themselves, its role in the Industrial Revolution would seem to be rather trivial. However interesting and prophetic the instances concerning chemical manufacture, they can hardly be said to have been decisive for the growth of the British textile industry by 1800, or even by 1825. Watt's separate condenser, despite the vast attention it has always attracted, was not essential to this or any other aspect of eighteenth-century industry. Supposing that at least 1,500 steam engines were at work in Britain by 1800, when Watt's patent expired, no more than a third had come from the Soho works; the majority were Newcomen engines or even Savery pumps supplying water to mill-wheels.[22] And the total power obtained from steam engines was but a fraction of that obtained from water-wheels and windmills over the whole country (the largest water-wheel erected in the British Isles – for mine drainage – at Laxey, in the Isle of Man, was built as late as 1854). Moreover, it is indisputable that Watt's later improvements of the steam engine – the sun-and-

---

[21] Obviously the falsity of Wedgwood's phlogistic hypothesis has nothing to do with my argument. See *Chymia*, 5, 1959, 189.

[22] J. R. Harris in *History*, LII, 1967, 133–48; A. E. Musson and E. Robinson in *Science and Technology in the Industrial Revolution* (Manchester U.P., 1969), pp. 393–426. The first use of steam for spinning in Lancashire (atmospheric engine with water-wheel) was in 1783; Boulton and Watt erected a rotative engine in Nottinghamshire in 1785/6 (R. L. Hills, *Power in the Industrial Revolution*, Manchester U.P., 1970, pp. 136, 152ff. This writer puts recognition of the advantages of the Watt engine by the textile industry as late as 1790).

planet gear, double-action, the parallel motion and so forth – owed everything to Watt's mechanical ingenuity and nothing to science. Even if the myth of the separate condenser were true its significance would be psychological rather than economic.

The purposefully useful enterprises of scientists in the eighteenth century seem, then, like those of their seventeenth-century predecessors, to have been largely misdirected, though exact as far as their own age was concerned; at any rate the work of Amontons and Coulomb on friction, of Euler and Coulomb on structures, was quietly ignored by their engineering contemporaries.[23] Even Sadi Carnot's study of the heat engine (1824) – the classical instance of technological progress presenting a new problem to science – was long in bearing practical fruit. In earlier cases where scientists had actually abandoned technical projects as futile (Huygens the marine chronometer, Huygens and Papin the atmospheric engine), their hopes were in the end realized by practical mechanics employing different principles from theirs.[24] There is a logical sense in which a formally correct scientific analysis must be true, but scientists could, and can, often be wrong in matters of practice, judgement and experience. The assumptions which had necessarily to be made in applying mechanics or chemistry to the technical issue commonly, in the eighteenth century, enforced an excessive simplification rendering the formally accurate answer of little mundane value. Since all real fluids are viscous, for example, theorems true of ideal fluids may prove deceptive in engineering practice. One must, then, conclude that the 'application' of science to technology bears some more general connotation than that examined so far, at least in the eighteenth century; it must denote, not the exploitation for particular purposes of general scientific truths or even particular discoveries, but some more diffuse activity: the adoption of scientific method, or familiarity with the general context of science, the practice of analysing problems in a rational or experimental manner, or even the development of an acquaintance among scientists, or some combination of these characteristics. The historical problem then (in face of such an indefinite conception of applied science) is to determine the functional significance

[23] Of course I do not mean to include here the practical and analytical work of such men as Macquer who were responsible for the direction of manufactories.

[24] It has never been proved, however, that the first steam engineers, Savery and Newcomen, were cognizant of the experiments of Huygens and Papin from 1675 onwards.

for technology of any one of these characteristics considered singly. For otherwise the whole discussion remains nebulous.

It is certain, to take one point, that there was a considerable popularization of science in Britain during the eighteenth century, in which entrepreneurs, manufacturers and inventors participated as well as clergymen and poets. The movement took two chief forms: the circulation of popular books by authors as various as Benjamin Martin and James Ferguson, Oliver Goldsmith and John Wesley, and the foundation of local societies, of which the best known are, of course, the 'Lunar Society' (or rather club – it was never a society in the ordinary sense) and the Manchester Literary and Philosophical Society.[25] In addition, itinerant lecturers continued the tradition of popular scientific demonstration begun in London by Hauksbee and Desaguliers, while instrument-makers found a lively market for microscopes, telescopes, globes and other apparatus. Probably more people possessed a discursive knowledge of science in 1760 than in 1660, though it is impossible to measure the extent of self-education or to assess its significance objectively. Only a very few of the popular writers and society members were at all eminent in science: such men as Joseph Priestley in the Lunar Society, William Henry and later John Dalton in Manchester. Inventors, engineers and manufacturers read the books and joined the societies; some impressed themselves on contemporaries' recollections by their interest in scientific discussion and their command of what they had read. But to find a causal relationship between the popularity of science and innovation in technology in eighteenth-century Britain requires a degree of faith. Most of these same people were dissenters and northerners also; perhaps these correlations may be equally significant. Moreover, it is difficult to determine whether a private devotion to science among British manufacturers compensated for the defects of formal scientific education in Britain.[26] It was dormant in the English universities during the early

---

[25] Among recent publications, much detail may be found in R. E. Schofield, *The Lunar Society of Birmingham* (Clarendon Press, 1963) and in the volume by A. E. Musson and Eric Robinson so often cited in this essay, *Science and Technology in the Industrial Revolution* (Manchester U.P., 1969).

[26] R. E. Schofield quotes a letter from Wedgwood to Bentley (7 March 1779) in which he writes: 'Finding my young men often at a loss to form clear ideas of chemical affinities, solutions, compositions, decomposition and recomposition of particles of matter which they could not see . . .' he devised a pictorial symbolism representing chemical reactions. This indicates, presumably, an early experiment in industrial

Industrial Revolution, though not in those of Scotland. The Royal
Society was not remarkable for great vigour and enterprise. There were
no trade schools. There was no scientific bureaucracy. The Royal
Military Academy at Woolwich (1741) was insignificant, while the
Navy had no comparable institution at all.[27] The very popularity of
science ensured its treatment as an entertainment, as it was regarded by
Samuel Johnson and Joseph Wright of Derby. Taking science seriously
– in its academies and factories – seems to have been very much a
characteristic of religious dissent.

In France it was far otherwise. Since the time of Colbert technical
education had been encouraged; there were trade schools, schools of
design, schools of commerce; above all there was the *École des ponts et
chaussées* (officially organized in 1755). A French school of mines, in
imitation of earlier German mining schools, was founded in 1778.
There were *écoles royales militaires* and an *École d'Ingénieurs-Constructeurs
de Vaisseaux* (1769). Even the theorists of the virtues of manual
education were French.[28] France, too, produced the first great theorists
of technology (Amontons, Perronet, Coulomb, the Carnots *père et fils*);
and there Macquer and Lavoisier became high officials; Lazare Carnot,
the organizer of victory; and Laplace, son of poor Normandy peasants,
made his way to great eminence from a career begun as a teacher in
military schools.[29] Chemical industry was systematically developed
under the direction of the state, and scientists were employed to carry
out mineralogical and metallurgical surveys.[30] Like Britain, eighteenth-
century France provides ample evidence of the popularity of science,
from the *Encyclopédie* and Algarotti's *Newtonianisme pour les Dames* to
the formation of provincial societies.

If interest in science, sophisticated analysis of technical methods and
problems, and the provision of technical education were essential
concomitants of industrial innovation it does not seem that France was

education to supply the want of formal training in schools, as well as Wedgwood's
belief that chemical knowledge was useful to the conduct of his business. (See *Chymia*,
5, 1959, 193.)

[27] The first Chief Master of the R.M.A. was indeed Martin Folkes, also President of the
Royal Society; but he possessed no ascertainable distinction in natural science.
[28] F. B. Artz, *The Development of Technical Education in France, 1500–1850* (Cambridge,
Mass., 1966).
[29] D. I. Duveen and R. Hahn, *Isis*, 48, 1957, 416–27.
[30] Henry Guerlac, *Chymia*, 5, 1959, 73–112.

notably at a disadvantage as compared with the midlands and the north of England.[31] (And, indeed, development of the more 'scientific' types of manufacture such as the synthetic chemical processes did proceed about equally quickly there.) Of course, it might be argued that other aspects of the English industrial milieu, its traditions of intellectual, religious and financial independence, for example, had to be allied with scientific propensities and knowledge to make the latter effective. But if that argument is pursued it seems to imply that these other differences between England and France were more effective in bringing about an industrial revolution than a diffusion of science that was common to both countries.

Another line of thought, and to my mind a more promising one, would be to examine the role of 'applied science' considered as the rational and experimental study of techniques, without presupposing the injection of important elements of abstract theory, or of pure scientific discoveries. For I must certainly concur with Charles Gillispie's conclusion that 'if the question [of the relations between science and technology in the eighteenth century] be approached in any detail . . . it proves extraordinarily difficult to trace the course of any significant theoretical concept from abstract formulation to actual use in industrial operations'.[32] And we have seen already how difficult it was straightforwardly to 'apply' a seemingly useful experimental discovery in science. Sometimes, indeed, scientific clarification post-dated the technical improvements which it ought to have preceded: the elementary chemistry of the differences between cast iron, steel and wrought iron (that is, the decreasing proportion of carbon in each) was established by Torbern Bergman (Sweden) and Guyton de Morveau and C. L. Berthollet (France) in the 1780s; this was the problem that had defeated Réaumur in 1722.[33] Meanwhile, in complete ignorance of chemistry, the Darbys at Coalbrookdale, Henry Cort and Benjamin Huntsman had respectively revolutionized the production of cast iron, wrought iron and steel. Moreover, if one supposes that what expert

---

[31] The point could be extended to practical mechanical skill also, with such obvious exceptions as the English lead in steam-engine building and (after 1770 and for a shorter period) in making textile machinery. Apart from the evidence of the *Encyclopédie*, the names of Vaucanson, Senot, Jacquard and Bréguet are in themselves demonstrative of the fact that excellence in mechanical craftsmanship was not confined to England.

[32] *Isis*, 48, 1957, 399.

[33] C. S. Smith, *Sources for the History of the Science of Steel, 1532–1786* (M.I.T. Press, 1968).

and scientifically literate technical men applied to problems of engineering and manufacture was not the content of science but a new rational methodology, the essential logical distinction already noted can be preserved; for the technical men directed their methodology to gain practical objectives, and judged its utility by its success in gaining them, while scientists directed their methodologies to the acquisition of knowledge. I do not mean at all to imply by this argument that there was a unique method of discovery (Hooke's 'philosophical algebra'[34]) that was simply transferred from science to technology; such a conception would be naïve. In scientific activity the basic logical structure was of far greater significance, and abstract concepts played a more vital role; science, moreover, was innocent of the 'scaling-up' problems (from pilot plant to factory, from model to full-scale engineering) that bedevil technological innovation. Yet it seems that, in the attempts made during the eighteenth century to analyse technical procedures with the object of improving their efficiency, one may detect contemporary notions of what constituted scientific procedure in (a) attempts to classify technical processes logically, notably in the *Encyclopédie*; (b) the employment of systematic experimentation, usually involving models; and (c) the treatment of data quantitatively.

One may speak of this empirical technical research – I have in mind such examples as Desagulier's and Belidor's investigation of waterwheels, Smeaton's model experiments on water-mills and windmills, young James Watt's experiments with his model Newcomen engine, Wedgwood's investigations of glazes and clays, or the quantitative data accumulated by structural engineers like John Rennie and Telford – as *imitative* (in a loose sense) of empirical science; it was not dependent on science, it did not apply science and, though it might sometimes use the language of science, it lay outside its conceptual structure. (There was no abstract conception of a heat engine until Carnot formulated one in 1824, even then defectively.) As R. E. Schofield has remarked, 'experiments, however cleverly performed, are not science unless they are guided by some sort of theoretical structure. Many things have been discovered by a process of empirical testing, but empiricism is not science.'[35] Such a process has very often, however, been sufficient to promote technology. Though ancient assay-masters were not chemists, they had great technical mastery of methods for determining exactly

---

[34] See Mary B. Hesse in *Isis*, 57, 1966, 67–83.
[35] *Chymia*, 5, 1959, 181.

the proportions of precious metal in an alloy, and for separating the metals. Stephenson, Fairbairn and Hodgkinson, for example, found empirical tests on model structures sufficient for the building of the Britannia Bridge.

Experimentation was no new means to technological innovation. Despite the occurrence of happy accidents it is hard to see how millennial progress could have occurred without it. When Darby experimented with coke for smelting iron, or Paul with rollers for spinning cotton (an idea later taken up by Arkwright), they were continuing very ancient traditions of technical innovation. What seems to have happened before the end of the eighteenth century is that such amorphous experimentation had been occasionally given a more systematic form, and that very gradually the design of mills, bridges or chemical plants was put on a more rational base. When an historical account assumes or implies that technological practices must be *either* derived from scientific knowledge *or* merely guided by tradition, chance, experience or rule-of-thumb, it necessarily ignores the role of coherent design procedures, which in many crafts (the shipwright's is the obvious example) constituted the core of technical skill. However ill-founded these procedures on abstract and demonstrable principles, they succeeded (if with far less than maximum economy and efficiency), and scientific knowledge was unable to supplant most of them before the mid-nineteenth century. Neither the shipwright nor the wheelwright nor the millwright nor the clockmaker of the pre-Industrial Revolution age was governed by an inexpressible mystique, nor were they mere slavish copyists. Though less subject to changes of taste than joiners and cabinet-makers they were consciously, though unscientifically, cognizant of what constituted good design in their respective crafts.

Their successors in the Industrial Revolution, of whom some began to be known as engineers or mechanics, were men of the same stamp, though masters of more sophisticated methods. It is a great misunderstanding to suppose that if some of them were certainly more than inspired tinkerers, they were therefore applied scientists as though no other category were conceivable.[36] Men like Smeaton, Wedgwood, Watt, Telford, Trevithick, Stephenson, Rennie were above all great

---

[36] Unless, of course, the term 'applied scientist' is taken to have some special definition in the eighteenth-century context, quite distinct from that which it has possessed for the last century or so.

technical designers. Certainly they used more exact and sophisticated experimentation than their predecessors; they also relied far more on the analysis of quantitative data. But these novel characteristics – and their significance was still limited enough – should be regarded rather as incipient modifications of an ancient tradition, partly enforced by the desire to use new materials like cast and wrought iron, than as the effect of a revolution wrought within technology by an infusion of scientific theories and discoveries.

I do not believe that my arguments in favour of the historian's distinguishing between technical research and the improvement of technical design on the one hand, and science itself or its application to industry on the other, are inconsistent with contemporary statements such as Richard Watson's:

The uses of chemistry, not only in the medicinal but in every economical art are too extensive to be enumerated, and too notorious to want illustration; it may just be observed, that a variety of manufactures, by a proper application of chemical principles, might, probably, be wrought at a less expense, and executed in a better manner than they are at present. But to this improvement there are impediments on every hand, which cannot easily be overcome. . . . It cannot be questioned, that the arts of dyeing, painting, brewing, distilling, tanning, of making glass, enamels, porcelain, artificial stone, common salt, sal ammoniac, salt-petre, potash, sugar and a great variety of others, have received improvement from chemical inquiry, and are capable of receiving much more.[37]

Or James Watt's inclusion of the laws of mechanics, the laws of hydraulics and hydrostatics, the 'doctrine of heat and cold' and the volumes of water converted into steam at various pressures, among the requirements to be known by a steam engineer.[38] Or George Stephenson's encomium of Smeaton: 'The principles of mechanics were never so clearly exhibited as in his writings, more especially with respect to resistance, gravity, and the power of water and wind to turn mills. His mind was as clear as crystal, and his demonstrations will be found

---

[37] *Chemical Essays*, I (second ed., London, 1782), pp. 39–40, 42. This seems to me more of a programmatic statement, and less a positive statement of past achievement, than appears to be claimed by Musson and Robinson (op. cit. note 5), p. 169. It need hardly be insisted that the elementary chemistry or biochemistry of the manufactures listed by Watson was unknown when he wrote, and that none had originated from chemical science.

[38] Eric Robinson in *Notes and Records of the Royal Society*, 24, 1970, 224.

mathematically conclusive.'[39] For we must set these statements in their industrial context, as well as allowing for an eighteenth-century writer's use of language which is somewhat different from our own. The chemical manufacturer did not work with purified reagents in a laboratory. He worked with sand, sea-salt, clays, limestones, metallic compounds, and coal whose compositions were complex and far from fully understood. He had to learn, partly in the light of technical experience and partly with the aid of scraps of laboratory chemistry, so to treat these materials that they yielded the desired products, using again a few analytical tools from science but also much familiarity with how a preparation should feel when handled, or how its appearance changed during treatment. So smiths had long judged the heat of iron. Wedgwood's pyrometric experiments were one attempt to quantify such subjective appraisals, but their one field of application hardly impinged on the accumulated mass of craft experience.

Similarly engineering mechanics in the late eighteenth century owed hardly more than an indefinite concept of 'force' (commonly rendered as synonymous with 'power') to the progress of theoretical mechanics since the time of Huygens and Newton, nor were its mathematics the mathematics of science. Rational and experimental students of machines did not start from the laws of motion, though they knew them. They employed no clear concepts of momentum and kinetic energy. They began with information about the performance of actual machines, so as to compare one design with another (for example, an undershot wheel and an overshot wheel) or, like Smeaton, they made model experiments to obtain more precise comparative data.[40] They examined the relation between the speed of the wheel and that of the water moving it. They investigated the effect of employing blades of different shapes, or boilers of different design. They developed a unit of their own for measuring the power of machines, which did not yet enter into the scientific vocabulary. By such search for perfected machines, by their eagerness to build only according to the most rationally efficient design, by their willingness to exploit useful devices (as Watt took the centrifugal governor from the windmill and applied it to the

[39] Musson and Robinson (op. cit. note 5), p. 74.
[40] At about the same time (i.e. in 1770) a structural engineer, E. M. Gauthey, made the first tests of the strength of building-stone in connection with an actual construction (S. B. Hamilton in C. Singer, E. J. Holmyard, A. R. Hall and T. I. Williams, *A History of Technology*, IV, Clarendon Press, 1958, p. 480).

steam engine) they greatly hastened the evolution of techniques. If to seek for quantitative measure is science, then these men were scientific in their approach to engineering; but I do not believe that to be an adequate definition of science. And just as the metallurgists created new craft skills by learning to manage smelting or puddling, or the chemical manufacturers mastered their retorts, so a new kind of engineering was created in which quantitative use of tabulated empirical data and simple equations linking the variables of concern to the engineer became important. In a sense this new form of design was as empirical as the old shipwright's rules for shaping a hull, since no fundamental physical reasons could be given in most cases to explain why its principles were valid; they had been established by testing, measurement and analysis of the behaviour of actual, or model, machines and structures. Accordingly, no reason of principle could be stated in 1820 why a high-pressure steam engine should be more efficient than one working at low pressure; and Thomas Young could write of the Telford-Douglas proposal in 1798 to throw a single cast-iron arch across the Thames: 'It would be difficult to find greater discordance in the most heterodox professions of faith, or in the most capricious variation of tastes, than is exhibited in the responses of our most celebrated professors on almost every point [concerning this proposed bridge] submitted for their consideration.'[41] Quite frankly, the greatest English engineers were wholly unable to determine the stresses in such an elastic structure, though much of the basic theory had been set out many years before by Coulomb. They had inherited a sound though limited engineering tradition, and developed it much further, without borrowing from or themselves creating a rational science of structures.

Modern engineering, with its use of formulae and calculation, was indeed nascent in the late eighteenth century. But there was then, and to a less extent is still now, much more to progress in engineering than the application of science. Before a structure or a machine can be built it must be designed; so must the plant required for a chemical manufacture. And as Hugh Clausen has insistently remarked, design is as much an art as a science:

One sees articles dealing with the scientific, analytical, or experimental aspects [of engineering] illustrated by formulae, graphs, and tables . . . but the next thing one sees is a photograph of the finished article, which tells

[41] Quoted idem, ibid., p. 484.

you nothing – except what the thing looks like from the outside. Of the really difficult and important thing, the transformation of ideas into ironmongery, there is usually not a hint. It is apparently inferred that these things just happen by themselves – that there is nothing of real importance in them.[42]

The technicians and entrepreneurs of the early Industrial Revolution in Britain excelled in 'transforming ideas into ironmongery', a problem made all the more acute for them by their desire to adopt new materials and new methods. In order to make this double form of innovation effective they had to modify traditional design procedures, though not departing from them entirely. Moreover, they had to solve quickly the problems so created. They could not wait for a slow accumulation of experience in building with cast iron, such as architects had accumulated in masonry construction, nor could the steam engine, like the water-mill, evolve over the centuries. The opportunity for engineers to create large, fireproof buildings, more powerful engines and swifter transport could not be postponed. Hence they turned to calculation and experiment to discover what could be done, drawing in a minor way on the ideas of the physical sciences, but employing also their own experience and technical intuition of what constituted sound design. But the point is that it was engineers and inventors who experienced the opportunity, and who created the new design methods to exploit it. They did not find these methods ready made to their hand in science. And it was they who chose, according to their necessities, fragments of scientific theorization to fit into their new ideas of engineering design – it was not the force of scientific truth that compelled them to do this, but the desire to derive quickly a sound and economical answer to a design problem. The history of the Industrial Revolution in Britain shows amply how ready the technical innovators were to work out new ideas empirically when, as was then so often the case, science had little guidance to offer.

---

[42] Hugh Clausen, *Engineering Design – the Background and Basis of Contemporary Life* (Institution of Engineering Designers Lecture, 1958).

IV

On Knowing, and Knowing how to . . .

'Logic, as it is generally understood, is the organ with which we philosophize. But just as it may be possible for a craftsman to excel in making organs and yet not know how to play them, so one might be a great logician and yet still be inexpert in making use of logic . . .' So Galileo wrote in his *Dialogue concerning the two chief world systems: Ptolemaic and Copernican.*[1]

One might reasonably suppose that any critical discussion of the writing of the history of technology would require some precise conception of what it is to 'know' or 'to have knowledge of' a subject or technique. But to avoid ambiguities, let us first attempt to be precise about the word 'technology' itself, of which a dictionary definition reads: 1. a discourse or treatise on an art or arts; the scientific study of the practical or industrial arts; 2. by transference, the practical arts collectively. This definition suggests a basic duality of meaning. Either technology is an all-embracing word — the sum of arts and crafts — whose history extends back in time to the neolithic; or technology, like the French *technologie*, implies a high degree of intellectual sophistication applied to the arts and crafts, such as has perhaps only existed within the last few hundred years or at any rate could not have existed before the invention of writing. By speaking of arts and crafts (*arts et métiers*), our ancestors avoided the difficulty; but to write 'a history of the crafts' now would seem distinctly odd. The word 'technics', employed by Lewis Mumford, is not natural in English, nor does his usage have the same meaning as the common cognate form 'techniques' as in the sentence, 'the violinist's technique was superb'. Hence in English the word 'technology' has more and more been generalized to include both the arts and crafts and *technologie*. It was to make this clear that the editors of the Oxford *History of Technology* chose to identify technology with 'making and doing' things; it seemed as broad and as time-free a distinction as could easily be proposed.[2]

Therefore the history of technology is the history of *homo faber*, of man the maker and doer; we can begin this history with the first known artefacts, though it may have begun earlier, and carry it continuously forward to the nuclear age. So much is commonplace.

It is obviously easier, though not easy, to say what has been done or made than how it was done or made. For the whole period of prehistory the how of technology must rest upon inference and analogy; even after

Reprinted from *History of Technology*, 3 (1978), pp. 91-103, by permission of the publisher. © 1978 Mansell Publishing Ltd, London.

written records start, the bulk of material bearing upon technological matters long remains slight and insufficient. Though the historian will always examine attentively the records of technological processes and methods available from antiquity to modern times, he must still rely heavily for his understanding on what he can deduce either from the actual objects or materials produced by these methods or from the workshops, tools and so forth used in their production. The deduction of methods by study of workshops and tools rests very much upon knowledge of the way such similar tools as furnaces and moulds have been used in more modern technologies that are better known to us; one cannot of course be sure that an ancient smith would tackle a certain problem of forging a piece of iron as would a more modern smith, an element of uncertainty that has to be accepted. However, the postulation of ancient technical methods by analogy either with modern practices or with those of modern primitive peoples can be rendered more plausible both by an actual reconstruction of the method — for example iron ore may be smelted in a reconstructed Roman furnace — and by precise analytical examination of historical objects themselves, from which the mechanical processes, heat treatments and so forth to which they have been submitted can often be determined.

Clearly the historical records concerning technology become richer and richer through time so that by the nineteenth century elaborate reconstructions and analyses become less and less necessary; a competent engineer can tell whether a given bridge is made of cast-iron sections or wrought-iron plates without a metallographic examination of the metal, though that might still have its value. Yet even in the nineteenth century there is still much to be learnt about the details of the how of technology by those same methods that are sometimes the sole ones available for earlier periods.

However the historian is often led by the nature of the material evidence as well as by written accounts to ask still more intricate questions. To some the answers are almost trivially obvious. If we ask why the early builders of Western Asia in Mesopotamia and the Indus Valley used bitumen as a mortar and a plaster,[3] it is plausible to see the answer in the water-resisting quality of the material. If asked to explain the occurrence of carvel-built boats in one area of Europe and clinker-built boats in another, the historian postulates the continuance of two independent traditions of boat-building.[4] But when we consider design in boats or structures, this kind of question about the mental processes of the ancient builder or technologist becomes pressing and difficult to answer.[5] I hope it is unnecessary to argue the point that the construction of a Roman aqueduct, for example, required a degree of planning and preliminary organization that was not needed by a Roman potter before he made a pot. The quantity of water required had to be estimated; the quantity and quality of water available from springs or streams estimated; the route surveyed; the methods of overcoming natural obstacles and delivering the water at the point of supply at the proper head established; the size and slope of the channel proportioned to the quantity of water required; and so forth, even before the organization of materials and labour began. We may suppose that Roman engineers made mistakes, but it is hardly likely that they built aqueducts or roads by chance or routine. Because we do possess literary texts, including the relatively ample works of Vitruvius and Frontinus, we can reconstruct the Roman methods of design to a certain extent and even state that certain ideas or principles of design that the Romans held were erroneous, which without the literary evidence we could hardly do.

Here then is a case where literary evidence leads us to the thought of the ancient technologist, which only the existence of literary evidence could do. The historian might establish from measurements that the Romans habitually gave their aqueduct channels an inadequate fall, but the Roman ideas of fluid-flow can only be presented to us in abstract terms and in the absence of writing cannot be reconstructed from the physical remains. The same may be said of medieval cathedrals. While comparative investigation of the surviving buildings can provide much information about the building methods used, the planning of the heights of arches, the diameter of pillars, the provision of flying buttresses and so forth, historians have also learnt a great deal about the methods of design, that is to say the thoughts of the builders, from documentary records.

In the history of technology, I believe the following definitions may be offered: a first-order conjecture implies the reconstruction of a technique based on inference by an analogy or analysis or both, and a second-order conjecture means the attribution of a purpose, design principle or theory to the user of a technique in the absence of literary evidence. Thus a first order conjecture is the outline reconstruction of the methods which were used by Greek or Etruscan potters to produce decorated vessels employing contrasts between two or three colours; there is a complete absence of direct evidence for the firing methods used and for any rationale that may have governed them.[6] But a second order conjecture is our explanation of the existence of welded blades having steely edges and softer central portions when we attribute to the smith a realization that while sharpened edges must be hard, the blade as a whole must not be brittle; in other words, we are putting into the smith's mind ideas of hardness, softness and flexibility for which we have no direct evidence.

As my terminology indicates, second order conjectures involving purpose, design and abstract concepts must necessarily be far more frail than first-order conjectures. In the example of the composite blade weapon, it might alternatively be supposed that this method of forging a blade was worked out by an intuitive evolutionary process involving (like all discoveries properly so termed) an element of chance. It is not essential that we attribute to the ancient smith abstract ideas of hardness, softness and flexibility and an awareness of an association between these ideas and pieces of metal prepared in given ways. One might rather postulate that the smith was as ignorant of such abstractions as of the metallographic structure of the various pieces of iron and steel. But my sole concern at this point is to argue that if we do wish to assign such abstract ideas to the ancient smith — and equally to all other craftsmen — we do so in a highly conjectural, almost poetic, fashion. It would certainly be a very great error to attribute to ancient technologists any abstract ideas for which we have no warrant in early technical literature. For example, we know that the oldest traditions and literature of metallurgy attach a mysterious importance to the nature of the fluid in which hot steel is quenched and recommend more than the use of olive oil or water for the purpose. We may therefore reasonably suppose that the most ancient smiths would have had a ritualistic attitude towards this strange and critical technique of hardening.[7] If, on the other hand, our literary evidence showed that simple fluids were invariably used, we would have no evidence for a different attitude among the smiths.

At this point we may return to our initial problem of what is *knowing* in a technological context. The statement 'the ancient smith knew how to forge a composite blade' is consistent with the statement 'the ancient smith had no metallurgical knowledge'. These statements are strictly similar to the single sentence, 'that boy knows how to ride a bicycle skilfully but he is ignorant of mechanics', which occasions no surprise because we do not expect a bicycle rider to understand mechanics or human physiology. Knowing how to ride a bicycle consists in acquiring a certain set of neuro-muscular co-ordinations as does mastery of many technical skills such as throwing a clay pot, making a good saw cut and ploughing a straight furrow. In the old days the principal instructor in such things was the clip over the ear. But clearly craft consists of more than acquired neuro-muscular co-ordination; it involves experience, a trained eye and a proper sensitivity to the fitness of things that may almost be called aesthetic, and in certain situations even demands an intellectual knowledge that is distinct in kind from the 'knowing how to' used here. But when we say 'that girl knows how to play chess' we imply far more than an acquired co-ordination or even memory of the pieces, board and rules, for someone who knows how to play chess must at least have knowledge of the simplest strategies. But here by 'knowing how to' I mean mainly non-cerebral knowledge as in bicycling, swimming, dancing and other familiar non-craft activities.

Clearly the spectrum of knowledge from knowing how to intellectual or theoretical knowledge is very great. Let us elaborate Galileo's example of the spectrum associated with the organ. Its construction may begin with craftsmen who know how to plane wood, tan skins, smelt metals and so forth. If we transport ourselves back only a few centuries in time, their essential skills could neither be analysed nor even described in a meaningful way, any more than one can learn to ride a bicycle or swim by reading a book. However, the actual organ-builder, though he may be unable to play the instrument and be wholly ignorant of musical theory, must know what an organ is and all its parts in their appropriate sizes, assembly, fitting, rectification and so forth;

he can learn something of his art from books, but abstract knowledge alone would be far less serviceable than practical knowledge alone. Then there is the player; he may even be totally ignorant of the notation of music, but know how to play the organ by ear, which is yet another exercise in neuro-muscular co-ordination of which all performance in part consists. Or the player may be Johann Sebastian Bach. Yet the composer of music does not necessarily have to be, as were Bach, Mozart and Beethoven, a virtuoso performer on an instrument any more than he needs to have a knowledge of the science of acoustics. At the same time the physicist whose science is acoustics need not be able to play or compose or to construct a musical instrument.

Hence the statement, 'Jones has a knowledge of musical instruments', is meaningless by itself. It might mean that Jones is knowledgeable as a particular kind of antiquarian; that he is a good performer on violin, flute and piano; that he keeps a shop selling musical instruments; that he writes music for symphony orchestras; and so forth; the possible meaning could only be learnt from the context. The sentence, 'Jones knows how to cut a whistle', would be more precisely informative, but in turn has no implication for his ability to perform on the whistle, knowledge of harmony and counterpoint and so forth. Suppose we find in some Danish bog a miraculously preserved mesolithic whistle. This is equivalent to the statement, 'Jones knew how to cut a whistle'. Let us also suppose the whistle has finger-stops. We cannot be certain that its maker made music since the stops might have been used for signalling or instructing well-trained sheep-dogs, but let us consider the idea that this mesolithic whistle was used to make music. We still cannot have any notion of what sort of music it was, in what social context it was played and what it meant to player and listeners, and indeed are unable to penetrate in any way into the minds of those remote people from the mere discovery of this artefact, interesting as it would be.

So far as history before the last few centuries is concerned, it is futile to demand, as a recent writer has done, a history of technology as knowledge on the ground that it is wrong to 'deny technology a significant component of thought'.[8] It is worse than futile to proclaim such a programme without considering the rather elementary distinction between knowing how to and intellectual knowledge and without looking at the different connotations of the verb 'to think'. The poet may 'think himself transported to the distant fields of home' or I may think myself lucky, or think about Tom Thumb as Dr. Johnson recommended in certain situations, but if we are to introduce thought in a more important way into the history of technology, a more precise definition of the thinking is necessary.

We may confidently believe that from prehistory to the present day craftsmen have pondered not only over their problems and failures, but also over their successes, especially when the latter resulted from variations upon established techniques, whether accidental or not. The problem is that for the reason I have already endeavoured to explain the mental processes of early technologists are largely hidden from us; only documents can reveal them authoritatively, and then but partially. If we try to reconstruct the thoughts of such craftsmen we can only create a mirror reflection of what is in our own minds; even if what is in our own minds is borrowed from Vitruvius or Theophilus we still cannot cease to be ourselves. It is mere self-deception to suppose that one can recreate thought from things, as for example those who have sought to fabricate a palaeolithic religion based upon the evidence of cave-paintings or, perhaps, an iron age astronomy on the evidence of stone circles. In such cases we do not truly see our forbears, but only see an image of ourselves distorted according to our own fancies and unconscious urges.

As for knowing how to, this is a matter of fact or at any rate of first-order conjecture. We have learnt that the Sumerians knew how to solder metals; we infer that they also knew how to anneal metal as when raising a bowl with the hammer, for this seems implicit in their successful technique.[9] Indeed, all the ancient peoples knew how to do many things by methods which are far from clear to historians, though plausible hypotheses can be suggested, for their alignment of tunnels and roads over quite long distances.[10] Our ignorance arises not so much from inability to imagine any simple method by which the ancient technologist might have solved the problem as from inability to choose between a number of possible methods. Thus, at one level, the statement of the achievement of early technology is nearly equivalent to a statement of the knowledge of the ancient technologist: we say that he knew how to do this because he did it, which is, in fact, our contemporary test of deciding whether Brown knows how to do something. If we seek after some higher levels of cerebral activity, we may simply be deluding ourselves, either because knowing how is essentially a matter of acquired muscular co-ordination and experience or because we simply have no evidence to go on. This last point is particularly impressive in considering some of the pivotal technical innovations of the past. Let us take the water-mill and windmill, where the former was introduced into Western Europe over a thousand years earlier than the latter. We have a fair amount of information on the early history of both devices; we have Vitruvius's text concerning the water-mill; we can compare the windmill with its Asian precursor and with the watermill; but our information is far too slender to permit us to make even the least guess at the mechanical knowledge, design or experiments of the first inventor or inventors in either case. And indeed we are hardly better off with Newcomen's steam-engine.

What then of knowledge in technology which is other than knowing how? Firstly, without pressing the evidence too hard we may claim that some such knowledge is at least as old as the civilizations of Mesopotamia and Egypt, whose documents make it manifest to us, but it must have been a tiny constituent in all early technological activity, which was after all mainly primitive farming. Otto Neugebauer has written, 'one must simply realize that mathematics and astronomy had practically no effect on the realities of life in the ancient civilizations . . . [And in particular] the requirements for the applicability of mathematics to problems of engineering are such that ancient mathematics fell far short of any practical application'.[11] His judgement is capable of wider extension; the higher learning represented by writings on chemical topics for example must also have been of very limited application.[12] Nevertheless, the surviving technical literature grows richer through the Graeco-Roman period and then lapses for several centuries to revive again in both Islam and Western Europe; incidentally, I do not here consider the Far East. Secondly, it may be remarked that those who knew often also knew how; whatever the status of Vitruvius may have been, and Frontinus was certainly an administrator, Hero like al-Jazarī or Theophilus later surely wrote of techniques with which he was familiar in practice.[13] Thirdly, we must recognize that a fair proportion of all documentary material in technology, up to the Renaissance and later is concerned with design as an aesthetic, which does not concern me here, and design as a theory of proportion, for things are made of different sizes and rules have to be made accordingly.

Now the concept of rule is in this context extremely important for we are not here dealing with rules of technical practice but with an abstraction. A craftsman might thoroughly know how to build a ship without awareness of such rules, but knowledge of them would enable him to design whole series of ships. Moreover, such rules were not dictated by the immediate requirements of technique as are the technical rules — if the technical rules are ignored, soldered joints will fall apart, corners will not be square and so forth — but represent a rationale of technique corresponding to theory in scientific technology. However, we should be careful not to over-emphasize the significance of canons of design because of the outstanding interest of their appearance in our documentary sources; they are always known in a constructional context as in Vitruvius's *De Architectura* and later on in manuals of building and ship construction, but were never relevant to most ordinary trades or tradesmen. Actually a great part of all the technical literature before the eighteenth century is concerned with technical rules, recipes, gadgets, 'secrets', and of course pleas for the patronage of the great and wealthy. We should assign such technical literature to the realm of knowing how rather than to what we are seeking for here, that kind of higher knowledge which caused the word 'mystery' to acquire its current meaning.

Now it may justly be argued that the extant technological literature before 1700 does not give a completely fair picture of the character of pre-scientific technology, and moreover that this literature has never been systematically analysed in an attempt to trace the history of technology as knowledge rather than as knowing how. Students of Leonardo da Vinci would surely wish to emphasize the high sophistication of his technological studies as in the movement of water and his endeavour to marry technology and philosophy, and it might be argued that he can hardly be a unique figure. On the other hand, however, it is obvious that even the best technical writing of the sixteenth century such as that by Agricola and Biringuccio is practical and descriptive in character and rather notably deficient in generalizations and explanations. Dr. Alex Keller's study of the machine-books hardly alters this general picture;[14] their authors are wholly concerned with the particular and 'practical' (though really highly fanciful), pursuing their mechanical hobby-horses *ad nauseam* through endless permutations and demonstrating repeatedly their total failure to grasp the most elementary theoretical notions about the functioning of machines such as the desirability of minimizing friction. In fact, the overt literature of technology in the sixteenth century is very far from reflecting accurately the success of craftsmen in knowing how to achieve all kinds of impressive and delightful new results. The growth in size of ships and the power of their armaments, the increasing refinement of furniture-making, ceramics, textiles and other consumer crafts; the application of coal to industrial processes; the ever-extending uses of wind- and water-power; all these constitute but a few examples of growing technological competence which is still, as in antiquity, much better-known to us from extant objects or from incidental records than from self-conscious descriptive treatises. Agricola's *De re metallica* naturally reflects the mining progress of the past; no successor described the further stupendous achievement of the next one and a half centuries.

If the increasing scope and refinement of the knowing-how of technology has to be painfully traced by studying multifarious historical sources, where is the growth of the knowledge of technology to be found? That such knowledge does grow is surely beyond doubt: it is to be found in the *Mémoires* of the Académie Royale des Sciences, in the *Philosophical Transactions* and Smeaton's *Reports*, in the encyclopaedias, in the writings of French and British engineers, in the *Transactions* of local societies, in the curricula of technical schools and eventually in professional examinations. Long before the end of the nineteenth century a huge body of knowledge, much of it mathematical in form, had come into existence relating to the design and performance of all kinds of bridges, steam-engines, boilers, heavy electrical engineering equipment, submarine cables, chemical works and even gliders and aeroplanes. Here surely we can see at last technological knowledge (*technologie*) clearly separate in character from the age-old knowing how to, though of course, they are related at the industrial level. Whence did this new knowledge arise? Nearly all historians have regarded it as deriving more or less directly from science; some have regarded it as essentially applied science;[15] and others, probably more correctly, as partly applied and partly imitated science. For often the engineer found scientific data inadquate and scientific theories too cumbersome; therefore he had to institute experiments to complete the data and introduce simplifications to make the theories usable. But there are obvious and well-known reasons for regarding whole new industries such as the electrical industry as science-based, just as there are equally good reasons for regarding the new prime movers devised rather later by Diesel and Parsons as the fruit of sophisticated scientific ideas which ultimately originated in the thermodynamic concepts of the 1850s. As L.J. Henderson reputedly declared in a memorable phrase it is not in dispute that science owed a large debt to the heat-engine, but in the second half of the nineteenth century that debt was more than repaid. The evolution of the new energy converters proceeded in a very different way from that of the old Cornish engine, now gradually falling into disuse.[16] In short, if one is to characterize the differences between the level and kind of technology practised around 1820 and that practised around 1890, virtually every historian would single out the origin of new technological ideas in scientific work and the training of technologists in mathematics and science, so that they were equipped to apply both analytical and experimental skills in their own work. It is a commonplace of educational history that, in order to produce qualified engineers of this new type, the structure of technical education was remodelled and extended in most civilized countries, on the widely praised German model of technical educations.

Many students of the Industrial Revolution in Great Britain have supposed this transforming influence of science upon technology to begin as far back as the first half of the eighteenth century, supposing this influence to have been made fruitful, in turn, by the Scientific Revolution of the sixteenth and seventeenth centuries.[17] These historians regard scientific method and scientific knowledge as major contributors to the technological innovations that both preceded and with increasing frequency accompanied the industrial revolution. For fear of being misunderstood, let me at once add that these historians do not exclude the role of empirical inventiveness, nor do they overlook the importance of wholly non-technical, economic and social factors in the British 'take-off'. I have been a sceptical critic of this interpretation, although I have not disputed the view that the increasingly scientific character of technology in the nineteenth century signalizes a turning-point in the history of material civilization. Indeed, I also agree that much scientific investigation of a useful or applied character was carried out in the eighteenth century, notably by the French school of mathematicians, culminating in the work of Lazare Carnot and Coulomb.[18] What I doubt is the validity of such investigation as an explanation of the Industrial Revolution in Britain until well after 1800.[19]

Naturally, just as in the eighteenth century there were eminent technologists like Smeaton, Watt, Wedgwood and other members of the Lunar Society with strong scientific connections — men to whom some would perhaps think the label 'scientist', though anachronistic, fully appropriate — so in the next century the more scientific technology professed by a Wheatstone or a Siemens, a Parsons or a Marconi was by no means devoid of important empirical elements. Thus Marconi had to learn by experiment that radio waves did not necessarily possess optical properties, that antennae of certain shapes revealed marked directional properties, that short waves were reflected by the ionosphere and so forth. Further, since technology deals with men and materials, neither of which are constant in their behaviour, the rules and tables used by engineers often codified the results of experience rather than experiment or theory. I do not think anyone would maintain, therefore, that an engineering discipline based on physical science and mathematics which contained major elements of analysis and theory is the same as a department of physical science.

One recent author has departed strongly from the general consensus that the penetration of science into technology is a historical reality. Dr. Layton argues that 'if basic science is the source of all new technical knowledge, then technology itself produces no new knowledge, and the technologist's role becomes that of applying knowledge generated elsewhere'.[20] To declare that technology produced 'no new knowledge' in the nineteenth century would certainly seem to be an exaggeration; and perhaps no one has ever made such an extremely negative claim. What may properly be said is that basic knowledge tended increasingly to derive from science, while knowing how to apply it to useful ends was very much the province of the technologist. Can any one doubt that knowledge of electromagnetic phenomena came from Oersted and Faraday, while the knowing how to use it was built up by Wheatstone and a host of other telegraph inventors and engineers? Or that electromagnetic waves were discovered by Hertz while Marconi was the one person in his age who knew how to make them useful? The development of wireless telegraphy in fact serves as a discriminating instance of what may happen generally: as the development of the new technique proceeded, so further injections of scientific knowledge by such people as Lodge and Fleming enriched it, while the technique itself developed its own logic and experience so that it was certainly distinguishable from university physics.

To recognize this distinction between physics and electrical engineering, which adds experience and knowing how to a part of the scientific content of physics, is not to claim, as Dr. Layton seems to wish to do, that science and technology are separate bodies of thought differing from each other in significant ways. In my view, it is in basic ideas that science and modern technology most closely approximate to each other. Dr. Layton enlists in his support, perhaps incautiously, the testimony of Aristotle and Alexandre Koyré, whose point he quotes[21] and seems to misunderstand. Koyré argued that technology was bad science. Moreover he held, as is obviously true, that the rigour and exactitude of science do not belong to the world of technology, but even this statement is less true today than it was in 1948 when, for example, neither industrial nuclear power nor commercial supersonic flight had yet been attained. Koyré contrasted science's world of precision with the engineering world of 'near enough'. But it seems an exaggeration to claim that imprecision and compromise, inevitable to engineering design as they may be, constitute another world of thought. If a physicist expresses dimensions in ångströms while an engineer is content with hundredths of a millimetre, the velocity of a radio wave, for example, is the same for both as are the basic formulae and parameters.

One may reasonably assert, I think, that in broad terms science in the nineteenth century, and to a lesser extent earlier, contributed to technology: 1. mathematical analysis, extended from physical science to engineering; 2. the method of establishing facts by carefully controlled experiments; 3. knowledge of relevant natural laws such as those of thermodynamics or genetics; 4. acquaintance with new natural phenomena such as electromagnetism or catalysis. These contributions at least came to technology from science, and I cannot conceive of late nineteenth-century technology without them; accordingly, I think the areas of technology and industry into which these contributions entered are very different from areas lacking such transforming injections of new ideas and methods as in the wrought-iron industry. Knowledge of making wrought iron in a puddling furnace consists entirely in doing it; a man either knows how to do that hot and exhausting job at the furnace mouth with his own hands or he does not; no literary description, formulae or pyrometric measurements can ever be substituted for experience of the look and feel of the metal. Some element of this same experience, skill and familiarity with the feel and look of things may remain in all levels of technology, but gradually loses significance as the formulae, measurements, print-outs and so forth come in. The technologist is, moreover, always concerned with doing and making and is not concerned like the scientist with experimental and conceptual problems. It is not the same thing to have knowledge of the theory of structures as to know how to build an aeroplane fuselage.

Thus, there is admittedly a sense — though a very different sense from that prevailing in Aristotle's time — in which the distinction between knowing and knowing how to is still valid so far as technology is concerned; and the knowing part has come to be almost wholly scientific in content and character. If this were all Dr. Layton's claim I should accept it, but he claims more by saying that there are separate bodies of knowledge.[22] This seems to me wrong because there are overlapping bodies of knowledge. To a very great extent technologists and scientists are on common ground; the scientist moves from the common ground into an area of refinement, complexities and new research that the engineer does not need at the moment, while the engineer moves from the common ground into a world of drafting, computations, approximations and compromises between conflicting optima — not to mention finance — into which the scientist need not enter. The 'laws of science refer to nature and the rules of technology refer to human artifice',[23] writes Dr. Layton, but he apparently forgets that the 'rules of technology' can only be valid in so far as they are consistent with the laws of nature and that these 'rules' have been continuously adapted during the last few centuries to make full use of science's increasing understanding of nature's laws. Since Dr. Layton quotes Aristotle with approval, let me in return refute him with a quotation from that great critic of Aristotle, Galileo:

> It would be novel indeed if computations and ratios made in abstract numbers should not thereafter correspond to concrete gold and silver coins and merchandise. Do you know what does happen, Simplicio? Just as the computer who wants his calculations to deal with sugar, silk and wool must discount the boxes, bales and other packings, so the mathematical scientist, when he wants to recognize in the concrete the effects which he has proved in the abstract, must deduct the material hindrances, and if he is able to do so, I assure you that things are in no less agreement than arithmetical computations. The errors, then, be not in the abstractness in concreteness, not in geometry or physics, but in a calculator who does not know how to make a true accounting.[24]

The history of modern technology has been in part the mathematical analysis of the 'material hindrances' which do indeed cause technological problems to resist the over-simple theories which science before the nineteenth century was alone capable of producing.

Notes

1. Galileo Galilei, *Dialogue concerning the two chief world systems: Ptolemaic and Copernican*, trans. Stillman Drake, Berkeley and Los Angeles, Calif., 1953, p. 35.

2. See also Charles Singer, 'The happy scholar', *Newcomen Society transactions*, 29 (1953-5), 123-32 (Dickinson Memorial Lecture). It should hardly be necessary to add that in giving technology a universal connotation as all ways of making and doing there was no intention to correlate a technology with manual labour; for example, one way of making things is to use a computer in designing a bridge.

3. See for example C. Singer and others (eds.), *A history of technology*, 5 vols., Oxford, 1954-8, vol. 1, pp. 250-6 (hereafter cited as *Oxford history of technology*).

4. *Oxford history of technology*, vol. 2, p. 578.

5. See A. Rupert Hall, 'The genesis of engineering: theory and practice', *Society of Engineers journal*, 60 (1969), 17-31, which attempts to make a small contribution to the history of design.

6. *Oxford history of technology*, vol. 2, pp. 262-6.

7. *Oxford history of technology*, vol. 3, p. 35.

8. Edwin T. Layton Jr., 'Technology as knowledge', *Technology and culture*, 15 (1974), 31-41, p. 32.

9. *Oxford history of technology*, vol. 1, pp. 624-5, 636-7, 649-54.

10. June Goodfield and Stephen Toulmin, 'How was the tunnel of Eupalinus aligned?', *Isis*, 56 (1965), 46-55.

11. Otto Neugebauer, *The exact sciences in antiquity*, Copenhagen and Providence, N.J. 1951, pp. 71-2.

12. I have particularly in mind the mass of cuneiform material dealing with chemical arts; see J.R. Partington, *Origins and development of applied chemistry*, London, 1935 and the many articles published during the last twenty years by Dr. Martin Levey (*Isis*, 47 (1956), 287 lists seven early papers by him). Despite this wealth of evidence, it is impossible to believe that in a predominantly illiterate society the transmission of technical information by written texts can ever have been of massive importance.

13. Hero of Alexandria, *The pneumatics*, 1851, trans. Bennet Woodcroft, facsimile reprint ed. Marie Boas Hall, London, 1971; Al-Jazarī, *The book of knowledge of ingenious mechanical devices*, trans. Donald R. Hill, Dordrecht and Boston, Mass., 1974; Theophilus (called also Rugerus), *The various arts*, trans. C.R. Dodwell, London, 1961 and *On divers arts: the treatise of Theophilus*, trans. John G. Hawthorne and Cyril Stanley Smith, Chicago and London, 1963.

14. A.G. Keller, 'Early printed books of machines, 1569-1629', University of Cambridge unpublished Ph.D. thesis, 1967. Dr. Keller's broad conclusions are apparent in his *Theatre of machines*, London, 1964.

15. The words 'applied science' have proved unfortunate as conveying the idea that science provides a recipe or prescription for some device or process, which the technologist has only to realize on the manufacturing scale; it is universally recognized that this provides a highly unrealistic model of the relations between science and technology. However, the original emphasis of the expression was probably rather to draw attention to the application of scientific methods and attitudes in technology; in this sense Smeaton's investigations of the efficiency of wind- and water-mills were certainly a work of applied science (and the results were published by the Royal Society) although Smeaton borrowed no general theories, parameters or calculations from the science of his day in carrying out his enquiries.

16. D.S.L. Cardwell and Richard L. Hills, 'Thermodynamics and practical engineering in the nineteenth century' in A. Rupert Hall and Norman Smith (eds.), *History of technology: first annual volume, 1976*, London, 1976, pp. 1-20.

17. A.E. Musson and Eric Robinson, *Science and technology in the Industrial Revolution*, Manchester, 1969; A.E. Musson (ed.), *Science, technology and economic growth in the eighteenth century*, London, 1972; Arnold Pacey, *The maze of ingenuity: ideas and idealism in the development of technology*, London, 1974.

18. C.C. Gillispie, *Lazare Carnot savant*, Princeton, N.J., 1971; C.S. Gillmor, *Coulomb and the evolution of physics and engineering in eighteenth century France*, Princeton, N.J. 1971; Jacques Heyman, *Coulomb's memoir on statics: an essay in the history of civil engineering*, Cambridge, 1972. For the extraordinary vigour of France in 'useful chemistry' as well as applied mathematics, see Henry Guerlac, 'Some French antecedents of the Chemical Revolution', *Chymia*, 5 (1959), 73-112.

19. See a fuller discussion in A. Rupert Hall, 'What did the Industrial Revolution in Britain owe to science?' in Neil McKendrick (ed.), *Historical perspectives: studies in English thought and society in honour of J.H. Plumb*, London, 1974. It is curious that while such historians as Professor Musson and Dr. Pacey (see note 17 above) have criticized my ideas as expressed in 'Engineering and the Scientific Revolution', *Technology and Culture*, 2 (1961), 333-40, on the ground that I underestimated the role played by science in the development of modern technology, Dr. Layton (see note 8 above) criticizes the very same article for its over-emphasis of the scientific character of modern technology. To satisfy both sets of critics, it seems that I would now have to argue that technology developed by applying science in the eighteenth century, but did not progress in the same way in the nineteenth, which seems absurd.

20. Layton, 'Technology as knowledge', p. 34.

21. Alexandre Koyré, 'Du monde de l'à-peu-près a l'univers de précision', *Critique*, Paris, 4 (1948), 806-23. Compare Layton, p. 36.

22. Layton, 'Technology as knowledge', pp. 36-40.

23. Layton, 'Technology as knowledge', p. 40.

24. Galileo, *The two chief world systems*, trans. Drake, pp. 207-8.

V

Isaac Newton's Steamer

When, in 1957, Professor Thomas S. Kuhn kindly invited me to give a public lecture at the University of California, Berkeley, I took as my subject 'The Genesis of Newton's *Principia*', one less hackneyed at that time than it might appear now. Afterwards, in discussion, a member of the audience asked me if I would comment upon Sir Isaac Newton's steam carriage. Now I had never heard Newton's name linked with road vehicles of any kind, much less as a forerunner of Cugnot. I do not recall my reply. No doubt in the heat of youth I may have indicated that Newton was a thinker, not a tinker, and that if he watched a fire it was not for the purpose of boiling kettles. One might suppose, too, that the topography of Lincolnshire around Woolsthorpe is hardly suitable for negotiation by primitive mechanical vehicles: I doubt if even a Stanley steamer would have rushed lightly up the hill from Grantham where Newton once led his horse (or rather bridle, the horse having slipped away). And if the Lucasian Professor of Mathematics had been accustomed to scorch along King's Parade or the Backs in a cloud of oily vapour, we should doubtless have a historical record of this amazing fact.

Nevertheless, many years later I acquired a neat plastic model of the 'Carro di Newton' (Fig. 1) as reconstructed by the Museum of Science and Technology in Milan. I have not seen the reconstruction—still less do I know the authority for it—but the description of it provided with the model runs as follows (Fig. 2):

> Iron chassis bearing the fire of the boiler. 4 iron wheels on rigid axles. No turning gear. Copper boiler. Directional gear of the steam jet and control of the boiler-valve handled by the driver on the traditional seat.

The information is added that this vehicle was 'planned by Newton in 1860 [?1680] it represents the first attempt of a jet vehicle'.

It is almost needless to add that very little of Newton's writing on mechanics precedes the preparation of the *Principia* (August 1684 onwards). The best evidence of his interest in machines is found in his earliest notebooks and in anecdotes of his boyhood. Some of his notes are copied from such sources as John Bate's *Mysteries of Nature and Art*.[1] He would not have found self-propelled aeolipiles there. In far-distant China, however, according to Joseph Needham, a Jesuit missionary had made a model steam runabout as early as 1671. Needham found in an early eighteenth-century geographical writer the story that Father Philippe-Marie Grimaldi, among the curiosities prepared for the entertainment of the Emperor of China at Pekin, had constructed a four-wheel toy, 2 ft

Reprinted from *History of Technology*, 10 (1985), pp. 17-29, by permission of the publisher. © 1985 Mansell Publishing Ltd, London.

Figure 1.   Newton's cart (about 10 cm long in this model) as depicted by the
Museum of Science and Technology in Milan.

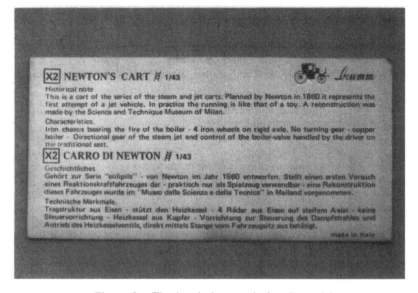

Figure 2.   The description attached to the model.

long, on which was mounted an aeolipile heated by charcoal. The steam-jet played on a vaned wheel, which drove a train leading down to the axle. The front axle of the toy was steerable, so that it ran in circles on a floor or table. A model of this device too—whose obvious antecedent is provided by Branca, not to say Guido da Vigevano—exists in the Museo Nazionale della Scienza e della Technica, Milan, and is illustrated by Needham. Of course, its details are just as imaginary as those of the 'carro di Newton'.[2] Evidently Grimaldi's toy was unlike Newton's in that it did not employ the force of reaction first exploited by Hero—it is the impact of steam on a 'windmill' that provides the moving force.

A closer reference, and one that may well provide the foundation for the story of Sir Isaac Newton's steam-carriage, was pointed out to me by an Italian historian of science and technology, my friend Dr Giorgio Dragoni of the University of Bologna. It is found in the *Physices Elementa Mathematica experimentis confirmata* of W.J. 'sGravesande (Leyden, 1720-1), Book III, Chapter X. The contemporaneous English version of this much-printed work, by J.T. Desaguliers, is almost unintelligible at this point, though his illustration is clear enough, so I will give my own translation from the Latin. In Experiment 4, 'sGravesande has described the aeolipile made of brass, how it is to be partially filled by heating it (dry) so that the enclosed air expands, then plunging it into water. When the water within is caused to boil violently, a strong blast of steam will issue from the hole (1/20th inch diameter). The following experiment shows the effect of elastic vapour (steam) more plainly.

The brass sphere of the aeolipile, 4 inches in diameter, is made thicker than in the last experiment. It is furnished with little wheels, as in Fig. 3. On the upper part of the brass sphere, a square brass steam-chest is soldered, open on one side. Parallel to this open side of the chest, dividing into two parts, is a transverse division (also of brass) pierced centrally by a one-eighth inch hole. The closed or inner half of the chest communicates with the sphere (so that steam in the sphere can rush out of the hole in the division of the chest, and emerge through the open side as a jet). To close the hole, a wedge-shaped piece of copper is used, which passes through slots in the sides of the rear or open part of the chest so as to butt up against the dividing wall.

To operate, the aeolipile is lifted off its trolley and partially filled in the usual way. The copper wedge is then inserted, and tapped tight with a hammer, to prevent the emergence of steam. After the sphere has been heated on the fire and a good head of steam built up, it is placed back on the trolley in proper position and the wedge removed. The steam-jet bursts out and the trolley rolls forward by reaction.

The steam, violently compressed, seeks with equal force to expand in all directions; these opposite pressures [against the wall of the sphere] mutually cancel each other out; but where the hole is open the steam that emerges is not compressed. Therefore, in one direction a certain amount of pressure is taken off and the opposite pressure prevails: the

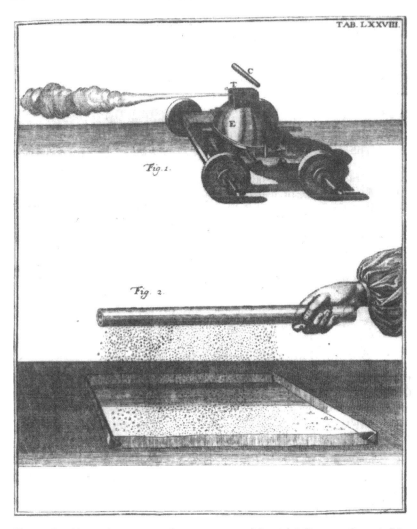

Figure 3. Newton's steam-carriage as presented by (a) 'sGravesande and (b) J. T. Desaguliers (opposite).

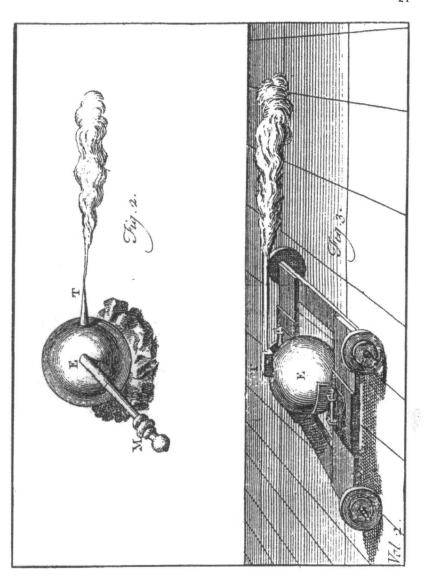

V

globe moves. The [flight of] rockets may be explained in the same way.

'sGravesande does not specifically invoke Newton's Third Law; indeed, the general topic here is *heat* and the dilatation (expansion) of bodies caused by it, but the meaning is clear enough. And here we certainly have locomotion produced by a jet of steam.[3]

Assuming that the original inspiration came from Newton, does any other early experimental Newtonian reflect the idea of propulsion by reaction? Desaguliers himself, in his early *Physico-Mechanical Lectures* (1717),[4] did not. Curiously, Desaguliers failed to insert the Third Law in its correct place after Laws I and II, and was forced to express and exemplify it in his *Addenda* (p. 79); his examples (e.g. the recoil of cannon) do not include the aeolipile. However, in a former page of miscellaneous experiments (p. 76) there appears the following:

> 6. The Force of the Steam of the heated Water, and the Centrifugal Force of a Body turning upon its Axis, together with the Reason of the rising of a Rocket into the Air; are shewn by the Experiments on the Aeolipile.

The breathless prose with superfluous definite articles is far from limpid, but it may be that Desaguliers showed Hero's aeolipile, which demonstrates (like the rocket) the force of reaction, but not locomotion. Desaguliers' major work (1734, 1744) adds nothing to this. Neither Keill nor Whiston have anything to the point, though the former devoted the whole of Lecture XII in his *Introduction to Natural Philosophy* to the Third Law.[5] In fact, eighteenth-century allusions to propulsion by reaction and the example of the sky-rocket seem curiously hard to find; the Third Law was generally illustrated by simple mechanical examples, as by Newton himself in the *Principia*, though it was of course known that the recoil of a gun is an example of this Law.

The reader of all editions of Newton's *Mathematical Principles of Natural Philosophy* subsequent to the first (1687) would suppose that Newton himself had done no more, for his enunciation of the Third Law is followed by very elementary instances of its application. Rather strangely, however, Newton reverted to the question of reaction in Book II, Proposition 37 (of the first edition only) which deals with the velocity of water-jets:

> And further if a [tall] vessel [filled with a fluid] be suspended by a very long thread from a nail, like a pendulum, when water flows out from it in any horizontal direction the vessel will always recoil from the perpendicular in the opposite direction. And the cause of the motion of those darts [rockets] that are filled with damp gunpowder is the same: gradually expelling the material through a hole in the form of a flame, they recoil in the opposite direction from the flame and are violently propelled away from it.[6]

This passage was omitted—fairly enough—from all late editions and translations of the *Principia* because, in them, the equivalent proposition is completely recast. One may regret that Newton did not transfer his remarkable statement about reaction propulsion to a more suitable position in his revised text, leaving his correct apprehension and application of the principle clear to posterity.

And whereas there are some earlier descriptions of European rockets—in Giovanni da Fontana, fifteenth century, for example—as a practical device there is nothing significant to report before Congreve's experiments at the end of the eighteenth century.

On the other hand, 'reaction' was fundamental to the operation of a family of water-wheels always referred to quite simply as 'reaction wheels'. It is conceivable that the working principle of this machine was derived from the Archimedean screw by the simple procedure of reversing the latter's mode of action, something which theoretically, if not historically, can be argued for a number of hydraulic machines, undershot water-wheel and scoop-wheel for instance, or Francis turbine and centrifugal pump.

The reaction water-wheel makes its earliest appearance in Giovanni Branca's *Le Machine* of 1628 and his most complex version is shown in Fig. 4. This bizarre contraption is notable for the fact that the directions of rotation of the 'cylinders' are not of the same mind as the gearing. But the reaction principle is clear enough. How much the reaction wheel was a part of the 'jet-propulsion' scene in Isaac Newton's time depends on how well the device was known, even on paper, and this is very far from certain; and on how closely the performance of water-jets was associated with that of steam jets. The affinity need not have been obvious.

Later on the reaction wheel takes its place fully in the water-power repertoire. In *A Course of Experimental Philosophy* of 1734, J. T. Desaguliers describes the curious twin-jet device, akin to a lawn sprinkler, shown in Fig. 5, which he attributes to 'the learned and ingenious Dr Barker'. Barker's Mill, as it is known, is the first of a sequence of eighteenth-century reaction wheels whose development in the hands of such as Leonard Euler was destined to contribute, in due course, to the emergence of the water turbine.[7]

The somewhat remote documentary antecedent to 'Newton's steamer', definitely relating Isaac Newton to the principle of reaction propulsion, is provided by the following, to which my attention was kindly drawn by Dr D. T. Whiteside; it is a sheet containing a double holograph draft:[8]

Sr

Sixty thousand pounds per an[num] must be paid six years and a half

[*Then at a different time*]

If the Pump proposed by Dr Papin can spout out 400 pounds of water every other second wth ye swiftness of 128 Paris feet in a second, it will spout it up 102 yards high or cast it to ye distance of 204 yards upon a level & do this 30 times in a minute. Whether this can be done can be known only by experience: & if it can be done

V

Figure 4. One, and the most complicated, of a number of ideas for 'reaction'
wheels among the designs of Giovanni Branca.

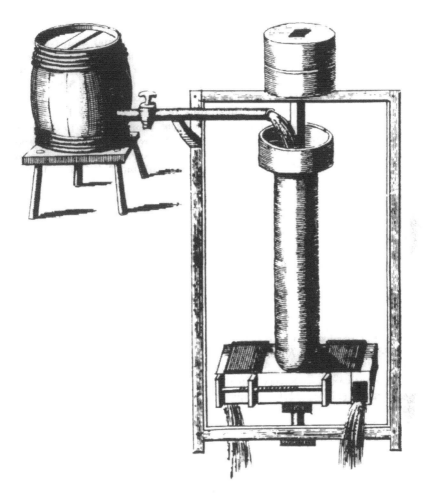

Figure 5. Barker's Mill (from J. T. Desaguliers' *A Course of Experimental Philosophy*).

I do not see but that such a pump may be applied to several good uses as to the draining of water out of trenches Morasses,[9] mines &c in difficult cases, making artificial fountains in Gardens, & to the towing & moving of ships & Galleys[10] by the recoil of the engine & force of the stream duly applied. But ye forces & uses of ye Engin must be learnd gradually by trying the simplest & cheapest experiments first & reasoning from these experimts.

When did Newton sketch these lines? Papin died about 1712 (in England); Savery's *Miner's Friend* had been published in 1698. Thus it would be a fair guess that Papin's pump and Newton's comment on it must lie between these dates. Moreover, Papin returned to England from Kassel in 1707, the year also of his publication of *Ars nova ad aquam ignis adminiculo efficacissime elevandam* ('A New Art for raising water most efficiently by the aid of fire') which, it seems, described an 'improvement' of Savery's engine.[11] Thus the date of Newton's comment is likely to lie between 1707 and 1712. A reference to the biography of Denis Papin by Louis de la Saussaye, *La Vie et les Ouvrages de Denis Papin*, takes us closer to the heart of the matter. For this confused and unreliable work indicates that Papin made a proposal to the Royal Society in London to construct a boat of 80 tons 'to be rowed by oars, moved with heat', on 11 February 1708.[12] By 'oars', here Papin indubitably meant a continuously revolving paddle-wheel—the idea going back to the anonymous *De rebus bellicis*—and according to Weld this scheme went back to a contribution from Papin appearing in the *Acta Eruditorum* for 1690.[13] As is well known, Papin had advanced Christiaan Huygens's invention of a gunpowder-powered vacuum engine by substituting water for gunpowder within the cylinder, rarefied into steam by heat applied outside. The oscillatory movement of a piston thus obtained was to move paddle-wheels—or any rotating shaft—by means of ratchets.

Now this, obviously, was a pre-Savery device and had nothing to do with the pumping of water or Papin's post-Savery engine for that purpose. And the 'oars' seem to have no connection with Newton's note. However, Weld's *History of the Royal Society* also records, from the *Register-Book* (Vol. IX, p. 108), a request from Papin that the Royal Society sponsor a comparative test of Savery's steam-pump and his own improved version; ultimately (21 April 1708), Papin was to claim ability to force a jet to the height of 300 ft.[14] In the fuller communication (of March 1708), Papin offered

> with all dutiful respect, to make here an Engine, after the same manner that has been practised at Cassell [in Germany], and to fit it so that it may be applied for the moving of ships.

Here, it is clear, Papin means to apply his post-Savery steam-pump to propulsion. How? There is no hint in his paper of the reaction principle. At Kassel, Papin had used his engine to provide a stream of water to turn a corn-mill, much after the manner of later times when a Savery engine was used to pump water behind the wheel of a blast-engine for an iron-works. It is therefore conceivable that he meant to use fire to turn a water-wheel which would be coupled to paddle-wheels for propulsion, a cumbersome but not impossible scheme, hardly stranger than the plan of Beau de Rochas mentioned elsewhere in this volume to force a vessel against a current by pulling on its own bootstraps, as it were.

Newton, as celebrated President of the Society, was asked to state his opinion of Papin's invention; both Weld and (from him) De la Saussaye

V

quote the passage, from the Royal Society minutes, which Dr Whiteside found in a draft in Cambridge. Newton, it seems to me, misinterpreted Papin's proposal and himself introduced the idea that the boat was to be propelled by reaction. I take it that he supposed the reactive effect would increase in proportion to the mass ejected (200 lb of water per second) and its velocity of ejection (128 Paris ft/sec). Hence his computation (taking one Paris foot as equal to 1·093 English feet,[15] and g as 32 English ft/sec²) that a jet of velocity 128 Paris feet per second reaches a vertical height of 306 English feet, and so (at 45° elevation) has a maximum 'range' of 612 English feet.

Thus it may be that Newton invented reaction propulsion almost by inadvertence. Certainly it would seem to have nothing to do with naturalists' observations that squids propel themselves by throwing out a jet; in fact, a quick search suggests that this method of motion in animals was known to neither Aristotle nor William Harvey.[16] In my opinion it is likely that Newton—who probably never set eyes on the sea—was wholly ignorant of the zoological fact.

It may be inferred that Newton did not necessarily accept as experimentally valid Papin's performance data, which he quotes. In particular, the historian (if not Newton) may well doubt Papin's claim to raise 200 lb of water through 300 ft each second; that is to lift six tons per minute, requiring some 100 horsepower! However, if we accept Papin's data hypothetically and apply the usual formula, the thrust obtained would be about 850 pounds, presumably enough to move the vessel carrying the engine. Adopting a more likely—still generous—capacity for the engine of 5 horsepower correspondingly reduces the thrust available by a factor of twenty to the less impressive figure of $42\frac{1}{2}$ lb.

Thus a little play with figures suggests—what one would guess *a priori*— that if one's only heat engine is a Savery pump, the jet-propulsion principle is not likely to serve as an efficient means of applying the engine to locomotion, at sea or on land. Nevertheless, this very propulsion method is said to have been employed by the American James Rumsey (1743–92) in a steam-boat demonstrated successfully on the Potomac (1787) and the Thames (1793);[17] I infer, however, that Rumsey made use of a Newcomen, rather than a Savery, pump.

No evidence has turned up so far to indicate that Newton, having inadvertently perhaps set his foot on the road to the Moon, ever thought of devising a road vehicle, or of employing steam as a way of demonstrating the Third Law by reaction propulsion. It is not inconceivable that he passed such an idea to 'sGravesande, who had spent most of the year 1715 in England. That is speculation, and until further evidence turns up we have to regard 'sGravesande as the first to apply steam power to land transport. Perhaps someone else can say whether he really did so successfully; whether, that is, the aeolipile on wheels can propel itself successfully in a demonstration.

Notes

1. E. N. da C. Andrade, in *Nature*, Vol. 135, 1935, p. 360; R. S. Westfall, *Never at Rest, A Biography of Isaac Newton*, Cambridge, 1980, p. 61. Bate's book received a third issue in 1654, but Newton is not known to have owned any copy.

2. Joseph Needham, *Science and Civilization in China*, Vol. 4(2), Cambridge, 1965, pp. 225–8, Fig. 472. The source is J. B. du Halde, *Description Geographique ... de l'Empire de la Chine*, Paris, 1739, Vol. III, p. 270. *Idem.*, *Clerks and Craftsmen in China and the West*, Cambridge, 1970, pp. 141–4. Needham does not mention Isaac Newton in this context.

3. In the arrangement of *Mathematical Elements of Natural Philosophy ... Translated into English by J. T. Desaguliers*, 4th edn, London, 1731, this section appears as Book III, Chapter IV, Expt. 4 (Vol. II, pp. 21–2).

4. J. T. Desaguliers, *Physico-Mechanical Lectures or, an Account of What is Explain'd and Demonstrated in the Course of Mechanical and Experimental Philosophy*, London, 1717.

5. John Keill, *Introduction to Natural Philosophy*, London, 1720 (the first edition in Latin was 1702).

6. Isaac Newton, *Philosophiae naturalis principia mathematica*, London, 1687, p. 332: 'Et propterea si vas, ad modum corporis penduli, filo praelongo a clavo suspendatur, hoc, si aqua in plagam quamvis secundum lineam horizontalem effluit, recedet semper a perpendiculo in plagam contrariam. Et par est ratio motus pilarum, quae Pulvere tormentario madefacto implentur, &, materia in flammam per foramen paulatim expirante, recedunt a regione flamma & in partem contrariam cum impetu feruntur.'

7. For a fuller discussion of reaction wheels and their place in water power's history, see Norman A. F. Smith, 'The Origins of the Water Turbine and the Invention of its Name' in *History of Technology*, Vol. 2, 1977, pp. 237–43.

8. I first learned of this interesting draft, Cambridge University Library MS Add. 3964 (B)(8), fol. 12, from Dr D. T. Whiteside. I am much obliged to Peter Gautrey of that Library for providing a photocopy of it. The draft was noted in A. Rupert Hall and Laura Tilling, *The Correspondence of Isaac Newton*, Vol. VII, Cambridge, 1977, p. 489.

9. The order of Newton's interlineations is far from clear.

10. Newton first wrote, then deleted: '& giving motion to gallies without the help of oars', which confirms that the propulsion-pump was powered neither by muscles nor by water flow.

11. Patricia P. MacLachlan, article 'Papin' in *Dictionary of Scientific Biography*, Vol. X, Charles Scribner's Sons, New York, p. 293 (a). I have not seen the Latin *Ars nova*, but a facsimile (Herman, Paris, 1914) of *Nouvelle Maniere pour lever l'Eau par la Force du Feu*, Paris, 1707. Papin's improvement to the original Savery engine consisted of interposing a hollow piston between the steam and the water in the working cylinder; in fact this piston floated on the water. The piston was meant to act as a heat-insulator between the admitted steam and the water which, as in Savery's engine, was forced up into the cylinder by atmospheric pressure when the steam was condensed. The object was to reduce condensation of the steam admitted under pressure to expel the water up the delivery pipe. The ovoid boiler of Papin's engine is said to have axes of 20 × 26 inches, while the pump-chamber or working cylinder has a bore of 20 inches also and a piston stroke of 16 inches. Papin reckons that a full stroke can be made every 2 seconds, and that each will eject 200 *livres* of water. The height of the lift he reckoned at 40 feet. This

performance is roughly equivalent to seven horsepower. It is much less than that indicated by the figures recorded by Newton, yet even so it is quite clear that the data are calculated by Papin, not based on experience.

12. L. de la Saussaye et A. Péan, *La Vie at les Ouvrages de Denis Papin*, Vol. I, Paris, and Blois, 1869 (all published), p. 240. This writer was greatly confused by the fact, of which he was unaware, that the English in adhering to the old calendar continued the year 1707 to 25 March.

13. C. R. Weld, *History of the Royal Society*, Vol. I, London, 1848, p. 382; De la Saussaye, as note 12, p. 263.

14. De la Saussaye, as note 12, p. 243.

15. This ratio slightly exaggerates the true difference between the two measures.

16. The squid's jet-propulsion is certainly included neither in Aristotle's *De incessu animalium* nor in Harvey's *De motu locali animalium* [1627], ed. and trans. by Gweneth Whitteridge, Cambridge, 1958. Aristotle was very interested in the cephalopods, not least the ejection of 'ink' by some species for purposes of concealment.

17. Charles Singer *et al.*, *A History of Technology*, Vol. V, Oxford, 1958, pp. 142, 152.

VI

HOMO FABRICATOR: A NEW SPECIES *

It is perhaps the most extraordinary coincidence in the whole of modern times that the revolution in Europe's population history was contemporaneous with, or to be more exact slightly preceded in time, the revolution in Europe's industrial production. The demographic revolution and the industrial revolution were both clearly happening before the end of the eighteenth century. Where would the hands for industry as well as agriculture have come from, had Europe's population not steadily increased? And how would the growth of industrial cities have been possible, if to typhus and cholera had been added smallpox and plague? It is indeed, if ever one can find such a thing, an example of the mysteries of Providence that Europe's population began to increase at just that moment when industry was capable of giving sustenance to large numbers, and that death-rates began to decline just as people began to shift from the countryside to far more unhealthy cities.

There is an enigma here which has never been fully examined. We do not know all that clearly why the first industrial revolution occurred in Europe; still less can we explain its transition from a population history of violent fluctuation with very slow secular growth, to one of rapid and continued expansion of the population. Still less, therefore, are the interrelations between these changes

* An earlier version of this paper was presented at the 2nd Course « Work and Health » of the *International School of the History of Biological Sciences*, Ischia (Naples) June 28th-July 12th, 1980.

Reprinted from *History and Philosophy of the Life Sciences*, 2 (1980), pp. 193-214, by permission of the publisher. © 1980 Leo S. Olschki, Florence.

clear. All one can say with much confidence is that they seem to have been, initially, independent one from the other. The eighteenth century increase of population cannot be correlated with geography, intense economic activity, or relative progress; it occurred in Sicily and Sweden as well as in Britain, in Brabant and most markedly in remote, agrarian Ireland. The numbers of people remained most nearly stable in France, the centre and cynosure of European civilization where, most observers would have judged in 1750, technology was more sophisticated and industrial arts more refined than in any other part of the continent. It is of course obvious that the « demographic revolution » in Ireland and in continental Europe was in no sense related to steam-engines and spinning-machines, innovations which before 1800 had made little impression outside Britain.

Conversely, one can show that industrial prosperity had little to do with human fertility and survival, at least before 1800, and very little, outside Britain, before (say) 1830. In time the opportunity provided by the textile industries for even little children to earn their own bread no doubt assisted their survival, just as the cheapness of child labour assisted the survival of industrial innovations. But all this relates to the development of a change that had begun for other reasons, long before; the population that grew in Europe was an agricultural population, and the reasons suggested for its growth, in Ireland for example, relate exclusively to the conditions of land-tenure, food production and so forth, such as affect both the nutrition of a people and its life-expectation.

In this article I shall concentrate on Britain because that is the country in Europe I know best, because it is the country in which industrialization began, and because it is that in which urbanization proceded most rapidly and completely. Even by 1850 more than half the population of England was living in towns and cities, less than half farmed the land. You will be familiar with the relatively swift transition – in a couple of generations at most – that has occurred in all developed countries from the historic, agrarian, food-producing economy where perhaps one-twentieth of the population is engaged in the production of inedible commodities, to the industrial economy where one-twentieth of the population or less is engaged in the production of food. This transition, occurring rather slowly in England in the first half of the nineteenth century, next in Germany beginning about 1830, then later in France and Italy, accelerating in the USA from about 1865, and finally taking place rapidly and forcibly in the Soviet Union after 1917, has of course affected different peoples in different ways, as has the demographic revolution also. Traditional peasant farming survived longer in Spain, Italy and France than it did in Britain, Holland, much of Germany, and the Soviet Union. Geography and social forces play their parts in effecting regional differences. But the overall pattern is the same, and it is of the part of this pattern that touches the lives of individuals that I wish to speak.

The most obvious social phenomenon, in the eighteenth century as in the twentieth, is the crowding of country people into towns. We have seen recently the enormous growth of cities in India, Africa and S. America, just as the villages of Lancashire grew into ugly textile towns in the eighteenth century. Manchester, already the capital of the northwest of England before it was invaded by the factory system, increased from about 10,000 inhabitants in the early 18th century to 95,000 two generations later; villages like Oldham or Bolton increased with the building of mills from a few hundred inhabitants to between 10 or 20,000 within a generation. Similarly in coalmining and metallurgical regions crude streets sprawled across the fields. In the early nineteenth century the number of city-dwellers in Britain was increasing by over a million in each decade; in the period 1841-1851 by almost two million. Country people pressed into the towns in search of employment, a movement reinforced by the concurrent changes in land-holding and land-management which largely brought about the destruction of the class of small peasant tenants. The movement of people brought about a redistribution of the English population in particular; historically, the south with its flatter, kinder landscape and more propitious climate had been the most thickly peopled region, but as mining and manufacturing industry developed in the northern area, with its dense conurbations taking shape in South Lancashire, the West Riding of Yorkshire, and the Northumberland coast, the balance of the population shifted in the nineteenth century towards the north, and this in the twentieth has proved an imbalance. A comparable recent phenomenon is the transfer of the rural people of the

Mezzogiorno of Italy to such urban centres of the north as Milan and Turin.

As in the mushroom cities of our own day, the living conditions of the new urban proletariat were probably wretched from the first. We have little historical evidence about, for example, the rapidly swelling industrial centres like Birmingham, Manchester, Newcastle and their satellites in the eighteenth century. We may be sure that no legal and administrative machinery existed to control housing, or to provide such services as roads, sewers and water-supply. Even before 1800 some of the housing required for the large industrial labour-force was provided by the employers; a few large firms were conscientious and public-spirited, not only providing sound housing but caring for the education, religion, and health of their employers. Some early model towns, built round a textile mill or group of mills for example, still exist and still provide reasonable living conditions. The general level of provision was worse or non-existent. Certainly in the Lancashire towns of the 1790s the evils of over-crowding and bad housing were notorious, with one or more families huddled into single rooms, often without proper furniture or bedding, the very worst in squalid cellars almost without light or ventilation and suffering from perpetual damp and cold. Towards the middle of next century the public health reformers collected a mountain of evidence concerning the common occurrence of such conditions among the urban poor. Thus Edwin Chadwick asserted that the bad areas of British cities offered a worse spectacle to the eye and nose than did eighteenth century prisons:

> Prisons were formerly distinguished for their filth and their bad ventilation; but the descriptions given by Howard of the worst prisons he visited in England (which he says were among the worst he had seen in Europe) were exceeded in every wynd [court] in Edinburgh and Glasgow inspected by Dr Arnott and myself. More filth, worse physical suffering and moral disorder than Howard describes are to be found amongst the cellar population of the working people of Liverpool, Manchester or Leeds and in large portions of the Metropolis.[1]

---

[1] E. CHADWICK, *Report on the Sanitary Condition of the Labouring Populations of Great Britain*, London, 1842.

And he adds that sometimes the sick were committed to prison, as offering them better food, lodging and hope of recovery. Friedrich Engels writes of housing in Lancashire as one might today of squatter settlements in Africa or South America. Chadwick said that the towns were « in respect of cleanliness almost as bad as an encamped horde, or indisciplined soldiery ». Roads were unpaved, there was no piped water, streams and rivers were choked with rubbish and filth.

I need not go on, because all this is well known, and can still be found – usually, in different places – in our own world. Nor should we imagine that such conditions for the poor in cities were new; they had existed for centuries, and whether we turn to Hogarth or Villon or a host of other sources it is obvious that the wretchedness of urban squalor was in itself not the creation of the industrial revolution: rather what was new in Britain was the extension of such conditions to very many cities and even small towns, and the fact that the cellar-dwellers were not street-beggars and petty criminals but immigrants to the town from the countryside. It was well known that these last were more subject to the diseases flourishing in association with evil living conditions than was the old urban proletariat. To many, like William Cobbett, it seemed that enclosures and industry were pauperising the once healthy agrarian mass of the British population.

To what extent this was true has been much debated. Some historians, like the Hammonds, have written of the degradation of the urban worker as though it represented a descent from his former more idyllic conditions as a peasant. This is a picture supported to some extent anyway by contemporary literary evidence. Economic historians, on the other hand, have drawn attention to the statistical evidence that real wages tended to rise in Britain during the early industrial revelation period, though the later stages of the war with Napoleon, and its aftermath of depression, saw this trend reversed. The historian must also remember that besides the very large numbers of the urban poor, whose dreadful conditions so impressed social reformers, there were (as always) large numbers of urban workers living – in terms of the day – in decent conditions. A skilled or semi-skilled worker, in regular employment, and with perhaps two or three children also bringing wages from the mill, could enjoy some prosperity if prudent and sober. Josiah Wedgwood, an enlightened employer, spoke of

the workmen earning near double their former wages, their houses mostly new and comfortable, and the lands, roads, and every other circumstance bearing evident marks of the most pleasing and rapid improvements.[2]

We know from the Manchester novelist Elizabeth Gaskell that a millhand could live in rough comfort even in a poor cottage on a nasty court; coal was cheap in the North so that good fires could be kept up; the living room could contain adequate if simple furniture, a few ornaments, check curtains, oilcloth on the floor and a bare sufficiency of utensils for eating.

On the other hand, the historian must not exaggerate the merits of the rural life which the town worker had left; he had left the country because conditions seemed desperate for him there, just as happens today. The late 18th and early 19th century was a particularly bad time for the English agricultural worker, the only time of violence in our recent history. A thatched cottage is not necessarily more healthy than a terraced house, nor were the conditions of farm work in all weathers necessarily better than those in a mill or factory. But they were certainly very different, and this I wish to explore a little.

One of the less obvious changes covered by the often-criticised expression « industrial revolution » is that in work discipline, which to some extent reflects the change of the work-task itself. Agriculture imposes a variety of tasks, some constant, like care and feeding of animals, milking and so on, some varying with the season of the year. There is rarely an unvarying task continued indefinitely hour after hour, day after day. The agricultural worker might not possess all the different skills of horse-keeper, cowman, shepherd, fruit grower and so on but he certainly possessed a range of skills which he employed at various times in the seasonal round, from ploughing to hedging to scything grass and so on. Even in regions of mono-culture, say of the vine, a range of different tasks constitute the yearly round. Similarly a rural craftsman like the smith exercised different

---

[2] N. McKendrick, « Josiah Wedgwood and Factory Discipline », *Hist. Jour.*, 4, 1961.

aspects of his skill in dealing with different tasks, from shoeing horses to making a pair of gates. As we all know, one tendency in manufacture organization, much older than the industrial revolution, was towards repetitive specialisation. This had been insisted on by Adam Smith, for example. It had long been the case that no one man ever made all the parts of a clock, or a gun. There were specialist metal-workers who made wheels, plates, arbors, hands, springs and so forth, each a different sub-trade, other specialists who made dials and cases. The «maker» was really an assembler. One aspect of the industrial revolution was simply that it collected different, connected sub-trades under one roof and one supervisory eye. Thus, in the nature of manufacture and even before the gradual spread of factory production, the urban worker was condemned to a more unvaried repetitive task in many cases than the agricultural labourer. His skilfulness in performing this one task increased, so that the work was better done than before, but the range of skills was reduced.

It was the deliberate policy of the progressive manufacturers of the eighteenth century both to organise the flow of production to avoid the useless carrying of materials and goods to and fro, delays, and obstructions, and to break production down into tasks, each of them carried out by a specialist. So the potter Josiah Wedgwood at Etruria, Staffordshire, in the 1760s followed «the scheme of keeping each workshop separate, which I have much set my heart on».

His designs aimed at a conveyor belt progress through the works: the Kiln room succeeded the painting room, the account room the kiln room, and the ware room the account room, so that there was a smooth progression from the ware being painted, to being fired, to being entered into the books, to being stored. Yet each process remained quite separate.[3]

Wedgwood preferred division of labour and the disciplined ordering of manufacture because – as Adam Smith stressed – it increased perfection and profit at the same time. He believed that «the same hands cannot make *fine* and *coarse*, *expensive* and *cheap* articles so as

---

[3] *Ibid.*, 32.

to turn to any good account to the Master». A worker had to be trained to a particular task in a particular type of manufacture for continual employment: «there is no such thing as making now and then a few of any article to have them tolerable». If a particular skill was no longer in demand, in accordance with the exigencies of the market, then of course those who possessed it were dismissed.

It was a consequence of this that each worker had to submit to a rigorous process of training. Wedgwood was very conscious that in forming his «new model» pottery he could not expect to recruit a labour-force in Staffordshire adequate to his standards of elegance and refinement in the pottery: the vase-makers, painters, glazers and so on had to be precisely taught their business. Starting with «hands who have never attempted anything beyond Huts and Windmills upon Dutch tile at three halfpence a dozen» Wedgwood had to produce artists who could paint «the most beautiful Landscapes with Gothic Ruins, Grecian Temples & the most Elegant Buildings» for his new and fashionable clientele, such as the Empress Catherine the Great whose great service numbered over a thousand pieces. Mathew Boulton likewise wrote of «training up plain country lads into good workmen».

If artistic excellence enforced specialisation, so did the use of production machines. The human hand is the most infinitely adaptable of agents: the prisoner of war with a knife and bits of glass and stone made the most exquisite model ships out of mutton bones. But machines are much more restricted in their capabilities. A woman can spin a thread with no more than a stick and a lump of clay: machine-made thread passes through five or more machines. So with wood-working: to manufacture an article, planers, drilling-machines, slotters, shapers, polishing machines and so on may be required in due series. If the worker becomes a machine-minder and tender, he also becomes just one piece in a very large puzzle.

In agriculture, the peasant or labourer often works alone and at his own pace. He has to satisfy his employer or his own conscience according to the conditions, state of the soil, weather and so on. In the factory or workshop he has to go at the speed of the sequence in which he is involved. This was one of the greatest changes in the new manufacture: it imposed punctuality, fixed hours, regularity of labour at a set pace, unremitting attention to the job, avoidance of waste and above all sobriety. In skilled work, whether of the delicate sort like painting a dinnerplate or the rougher sort like puddling iron, if it were not done just right all the labour would be wasted and the master's profit disappear. The clock replaced the sun, and repetitive discipline was substituted for the freer and more easy going ways of earlier days. Work in the pre-industrial age was hard, but it had been punctuated by fairs and wakes, market-days, celebrations at weddings and on Saint's Days, village football and harvest suppers. There was a kind of rhythm of labour and relaxation partly imposed by reason and climate. Relaxation was associated with drinking, sometimes a kind of tolerated violence, a loosening of the standards of ordinary behaviour – having a good time, in other words. In the industrial town all this disappeared. For one thing the social cohesion and solidity of tradition of the village no longer existed. For another the employers would not stand for drunkeness, or irregularity at work – holidays only came when trade was bad, at the employer's command. The celebrations, sports and ancient jovial customs of the English countryside died away for ever, unregretted by Methodists and moralists, and happily forgotten by the pharasaical middle class.

Historians have discussed the managerial problem at some length: as mines, mills and factories increased in size it was necessary to find managers, overseers, foremen, inspectors and so on who could organize the work in detail, and above all enforce discipline. Such a group of « executives » did not exist and had to be created. Fear was the main weapon against the men. Wedgwood wrote with wry pride that « my name has been made such a scarecrow to them, that the poor fellows are frightened out of their wits when they hear of Mr W. coming to town, & I perceive upon our first meeting they look as if they saw the Devil ». Management was unquestionable authority and stern rectitude; the labour force was by definition weak, immoral, lazy, and easily misled. Most employers, like Wedgwood, and like the managers of large households, relied on a system of written rules, breaches of which were punished by fines or dismissal. These covered not only the technical work of each kind of operative but his general conduct while at work. Down to the twentieth century employers forbade talking and singing at work. In Wedgwood's works as in others later uncleanliness, fighting, drinking, waste of materials, throwing things and above all unpunctuality were punished by fines. Strenuous efforts were made by all the new managers to get their workers in to time – hence the introduction of the factory hooter or bell, and the « knocker-up » of Lancashire. Good timekeeping was rewarded, lateness at work punished. The clock, later the automatic time-clock, ruled all.

Wedgwood's Etruria pottery works in the 1770's already exemplified most of the principles of « new management » of the workers, on which (it is evident) the constant vigilance of « the Master » himself was the essential driving-force. It was Wedgwood who broke badly-made pots, and who hobbled through his work, on crutches when his leg was broken in an accident.[4] From Boulton and Watt's new foundry of 1796, some twenty years later, further details of the enshrouding of the worker in rules and regulations emerge. In the new Soho Foundry, machines and processes were meticulously listed, as also the speeds of the machines and the order of flow of work through the shop. It codified

all the production processes and fixed a definite standard regulation on the part of the management, thus relieving the workmen of the larger part of their independence and individual responsability.

Each workman had a fixed, standard job assigned to him. To encourage productivity, Boulton and Watt devised a bonus payment scheme; in the case of boring cylinders for steam-engines, for example, a formula was used to find the normal time for the job, in days, increasing with the magnitude of the cylinder; the team of men was rewarded by a payment of five pence extra bonus for each day saved out of this « norm » by quicker working.[5]

I have mentioned these various points because they have led historians of industry to conclude that in Britain, in the late 18th century, began the « subordination of man to the machine » which continued and became more emphatic during the nineteenth century.

---

    [4] *Ibid.*, 43-45.

    [5] L. URWICK and E. F. L. BRECH, *The Making of Scientific Management*, London, 1946, II, 28-29, 32-33.

Technical considerations – that is, those aimed at efficiency of production and maximisation of profits – overrode all others. As one book put it forty years ago,

The generation of today has inherited the evil social consequences of that warped and narrow outlook – the squalid tracts of old industrial towns, the deteriorating conditions of life for large sections of the population, the inertia of intellect and of moral initiative in many of the younger people of the poorer disctricts.[6]

The feeling behind all this is first that the new system of manufacture regimented the worker, and degraded him to the condition of being a slave to the machine – the old Fritz Lang film *Metropolis* is a fantasy of the extreme exaggeration of this industrial world. The point to which it leads is the utter dissociation of the mentality of the worker from his work; in the words of a distinguished English novelist

The nature of the work is irrelevant [to the worker] [...] All that matters is that it shall produce the money necessary for one's needs, and that the conditions surrounding it shall be as tolerable as possible. It follows that in the circumstances there can be no serious question of *enjoying* work.[7]

In the last resort work can be almost a matter of conditioned reflexes, with the conscious mind devoted to wholly different subjects – the muzak, gossip with one's mates, Walter Mitty dreams or private worries. The integration of personality consistent with the old system, in which a man's work might also be his life as it still is today with such people as artists, academics, and writers, and in which labour and play merged together without the sense of « the bosses' time » and « my time » vanished altogether.

Loss of personality, responsability and ambition, reduction to an most convict condition of mindless toil – even in pleasant surroundaings – is of course also associated with the growth in numbers of employers in individual firms. The most celebrated early modern

---

[6] *Ibid.*, 170.

[7] Nigel Balchin in D. Cleghorn Thomas (ed.), *Management, Labour and Community*, London, 1957, 137-138.

study of industrial psychology, the Hawthorne experiment, was tried on the 30,000 employees of the Western Electricity Company of Chicago. Perhaps such large numbers did not work for a single organisation, and certainly not on a single site, in the 19th century. Already, however, before 1800 in Britain a number of units employed over 1,000, among them four Cornish tin mines, and a number of large quarries; the Carron Iron Co. in Scotland employed over 2,000 by 1790 in its very large complex of furnaces and workshops.[8] In South Wales some of the ironworks were nearly as big. In the textile industry Robert Owen's Lanark Mills was the first really large enterprise but again by about 1810 a dozen or so factories employed a thousand or more people – many of them women and children. Moreover, the replacement rate was extremely high, and the labour mobile. The old, secure relationship of farm and household could not exist under these conditions.

The proportion of children was very large, partly because at this time of population growth half the British people were less than 18 years old. Children were cheaper labour than adults, and more adaptable. Labour discipline was brutally enforced by some employers; others paid attention to the children's need for shelter, food, religion and education. A typical recollection of the worst conditions is this of 1857:

> In stench, in heated rooms, amid the constant whirling of a thousand wheels, little fingers and little feet were kept in ceaseless action, forced into unnatural activity by blows from the heavy hands and feet of the merciless overlooker, and the infliction of bodily pain by instruments of punishment invented by the sharpened ingenuity of insatiable selfishness.[9]

In the famous novel *North and South* by Mrs. Gaskell, by no means a sentimental detractor of Lancashire industry, Bessie acquired the pneumonic affliction that destroys her in the carding-room where the air was full of cotton-dust – no improbable event. Many children

---

[8] SIDNEY POLLARD, *The Genesis of Modern Management*, London, 1965, 71, 76, 91 ff.

[9] SAMUEL KYDD, *The History of the Factory Movement* (1857) quoted in N. McKENDRICK, « Home Demand and Economic Growth », in *Historical Perspectives: Studies in English Thought and Society in honour of J. H. Plumb*, London, 1974, 159.

became diseased or deformed by early exposure to unhealthy conditions of work. In many factories a third of the labour force was less than 18 years old; in textile mills usually a half to three-quarters. In cotton mills they started to work at the age of 9 or 10, in silk mills and coal-mines as early as five years old. Many children, and adults too, were recruited into industry forcibly from poor-houses. Paupers, even children, were not necessarily cheaper to employ, but they were available; in the north west many Irish were employed for the same reason, and they came over to Liverpool in droves.

Of course child labour was no novelty. Small children had always helped with the farm work and cottage industry, and were not necessarily better treated there than in factories. Moralists had long insisted on the virtue of finding a useful occupation for tiny children, to save them from « the sins and pitfalls of sloth and idleness ». Long before the Industrial Revolution the writer Defoe (author of Robinson Crusoe) had praised the cottage textile industry of Yorkshire because the youngest child could earn its own bread. But there is a kind of horror about the thought of these infants amid the sprinning-mules and power-looms which apologetics cannot eradicate, even though it is true that (by the early nineteenth century) parents eagerly drove their children into the mills, to increase family earnings. Many young people, girls included, preferred work in the mills to domestic service - still, and for generations to come, the principal occupation open to poor women. To refer again to Mrs Gaskell's *North and South*, the clergyman's daughter removed to Manchester cannot find a servant-girl to help with the housework because of « the difficulty of meeting with anyone in a manufacturing town who did not prefer the better wages and greater independence of working in a mill ». The mill-girls had a society of their own, much freedom, and money to spend as they chose.

They constituted one group who certainly had a greater spendable income as a result of the industrial revolution. Before the textile mills, pot-banks and so on offered jobs to working-class girls, they received little more than rough board and lodging and cast-off garments for their labours. Mill-girls obviously had to contribute to family expenses, but they still had some cash left for clothes and baubles to wear on Sundays and holidays - even gloves like ladies.

We know this because in the early nineteenth century moralists began to declaim against them for their love of finery and dressing like their betters. « England surpasses all the other nations of Europe in the luxury of dress and apparel, and the luxury is increasing daily », wrote a traveller in 1790; even the poor wear silk stockings, adds another, so that the lord and the labourer are almost indistinguishable in dress.[10] Of course, complaints against the luxuriousness of social inferiors have always been common, and are not found in Britain only after the industrial revolution, but they fit in with other evidence of increasing prosperity among craftsmen, skilled and semi-skilled workers. When a farm labourer earned perhaps five shillings a week, and the maintenance of a pauper child in a factory cost four shillings a week (or less), a moderately skilled man in industry might earn 15s. a week, and his whole family twice as much, or even more. If we rather optimistically put the regular earnings of such a prosperous working family at about £ 70 per year, it was very well above the subsistence level and entering upon modest comfort. It has indeed been argued that the greater prosperity of the upper working class – which is reflected among other things in more widespread literacy and the increasing sale of popular books and newspapers – provided much of the new market for British consumer goods – clothing of course, but also such Birmingham made articles as buttons and buckles, japanned ware in the home, Staffordshire pottery (replacing pewter and wood) and furnishings. However, one should not forget that the export trade, which one may presume profited owners more than workers, largely contributed to Britain's rising industrial prosperity at this time: Wedgwood, for example, exported some 80 per cent of his manufacture, while in the cotton industry exports increased from virtually nothing in 1760 to a value of over £ 18 million in 1833.

So far, I have tried to analyse the changes in human life and labour that accompanied the transition to urban manufacture as the economic basis for Britain in the Industrial Revolution period. It was a transition that was to occur in varying degrees and at different times everywhere. We have seen that though in any romantic view country life must

---

[10] MCKENDRICK, *ibid.*, 193, quoting K. P. MORITZ, *Journeys of a German in England in 1782* (1965) and J. W. VON ARCHENHOLZ, *A Picture of England* (1971).

be idyllic compared with life in a grimy industrial city, the social reality as regards the late 18th and early 19th century is less than easy to establish. The fact that people then as now moved to the towns in search of work and wages indicated that both were lacking in agriculture. It has been pointed out that those who were led, or misled, into trying their fortunes in industry could not easily retreat to the countryside; the unsuccessful urban proletariate was as it were trapped in the urban slum. But equally we have evidence that agricultural and domestic wages were pushed up in the neighbourhood of industrial employment, indicating a beneficial effect on living standards (as measured by money wages). The plain fact is that the romantic view of the ploughman and the lowing herd winding slowly o'er the lea has never yet, in any country, held the agricultural labourer fast to the land, though the more successful have sometimes returned to the land.

I wish now to turn to the specific question of health, which will prove I fear just as doubtful as the social issues just discussed. The basic fact is that (as I wrote before) the British population continued to grow rapidly as more and more people worked in industry: therefore, by this crude test, the population was « healthy », healthier than it had been when more strongly rural in the seventeenth century. Yet a great deal of the increase, during the early 19th century as before, took place in the countryside. The towns were death-traps, destroyers of people, as they had always been. It has been argued that the hard labouring work was done in English cities by burly Irish peasants – the « navvies » of later days – because the English city-poor were too puny, weak and unhealthy to undertake such physical exertion. This explanation is possibly too simple So, in my view, have been most accounts of the 18th century population explosion and its relation to medicine. Here is a statement encountered almost at random, referring to Britain of course:

Until the agricultural improvements of the 18th century and Jenner's discovery of the smallpox vaccine began to affect the even balance between life and death, it was to be expected that the majority of newborn children would not live past their fifth birthday.[11]

---

[11] I. F. CLARKE, *The Pattern of Expectation, 1644-2001*, London, 1979, 3, cited as an

The actuarial fact is true enough, the explanation offered is nonsense. It would not apply to Ireland and continental Europe, where population increase also occurred. Neither agricultural improvement nor vaccination had any effect early in the century, when the population increase began. Agricultural improvement, so-called, had a depressive effect on the English rural working population – on this all are agreed. Vaccination cannot rapidly have been applied to very large numbers of people and so-forth.

I think myself that perhaps a slight softening of the climate, some improvement in diet, the growing economic prosperity of Europe due to her domination of the Near East and most of the rest of the globe, and the more or less complete disappearance of bubonic plague are all factors that may have contributed to greater longevity, especially in children. But since we know that the thriving of population occurred in the country rather than the towns, we need not in considering the health of urban man concern ourselves much with the reasons for it.

Cities were unhealthy because the congestion of people assisted the spread of diseases transmitted from person to person, like tuberculosis; because bad water and absence of sewerage promoted the spread of water-borne diseases like typhoid and cholera; because poverty and filth and overcrowding favoured the existence of insect disease-vectors like the typhus-carrying louse. Further, town-dwellers usually were worse off for food. The more fortunate farm labourer could eat ample vegetables, and produce for his family perhaps eggs, pig-meat and milk, often bought cheaply from the farm. The town-dweller must eat food from the shops, and this was an age when, in Britain at least, food was adulterated as never before or since. Chalk was added to flour and milk, milk diluted with water, tea (now the Englishman's universal drink, in place of beer) mixed with the sweepings of the floor from the warehouses, sugar was « extended » by adding the poisonous sugar of lead, cheese and sweets coloured with red lead. Dreadful things were used as preservatives or to disguise the taste and smell of decomposing food. In fact the

example, merely. Most lay historians would probably cite « medical advances » as a main cause of population growth in the 18th century, in a wholly anachronistic manner.

beginnings of the food industry in Britain were totally shameful, a condition only put right fairly late in the nineteenth century. Food purity could be (and ultimately was) protected by legislation in all civilized countries, enforced by a proper inspectorate and prosecution in the courts of law, just as with criminality in respect of weights and measures; law and scientific analysis thus took the place of custom and the traditional practices of the market-place – with, it must be said, a further increment to public health, for once again we must be careful not to idealize the careless methods of our ancestors whereby food was ignorantly exposed to contamination. With respect to water and sanitation, what was required was « social engineering » – the provision of pipes and drains, of filtration and sewage treatment plants. These engineering problems, some of which have medieval or older antecedents as with the ancient sewers of Rome and Paris, began to be generally treated in British, Continental and American cities from about 1840. An important factor was the recognition, from the 1850s onwards, of the transmission of that most terrifying disease of the 19th century, cholera, by contaminated drinking-water, when wells or reservoirs are polluted by infected sewage.[12] Cholera was endemic then as now (to a lesser extent) in Asia; periodically – as in 1830-1832, 1840-1849, 1863-1866, 1883, and 1891-1892 – great waves of infection and mortality spread from the East through Russia and the Mediterranean to Europe and North America. One of the last great city outbreaks of cholera, due to the distribution of untreated river water, occurred at Hamburg in 1892; the composer Tchaikovsky died of cholera during this same pandemic.

« Fevers » were of course always associated with slums and the poor; the known fact that epidemic disease was no respecter of persons or class barriers was one great social reason for cleaning up the cities, and improving the hygienic conditions of the proletariate.

Disease, maiming and the shortening of life also came, however, directly from industry. Particularly in the unregulated conditions of the early and mid nineteenth century many of the large British

---

[12] A famous English study was that of the 1854 outbreak associated with the well at Broad Street, Soho, London, by Dr John Snow.

industries were extremely dangerous. In mining alone the number of accidental deaths caused by explosions, asphyxiation, falls and so on was put at 1,000 per year. Railways – all forms of early steam-engine one may say – were liable to frequent accidents through scalding by steam, boiler explosions, and injury due to rapidly moving parts. All mills and factories were dangerous for similar reasons, since, in the early years, little or nothing was done to shield workers from the working parts of machines, and the shafting and belts that transmitted power. Power, which made modern industrial production possible, was probably the greatest cause of accident and injury to the workers.

The hazards of manufacture to health have long been recognised. From Egypt of 3,000 years ago the scribe writes:

I have seen the metal worker at his furnace. His fingers were like the hide of crocodiles and he stank worse than fish spawn.

The stone mason seeks work in every hard stone. When he has finished, his arms are destroyed and he is weary. When he sits down at dusk, his thighs and his back are broken.

The weaver in the workshop squats with his knees to his belly and breathes no fresh air. He bribes the doorkeeper with bread to let him see the light.[13]

The Graeco-Roman world recognised the same sort of familiar facts, while in the later West occupational disease began to be described by doctors from the Renaissance onwards. Notably the Italian physician Bernardino Ramazzini, of Carpi near Modena, published in 1700 *De morbis artificum* (« On the diseases of workers »).[14] Ramazzini (1633-1714), a man remarkably distinguished in his profession, noted a great medical truth: that when a physician enters a patient's house he should ask, in addition to the traditional Hippocratic questions about the state of the sufferer's body, « What is his occupation? ». That is, of course, he adds, when the patient is one of the common people. He tells us how he was first drawn to an interest in the spe-

---

[13] H. E. SIGERIST, *History of Medicine*, I, Oxford U.P., New York, 1951, 257-258.

[14] See B. RAMAZZINI, *Diseases of Workers*, trans. W. C. WRIGHT, intro. G. Rosen, repr., New York and London, 1940.

cificity of occupational disease by discovering that the men who were employed to empty privies and cesspits suffered from inflammation of the eyes leading to blindness, though no other part of the body was attacked. Altogether Ramazzini deals with the disorders of some 53 classes of workers, from the humble surgeons who were poisoned by the mercurial ointments rubbed into the sores of syphilitics to the lung disease of rag-pickers; from the ophthalmia of blacksmiths to the dismenorheia of laundresses. Ramazzini realised that many industrial vapours or steams – the mercury vapour produced in gilding, for example, or the steam of hot lye, or the carbon dioxide produced by alcoholic fermentation – are extremely harmful to the workers, as also is the pneumoconiosis brought on by constantly inhaling any kind of dust, as in flour-mills, stone-cutting, mining and so on. At the same time he somewhat confused the issue by treating in the same category the troubles that arise from defective posture, as with the cramped cross-legged position of tailors or the varicose veins suffered by footmen or the noblemen everlastingly standing at the Court of Spain.

Thus we see that occupational disorders by no means began with the industrial revolution, and on Ramazzini's evidence it might be expected that *any* labour is fraught with peril to health – the farmworker may well suffer from hernia, or (commonly) rheumatism, or even contract the animal diseases of anthrax, brucellosis or glanders. But in Ramazzini's day the number of workers to be affected by industrial disease was small. What was new about the urban diseases of the industrial revolution was the widespread incidence of some, not least the common fevers, and the inevitability of others. If a man worked long enough dipping pots in lead-glaze he was bound to be poisoned by the metal, and show the symptoms of potters' palsy. Similarly match-girls who stayed any length of time in the job could not escape poisoning by the white phosphorus they used. The very regularity of labour, the confinement of the worker to a single task, virtually ensured that if it were in the least unhealthy the worker would become afflicted. Machinery itself rather increased the hazards, than relieved the worker of danger, not only because of the chance of mechanical injury but because fast-moving machines themselves increase the production of dust, notably in coal-mining, quarrying and tunnelling. Trades like that of the cutler – making knives and sharp-edged tools – were dangerous because of the inhaled metallic particles.

I need go not into many details: you may find them, for Britain, in *Dangerous Trades*, edited by Thomas Oliver (1902). Let it be sufficient to assure you that the experience of the 19th century was that agricultural employment was several times more healthy – in terms of longevity – than craft or industrial employment. Some occupations associated with high mortality, like those of the plumber and publican (beer- and wine-seller) were old, others like that of the phosphorus matchmakers quite new.

The widespread diseases linked with industrial development are those caused by poor nutrition, congestion and air- and water-pollution. I suppose that defective food does not constitute a disease properly speaking – though we do speak of vitamin-deficiency diseases – but its effects on longevity (especially of infants) and on the general health of populations are notorious. One way in which the employment of married women in textile-manufacture, for example, was deleterious was in the neglect of babies and infants it provoked; if the infants were not regularly breast-fed by their mothers, under 19th century conditions, their likelihood of survival was low; regular work at the mill prevented breast-feeding in a way in which field work (one supposes) had not. Yet I would not argue that in the long run industrialization has produced a strong and universal adverse effect on nutrition; locally and temporally no doubt it did so. In the end, and with the aid of policies, legislation and campaigns directed towards the adoption of better food-habits, especially in relation to young children, our industrial society has effected a reduction of mortality, an improvement in child health, a marked increase of human stature in all prosperous industrial countries, and a reduction in the age of puberty. In the long run industrial society has emerged as healthier and stronger on the whole than any agrarian community of the past.

Some diseases spread by the congestion almost inevitable in industrial cities have been cured by improved housing and hygiene, for example tuberculosis, typhus, typhoid and diptheria; in general the fevers so destructive of childhood have yielded to hygiene and

innoculation prophylaxis. Respiratory ailments diffused by close
contact have proved far less amenable.

The widespread illness associated with industry has, of course,
largely been caused by coal-smoke, and more recently the fumes of
burnt oil fuel. The story is a familiar one and I need not go into
details. In Britain, the nuisance of coal-smoke is much older than
the industrial revolution; in the early days the smoke came from
domestic fires. Already in the sixteenth century London particularly
was burning large amounts of coal for household warming, as well
as for industry (like soap-boiling) and complaint about its thick,
noxious vapour soon followed. In the north of England and Scotland
too coal was being used for industrial puposes, like salt-evaporation.
The steam-engine provided another use for coal, and shortly before
the end of the 18th century began to bring power, and dark smoke,
to the factories and mills. After smoke had fouled many of the formerly
green valleys of Lancashire and Yorkshire, the extension of steam
railways from about 1830 brought smoke and dirt into the centres
of cities formerly relatively clear of atmospheric pollution. Despite
legislation against it, and the attempts of engineers to improve combustion so as to reduce the emission of smoke, the problem increased
throughout the 19th century. Needless to say, gas-works and electricity
generating stations, though providing clean sources of energy, themselves contributed to the foulness of the air.

Filthy air, polluted water, omnipresent falling soot and grime,
became normal features of industrial and capital cities, particularly
of centres of metallurgical industry such as Sheffield and Pittsburg.
Such was the ordinary condition of industrial life, worsening still
further in the twentieth century until strong remedial measures were
taken and technological change offered at least some partial relief,
though this has often consisted of change rather than cure.

The major effects on health of the enormous increase in the
consumption of fossil fuels during the last two hundred years have
been in relation to respiratory and lung disease, and rickets. The
incorporation of particles of carbon and silica into their respiratory
organs obviously does them no good, at the least the uptake of
oxygen is inhibited. The relation of rickets to industry is more indirect.
Rickets is a bone-disease, a failure in the bones to take up calcium,
caused by the body's inability to behave normally in the absence of

adequate sunlight. Without the energy of ultraviolet radiation the body cannot manufacture calciferol, so the bones suffer. In northern Europe, under urban conditions, the amount of solar radiation received is always low; the effect of coal-smoke in the atmosphere was to block the sunlight still further by a factor of ten per cent or more. The diminution was sufficient to cause rickets in the children of British cities – unless specific dietetic measures were taken – while country children might be perfectly normal. Cleaning up the air in the last generation has made an enormous difference, but it is still wise for children to be protected by additional doses of Vitamin D.

By removing coal-smoke and automobile fumes from the atmosphere, public health authorities have taken an important step towards the restoration of a « normal » environment for industrial man, second only in importance to the purification of water supplies. I said « normal », but of course man is not and cannot ever be a « normal » inhabitant of the planet in the way of other animal species. The pattern of his growth in infancy, his exigent need for clothing, fire and shelter at least put him in a place of his own in the natural world. Even the now-vanished way of life of the Eskimo or Australian aborigine necessitated the creation of a « micro-environment » in which alone he and his family could survive, by man himself. The industrial environment, the extreme of abnormality, is but a vast exaggeration of the same thing – not indeed created for the sake of those who live in it, but arising as an incidental of the industrial production to which their lives are tied. In the industrial city man has survived, where birds, animals, trees and even grass have perished. Man sails his boats for pleasure on waters in which no fish can live; industrial man has accepted the hazards which defeat nature, though at great cost to himself. To have become industrial man, to have found remedies for the plagues and poisons that decimated cities in the past, is a feature of human evolution even though, romantically, we may find such a way of life detestable and seek with all means to avoid it for ourselves as individuals. So long as the numbers of our species are as large as they are, and the wants of each member in terms of food, commodities and transport remain as insatiable as they are in the Western world, we have to remain industrial men, *homines fabricatores,* and reduce or cope with the evil aspects of industrial life as best we can.

VII

THE ROYAL SOCIETY OF ARTS:
TWO CENTURIES OF PROGRESS
IN SCIENCE AND TECHNOLOGY

ON 22nd March 1754 a number of 'Noblemen, Clergy, Gentlemen and Merchants' met at a Coffee-House in Henrietta Street to organize a 'Society for the Encouragement of Arts, Manufactures and Commerce'; they were a group of men 'actuated by the most liberal motives (the benefit of mankind in general and of their country in particular)'. Among those present who were to be particularly involved were William Shipley, teacher of painting, who had first proposed such a society; Lord Folkestone, its chief early patron; Stephen Hales, biologist and humanitarian, and Henry Baker, microscopist and poet. The purpose of the new Society of Arts was clear

from the beginning: it was to offer monetary prizes to those who best attained certain stated objects, which were to be proposed annually. To make this clearer, let me quote Henry Baker's own report of the Society's progress, written less than two years after its inauguration:

'[We] find that the premiums. . . bestowed are likely to produce great advantages to the nation, by employing many hands, and saving annually large sums of money . . . By means of this society, a large mine of cobalt ore has been discovered . . . on the estate of Fr[ancis] Beauchamp Esq . . . Wherefore the assistance of all persons skilled in the fluxing and management of minerals,

Reprinted from *Journal of the Royal Society of Arts*, 122 (1974), pp. 641-655, by permission of the publisher.
© 1974 Royal Society for the Encouragement of Arts, Manufactures and Commerce, London.

or in other chemical operations . . . is requested and hoped for, to establish and perfect this manufacture.' Secondly, Baker writes, the red dyestuff madder has under the Society's encouragement been grown again in England, after long neglect, so that its importation cost £150,000 per annum. All who use the dye should encourage this home-grown product, he adds. Further, oil-tanned leather has been prepared here, such as is used for buff jerkins, again equal in quality to the import, and (in response to another premium) the largest copper vessels are now made coated with pure tin: so 'that for the future, none who value the health of their families, will use copper vessels un-tinned, or permit their pots, saucepans, or other kitchen vessels to be tinned with a mixture of lead, in the former unwholesome manner'.

Finally, the prizes offered for drawing had also succeeded, but Baker is careful to emphasize that the Society was not aiming at aesthetic proficiency in landscape and portraiture; no, he says, it realizes that skilled drawing is essential to many trades and conduces greatly to the improvement of manufacturers, as a rival nation, after encouraging this art, has discovered.[1]

From this we can see that the first object of the Society was not so much (as we might expect to-day) the encouragement of invention as of proficiency in established trades; not so much a new labour-saving technology as the employment of a myriad of skilled craftsmen – not artists only but (as Baker said) mechanics, joiners, upholsterers, coach-makers, weavers, smiths, milliners, embroid-erers, fan-painters and so on. This was good mercantilist doctrine, full of sound common-sense but not untinged with chauvinism, for the founders of the new Society clearly held that the finer and more productive British manufactures were, the richer Britain's exports would be; the richer *her* people, the poorer her neighbours. The balance of payments problem has long been with us.

I make this point because, although the Newcomen engine had by this time been pumping mines for forty years, and patents already existed for moderately successful carding and spinning machinery as well as Kay's flying-shuttle, the inauguration of the Society of Arts clearly belongs to the pre-Industrial Revolution age. In what follows I shall speak mainly of the Society in relation to the technological progress coterminous with its history; as for the fine arts, these

walls are more eloquent than I; *si monumen-tum requiristi, circumspicite.*

We may first note, then, that an interest in technological progress – one might say a belief in its virtues – is no product of the Industrial Revolution. Pride in the technical prowess of modern Europe is already evident in the sixteenth century, and the inventions of gunpowder, printing, and the magnetic compass were ready instances to hand, no European then realizing Chinese priority in these new arts. Francis Bacon had foreseen a favourable progression from increased knowledge to an extension of the 'empire of man over nature' and so towards the amelioration of the human condition. Later in the seventeenth century the Royal Society in its series of (abortive) permanent committees included one 'To consider and improve all mechanical inventions', another for 'Histories of Trades' and a third for agriculture. It is possible to exaggerate the Royal Society's concern for technology; nevertheless there was (in its early years) interest in new inventions and processes, and a more sustained attempt to codify and improve agricultural methods which was not without relevance for the introduction of the potato, the turnip and 'Norfolk husbandry'. Moving on to the eighteenth century one finds a literate or gentlemanly interest in technical progress almost everywhere – I mention only the splendid technical monographs and plates in the great *Encyclopédie* of Diderot and D'Alembert.

However, there is a difference between this pre-revolutionary age and ours that is significant. Before the Industrial Revolution the notion of technical progress carried with it strong humanitarian, charitable overtones; its object was to ease the toil of the plough-man wearily plodding home, not to convert him into a machine-minder, to employ the destitute and (particularly) to enhance medical care for the sick. Indeed, such hopes lingered long into the nineteenth century, and it was true that new industries furnished both wages and cheap goods to the poor, if only rarely the kind of social welfare prac-tised at New Lanark and Lowell Mills. It is my impression that at this early period people did not value speed of travel, for example, as an object to be sought at any cost, nor did they insist that whatever tech-nically can be done should be done, as we are inclined to do. In other words, techno-logical progress was envisaged as taking place within a known and established moral

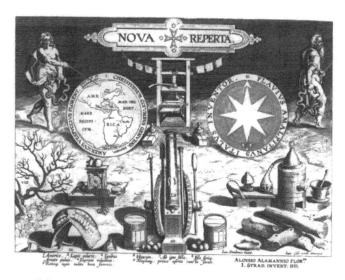

Title-page from Nova Reperta, *by Jan Stradanus, ?1590. Amongst the
technological discoveries represented are gunpowder, the printing press,
the iron mechanical clock, distillation and the silkworm*

and social order, for it was impossible to see then that it would bring about a total change in civilization compared with which the French Revolution was a mere village brawl. But of course confidence in the immutability of moral and social patterns had its drawbacks; on the one hand it made people insensitive to the actual harsh realities of the Industrial Revolution, and on the other it made them naïvely blind to the magnitude of the price which always, it seems, has to be paid in compensation for technical advancement. We have learnt that, like democracy, technology demands eternal vigilance; William Shipley and the Society of Arts, following after all a not too adventurous path, had no sense of trying to command the whirlwind.

If we seek to examine in a broad way the factors that have made for technical advance, other than the basic levels of scientific and craft knowledge, I think we shall discover three. The first is the human element that sent Watt from instrument-making into mechanical engineering, Joule from brewing into physics, Marconi into radio and Parsons into the development of the turbine. Secondly, there is a great complex of economic

and social factors that make a time 'ripe' for a particular advance, or perhaps works oppositely to favour lethargy, as seems to have happened in Britain during the later nineteenth century. And thirdly, there are the more deliberate stimulative or regulative influences exerted within a society. Does a particular community readily reward technical inventiveness or not? Does it freely allow, or does it regulate, innovation in certain fields like public transport, or the manufacture of foods and pharmaceuticals? Does the community educate people in the way that is necessary if some of them are to be capable of technological invention?

Now clearly when we turn to the rôle of a Society devoted to the encouragement of arts, manufacture and commerce we are concerned with this third category; we have to ask ourselves two questions: first, how has the Society of Arts sought to direct its encouragement, and second how effective has this been, in relation to the magnitude of social and economic pressures, for example? Obviously no prizes would establish a madder industry in England if the growth of madder were consistently unprofitable; nor, in spite of prizes of over £1100 given to encourage sericulture in the American colonies, was the Society successful in its attempt. (I would like to think that the Society's similar awards for viticulture had some connection with the now-flourishing wine-industry of the USA, but I doubt if it can be so.)

When we consider that the vast majority of the population worked on the land during the early decades of the Society's history, and that most of the members were land-owning gentry, it is not surprising to find that agricultural improvement was the subject of many awards; and in the opinion of Arthur Young and other observers the Society of Arts did contribute effectively in that way. Prizes were given for the cultivation of new fodder-crops such as Swedish turnips and the mangold-wurzel; for the improvement of land by draining and manuring, and for a variety of simple agricultural machines like chaff-cutters. Animal – and plant – breeding did not attract the Society's attention, but it is reckoned that something like twenty million trees were planted in new woodlands under the Society's encouragement. I think one may discern here its rôle in making a certain kind of activity fashionable; as it happened the wooden walls of England were outmoded by the time late eighteenth-century oaks grew to maturity, but the Society doubtless did something to alleviate the excessive deforestation of these islands.

The most familiar agricultural machine of the horse age, the reaper, came to us from the United States and spread only slowly. It is just possible that had the Society promoted certain Scottish inventions energetically, that of Cyrus McCormick (1831) might have been less successful. To its credit, the Society did offer an award for a practical reaper over many years (from 1774) but it refused the prize to John Common in 1812 and again to Patrick Bell in 1830 – on the second occasion maintaining that the device was well enough known already! However, in this case one should add that it seems doubtful whether either of these inventions was really as effective as the McCormick reaper which attracted so much attention at the great Exhibition of 1851.

The temporary activity of the Board of Agriculture and the foundation of the Royal Agricultural Society (1838), together with the shifting of the balance of weight in the economy, all tended to reduce the Society's concern for agriculture. How well did the Society of Arts adapt itself to a new world in which the preponderance of population and manufacture shifted decisively northwards, in which mine, factory and mill were dominant while the old crafts commenced their slow recession? As one might expect, one's immediate impression is that the Society was rather ineffective in promoting the industrialization of Britain; indeed, its first historian argued that the inventors and entrepreneurs who did this were so far in advance of their time that no one could have discerned the importance of their innovations.[2] This argument is neither historical nor rational, and is in any event too pessimistic. While a number of Samuel Smiles' engineering and industrial heroes had no connection with the Society – neither Crompton of the mule, nor James Watt of the steam-engine, nor Brindley of canals (for example) were members of the Society – many others were members; among them the great manufacturers Josiah Wedgwood of Etruria and Matthew Boulton of Soho; Henry Cort and Isaac Wilkinson, celebrated ironmasters; Joseph Bramah and the elder Brunel, famous for mechanical skill; the civil engineers Thomas Telford and Robert Mylne and indeed all three of the first Presidents of the Society of Civil Engineers were linked with the Society of Arts. Even Watt came at least once to a Committee. Hence the Society was in close touch with the new world of engineering and industry, and remained so into the next generation; it attached particular importance to some achievements – such as the Duke of Bridgewater's canals and Abraham Darby III's iron bridge at Coalbrookdale – by the award of medals.

Of course it is true that the Society of Arts can take no credit for the development of the iron industry in Britain, or that of the steam-engine, and little for the creation of the Lancashire textile industry. It may even be doubted whether the awards of prizes and medals would have had the least effect in strengthening enormous economic forces. If we examine the subjects of awards made by the Society at this time they seem to have been made for improvements in windmills, spinning-wheels, saw-milling, whale-fishing, canal-locks, pile-drivers, dredgers, carriages and so forth – things useful enough but familiar in the arsenal of technology since the Renaissance. It was perhaps natural that members of the Society had difficulty in foreseeing the future importance of certain inventions. When John Kay applied in 1765 for an award in respect of machinery he had devised for making wire cards (used in preparing fibres for spinning), the Committee was clearly unaware that Kay had been the inventor of the highly useful flying-shuttle some thirty years before. His application was allowed to lapse, but an award for such machinery was made later to John's son William Kay. The latter's brother Robert was equally unsuccessful in obtaining a prize from the Society for another useful invention, the 'dropping-box' which enables the weaver to change the shuttle, and hence the colour of the weft-thread, at will.[3] However, a good deal later, in 1806, awards were made for the successful establishment of power-loom weaving,[4] and in 1807 for improvements to the draw-loom employed by handloom weavers of figured fabrics, usually of silk. The Society also encouraged the development of the Jacquard loom (a mechanized version of the drawloom), which was, of course, a French invention.

As in our own day, technical forecasting proved extremely uncertain. While the Society could see the value of a better harpoon-gun to whale-fishers, or of improvements in the lace-industry and the semaphore telegraph, it missed the significance not only of modifications to existing machines such as Kay's dropping-box, but of whole branches of technology like the water-turbine later. The gentlemen of the Society of Arts called in expert advice on occasions (I have mentioned Watt's name in this context) and are certainly not to be blamed for imperfect foresight; we ourselves may well find, unfortunately, that *Concorde* proves to be the *Great Eastern* of the twentieth century, and cannot really flatter ourselves on great prescience. If the Society of Arts did not always choose well, its mistakes were harmless ones.

There is another factor. Until 1845 the Society refused to give any recognition to a patented invention, and its awards were conditional on the recipient's not taking out a patent in the future. Indeed, though the Society had no objection to an inventor's making all possible pecuniary profit from his invention in free competition, its policy of laying all improvements open to unrestricted public use was opposed to the monopoly privileges granted by a patent. An inventor therefore had to weigh the likely profit of a patent against the value of an award from the Society, while the Society inhibited itself from distinguishing the many like Arkwright,

Cort or Watt whose achievements had been protected by patent – and by lawsuits to defend patents – long before the achievements became celebrated. An obvious example of conflict between the Society's policy and the patent system is provided by Joseph Bramah's preventing the award of a premium to a certain claimant, on the grounds that the machine in question infringed a patent of his own.[5] So that while the difficulty of foreseeing the future meant that some things of yet unknown merit went unrecognized, the provision about patents usually meant that the most obvious merit could not be recognized either.

Let me not set too gloomy a picture before you. There are many instances of the Society's showing an alertness to real manufacturing problems, and aiding their solution practically in its own way. For example, it was aware of the shortage and high price of that essential substance, soda, and early offered prizes for the production of home-produced barilla, for improved kelp, and finally (in 1783) for soda made from salt.[6] This was a prescient suggestion for the development of a new chemical industry that was to start its major growth some quarter-century later. Prizes were also offered for dyestuffs, varnishes, various minerals, glass of optical quality and so forth, but more significant were the awards made to such improvers of the infant gas-light industry as Samuel Clegg, inventor of the gas-holder and works gas-meter (1808). However, the main pioneers like Murdock and Winsor again went without recognition, perhaps because of patent protection.

It is my impression that such awards had negligible effects on *major* industrial changes. Gas lighting spread with rapidity because its usefulness was so great and obvious. Similarly, in the case of the manufacturer of synthetic soda, experiments had been pressed by James Keir, John Roebuck and others for more than ten years before the Society offered its prize, and a number of (abortive) patents for such a process were already in being by 1783. Moreover, although synthetic soda was certainly produced in Britain before Leblanc's process was worked here, it was this process (patented in France) that ultimately held the field.[7] Indeed, it is and was obvious that the economic prize to be won by synthesizing soda cheaply and practically was so vast, that no other incentive was needed. On the other hand, the Society's offer of a considerable sum (£100) for a

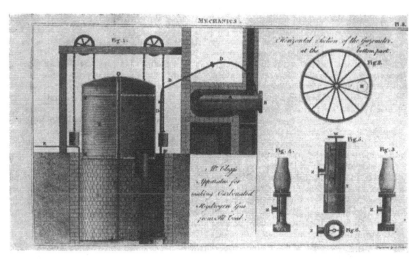

'*Mr. Clegg's Apparatus for making Carbonated Hydrogen Gas from Pit Coal*':
Transactions of the Society . . . , *Vol. XXVI (1808), Pl. 8*

county map (1759) is said to have stimulated the publication of the numerous maps of that kind that were published afterwards, though of course even then £100 would be but a small fraction of the cost. I must record also that the Society helped William Smith to print the first of all geological maps (1815).

Now I turn to an aspect of the Society's history during its first century which seems to me enormously creditable and useful – I mean its concern for industrial health and the preservation of life. As we all know, the humanitarian consequences of technological progress were not realized during the early stages of the Industrial Revolution. On the contrary, the working conditions of men, women and children deteriorated, dangerous machinery and chemicals became more commonplace, and new ones were constantly being introduced. But the Society's first major concern of this sort was with a process in use for many centuries, that of applying gold in a mercury amalgam to beautify base metal. No one had been troubled that the vapour inhaled when the mercury was driven off by heat produced a condition in the gilders also associated with Lewis Carroll's 'Mad Hatter', and early death. In 1771 the Society offered a prize for the prevention of this disease, which was soon awarded to one J. Hills who suggested the use of extraction ducts to remove the vapour, with or without glass screens. Similarly, the Society tried to protect workers who incurred silicosis by using grinding-wheels, the pottery dippers who were poisoned by the lead glaze with which they worked, the painters employing white lead paints, and above all the boy chimney-sweeps. It was an Act of Parliament coming into force in 1842 that finally made illegal the use of children to sweep down soot; but it may truly be claimed that it was this Society's eagerness to develop an effective mechanical chimney-sweeper, from 1796 onwards, that made the Act possible. Other awards were made by the Society to Henry Greathead, builder (if not inventor) of the first unsinkable lifeboat (1802)[8] and to Captain Manby who invented the first projectile apparatus for casting a line to a stranded vessel (1808); but both of these benefactors also received large sums from the public purse. In this area of activity the Society was, in its early years, devoting itself to objects which were not also urged by economic considerations of self-interest, and here its rôle in defining problems and recommending the search for solutions was significant.

Throughout the first seventy or eighty years of the Society's history which I have considered hitherto, it had endeavoured (largely by the simple method of offering monetary prizes and medals) to promote

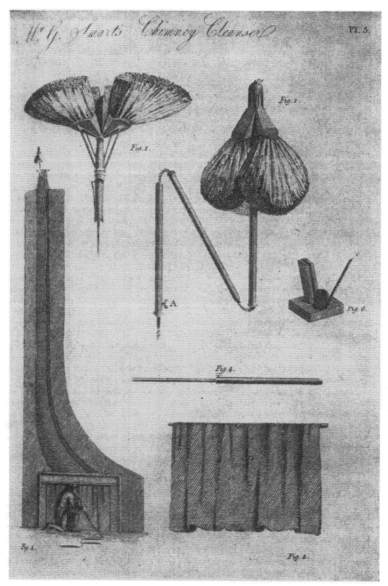

'*Mr. G. Smart's Chimney Cleaner*': Transactions of the
Society . . . , *Vol. XXIII (1805), Pl. 5*

technical or commercial innovation. Whether its particular objects of this sort were well-chosen or not, they were quite precise. In the next and most distinguished period of its history the Society largely abandoned both the method and the particular objectives, and its interest in the economic health of the nation was manifested in new and more diffuse ways. In the mid-Victorian age, the age of the Prince Consort and Sir Henry Cole, of the two Exhibitions of 1851 and 1862, and of thriving success in the Society itself, its function was rather communication, publicity and education.

At the start of this phase in 1845 the Society of Arts was almost in dissolution, a fate from which it was saved by reorganization and the shift of interest I have just characterized. The award of prizes and medals has never of course been abandoned, and there is no reason to doubt its continuing value and interest; but it had become (in its old form) irrelevant to the massively industrializing nation of the 1830s, at a time when no one could any longer guide effectively the enormously open field of technological invention. Arthur Aikin, the Society's Secretary from 1817 to 1839, clearly perceived about half-way through this period that a more valuable function for the Society would be the *communication* of knowledge of technical or commercial matters, and the discussion of papers, rather than prize awards, at the weekly meetings. Fifty years earlier it had already been hoped that the Society would serve 'to communicate the labours of the ingenious of one part of the British dominions to the inhabitants of the other', a plan which 'it was hoped would appear very eligible to those who have observed the slow progress with which useful knowledge is disseminated'.[9] Communication, education proselytization become henceforward the Society's objectives, rather than direct pecuniary stimulus. Thus, if you examine Vol. 51 of the Society's *Transactions*, for 1835-7, you will find useful and quite detailed accounts of Joseph Glynn's work on the drainage of the Fens by steampower, of an improved safety-lamp, of Marsh's test for arsenic, of an improved loom for weaving figured fabrics, and so forth, besides the usual shorter notes on mechanical contrivances. The Secretary himself contributed two lectures on 'sacchine substances' and Cornelius Varley one on eyepieces for optical instruments. This was very different from the early volumes, which were little more than lists.

From 1852 onwards, in its new *Journal*, the Society regularly published the twenty-four lectures delivered annually, much as the lectures are published to this day. However, there was at that time no planning of the topics of communications; as Trueman Wood puts it: '[The Society] never refused a hearing to anybody who had fresh information to give on any subject likely to be beneficial to human progress or human welfare.'[10] There were exceptions to this capricious approach; for example, at the suggestion of the Prince Consort a series of lectures by most distinguished authorities (Whewell, Playfair and so on) was arranged by the Society in connection with the Great Exhibition, in 1851-2; and after the receipt of the Cantor Bequest in 1862 it was possible for the Society's Council to arrange for systematic series of lectures on industrial, technical and hygienic topics, but the total effect was still highly miscellaneous.

Of the Society's chief public rôle in these years, its initiation of the two great International Exhibitions, I shall say nothing now because any proper attempt to measure their commercial and psychological significance would be inappropriate, and I doubt if it could be argued they really helped technological and scientific progress in this country, except in so far as they did something (though not enough) to alert British manufacturers to the vigour and originality of foreign competition. In general the national mood was one of complacency: the exhibitions proved that Britain was the workshop of the world, that the volume and the quality of her production was unrivalled, and her control both of markets and of natural resources beyond challenge. The Society of Arts rated high in its counsels at this time men of great engineering distinction such as Robert Stephenson, Sir William Fairbairn, John Scott Russell (its Secretary) and Sir William Siemens, but neither these nor others could foresee how different the international technological environment was to become, from its state in 1851. Of course, I do not mean that such men did not comprehend the recent rapid progress of technology; on the contrary, and the attempt to display the nature of this progress was made in publications associated with the two Exhibitions, for the latter of which the Society of Arts assumed responsibility. What is rather the case is that it was not clear in Britain that both the form and the foundations of technical development were to alter – that empirical inventiveness and mechanical skill were to count for less in the future, and scientific knowledge plus meticulous management were to account for a great deal more. In any event the International Exhibitions were occasions of commercial display, rather than of economic and technical analysis,

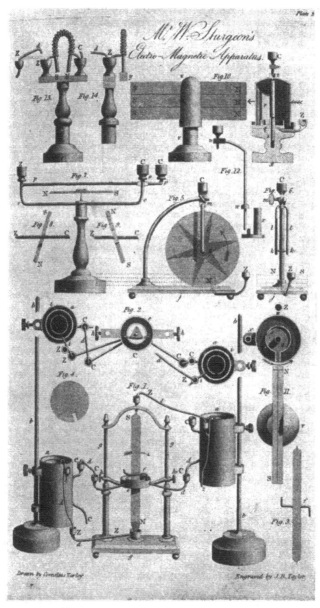

'*Mr. W. Sturgeon's Electro-Magnetic Apparatus*':
Transactions . . . , *Vol. XLIII (1825), Plate 3*

though something of the latter did occur.

Hence the Society had not only developed a close and prestigious link with the Court – reaffirmed when the Prince of Wales after a brief interval succeeded his father as President – and with other elements in what would now be called the Establishment, but was emphasizing its service to Commerce at some expense to its interest in Manufacture. Whereas Shipley had wished to train craftsmen, the Society of Arts now examined clerks and book-keepers, while from 1879 it handed over its concern for technical education to the newly formed City and Guilds of London Institute. Perhaps this was symptomatic of a national shift from the blue overall to the white collar. The annual exhibitions of new inventions, and the permanent museum of machines, had already been abandoned by the Society, partly as a consequence of the reform of Patent Laws in 1852, in a measure which the Society itself had helped to promote.

But I run ahead rather as to the details, for I must certainly not omit a very significant event in the Society's history, its association with the Mechanics' Institutes which began in 1848. As is well known, these Institutes, ancestors of many of our Technical Colleges and Polytechnics, came into existence during the second quarter of the nineteenth century, their object being the general education of working men as well as their technical training. This Society was instrumental first in creating a Union of the Institutes in July 1852, then in the following year organizing an exhibition of their work (which in turn led to the foundation of what was to become the Science Museum) and finally organizing a system of examinations for the Institutes, which also took place in 1853 for the first time.

The Union of Institutes, numbering 368 members as early as 1855, lasted only some thirty years – by which time much of the original vigour of the movement had gone – when the Society's examinations were opened to all comers, but the examination function has of course remained an important part of the Society's work up to the present time. These were the first examinations of proficiency in the usual school subjects – including science, mechanics and drawing – available to adults who were not University students; the examinations by the Government's Department of Science and Art were begun about the same time, but the University 'local' examinations began only in

1858, and these, of course, became school examinations. The first Education Act became law in 1870. However, the strength of the Society's interest in public examination moved rapidly towards commercial and clerical subjects. Book-keeping was examined almost from the beginning; shorthand was introduced in 1876, and - sign of a new age - typewriting in 1891. Science subjects were taken over by the Government, and the examinations in technological subjects (begun only in 1873) were handed over to the City and Guilds Institute in 1879. For two years, indeed, the Society abandoned all its examinations, but it resumed a restricted range in 1882 after a local government measure of 1890 enabled many more candidates to receive instruction for them. Apart from the usual school subjects the Society examined in book-keeping, shorthand, domestic and political economy, and sanitary knowledge. Most of the candidates were under twenty and some were mere children.

There is no doubt of the success of the Society's examinations in commercial subjects, for which in the year 1929 there were over 100,000 candidates, but I think it may be regretted that the Society relinquished so readily its formal connection with technological education. I cannot see that the proper education of engineers, mechanics, industrial chemists, surveyors, architects and designers was any less proper concern for the Society in the 1880s than the training of clerks and secretaries, or than the training of tradesmen had been in the 1750s. Nor can it be said that the training, examining and professional status of British technologists were all so well provided for in the 1880s, or for very many decades thereafter, that no further efforts by this Society would have been valuable. Quite on the contrary. In engineering practice and elsewhere the system of articled pupillage long prevailed, so that the number of graduate technologists remained pitifully small; and the vast majority of those working at the bench in industry was still trained only by the methods of apprenticeship, with their emphasis on 'practical' and often very narrow skills. As early as 1868, as a consequence of Lyon Playfair's report on the Paris Exhibition of the previous year, the Society had appointed a committee to report on technical education in this country.[11] At a time when an MP could maintain that 'where there was one in England there were twenty in France or Germany, who knew enough of scientific principles to be able to apply them to the common affairs of life'; and the authors of the committee's report could assume the necessity for a reform in Britain without argument, it was perhaps unfortunate that the same report accepted without question the superiority of the English system of pupillage for the gentry and apprenticeship for the working class, and so recommended against the institution of Polytechnics in this country even while actually recognizing their value to Continental countries. Once more British insularity triumphed, and while the Society's committee strongly emphasized the need for teaching pure science in schools up to the age of 17 or 18, it was willing (with the exception of the chemist A. W. Williamson) to cut off formal technical education at that point. It saw no need to teach formally such subjects as hydraulics, theory of structures, or thermodynamics.

Certainly within a few years a good deal was to be achieved by the Government and the City Companies, and perhaps it was natural for the Society to feel it had played its part. But, with hindsight, one sees that a great opportunity was missed, both in the report of 1868 and the renunciation of technical examinations in 1879. A few forceful men of vision, like Williamson, pushing from this Society for the cause of scientific training and indeed research in technological subjects, pushing for new attitudes in British industry, might possibly have achieved much during the 1880s and 90s, when this country was still wealthy and thrusting, which was in fact to be postponed to the depressed and cheerless times of the 1920s and 1930s.

This being said, let me not give the false impression that the Society was out of touch with technical and industrial progress; quite the opposite is true, and it is also true that during the last century or so the Society has invited speakers and awarded medals relating to pure science, which it did not do in its earlier history. Michael Faraday was third recipient of the Society's Albert Medal, and among the names of subsequent recipients are those of many eminent scientists, Pasteur and Helmholtz, Sir William Crookes, Lord Rayleigh and Mme. Curie, Lord Rutherford and Lord Adrian. But I think we should make a distinction between topics covered in single evenings or series of lectures in this room, sometimes several times over at intervals of years, and the definite efforts made by the Society towards particular ends. None of the special exhibitions arranged by the Society have had much connection with science or technology, apart perhaps from the first Photographic Exhibition of 1852, but I should not fail to mention the Society's continued concern for the health and nutrition of the British people. Although the domestic preservation of foods with the aid of sugar or salt is old, and canning was invented early in the nineteenth century, the modern long-distance transportation of such foods as meat and milk involved new techniques of preservation. The Society was involved about a hundred years ago in attempts to improve the milk supply of London, but more significantly promoted the search for new methods to preserve meat so that it could be brought from Argentina or Australia. The first attempt to bring 20 tons of refrigerated meat from Melbourne was unfortunately frustrated in 1872-3 owing to failure of the machinery; an attempt in 1879 succeeded and meanwhile ice-chilled meat had been imported from the USA, and tasted by members of the Society's Food Committee. Thus a very disagreeable trade in live cattle was slowly brought to an end, and a different trade with the antipodes opened up. It must be admitted that though the Society rightly foresaw the importance of refrigerated transport, the economic forces favouring attention to the problem were also extremely strong.

If foreign sources of food were essential to the vast industrial population of Great Britain, sanitation was no less so. Technological development produces urbanization, urbanization increases the density of sewage production, industry pours effluents into the drains and water-courses. A minor invention, that of the water-closet, notably increased the risk of disease organisms in sewage contaminating ground-water supplying drinking wells, as happened in the singular case of the Broad[wick] Street pump, investigated by Dr. John Snow. Not to dwell on the horrors; thirty years after the passage of the first Public Health Act in 1848, sewage disposal remained one of the great technical problems of Victorian cities, transport being the other. The Society of Arts was active in both fields, but particularly in the former, especially after the organization of its first conference on the Health and Sewage of Towns (9th-11th May 1876). To-day we take this whole issue for granted, but it was very different a hundred years ago, when even piped water was far from free of danger, when most towns had an average of only one privy of any sort for each ten inhabitants, and the night-soil cart was a regular feature of urban life. At this time a death-rate of 28 or more per thousand was not unusual in British cities.[12] The system of water-borne sewerage, even if it could have been suddenly introduced, was by no means universally approved; many authorities deplored the loss of important chemical elements, nitrogen especially, which ought to be restored to the land. But could dry sewage disposal be made both cheap and aesthetic? On the other hand if sewers were used, how was the massive quantity of effluent to be disposed of? Could it safely be used to fertilize agricultural land, and would it endanger subsoil water? Experience had proved that the direct passage of liquid sewage, even when diluted with large quantities of storm-water, into streams and rivers, produced ghastly results. Could treatment plants be practical and economic? And what about the industrial muck in the drains that plagued industrial cities like Coventry, Manchester and Glasgow, or those places where the sewers *now* opened directly into water-courses?

The Society of Arts canvassed all these problems and purported solutions to them in its public discussions over ten years, and indeed it has maintained an occasional concern for such issues down to the present day.[13] I omit further details; let it suffice to note that, as a result of its first Conference in 1876, the Society concluded that no single method of sewerage was universally satisfactory, that health was more important than financial cost, and that Parliament should provide for a better administrative supervision of house-drains and their connection to municipal sewers.[14]

From the great bulk of the Society's *Journal* reporting its manifold activities, I might have selected many more instances of the Society's humanitarian concern, from life-saving to the quality of the natural environment; but instead I chose a number of examples of technological development in the last hundred years, to see what I could find of them in the Society's history. The first is the water-turbine, a prime-mover developed particularly by French and American engineers; about 2 per cent of all the world's energy is derived from moving water to-day, but 25 per cent of all the world's electrical power is hydroelectric. Now I find scattered through the *Journal* a number of short notes on the exploitation of water-power resources in various parts of the

world, but that this Society paid no particular attention to such resources, or to the technology by which they are exploited, through the whole of the nineteenth century. The period from 1890 to 1920 was one of enormously dramatic development in hydro-electric power generation, witnessing the harnessing of Niagara and the building of large installations in the Alps, the Pyrenees, Canada, Scandinavia and so on. We had in this country, by then, only one large scheme, at Lochleven in Scotland, producing some 15MW. A correspondent writing to this Society in 1918 mentioned 'the prejudice, prevalent even among engineers', against waterpower;[15] and later speakers emphasized the lack of opportunity for education in hydro engineering in Britain; that the necessary plant was scarcely manufactured at all in this country was also recorded.[16] When such was the attitude of the British engineering profession and of British industry, conditioned by the traditional British belief that no energy that was not dug out of a mine was worthy of consideration, it is hardly surprising to find it reflected by the Royal Society of Arts. The First World War wrought a great change. It became clear both that we were moving heavily into oil and that eletricity (if not the most efficient form of energy conversion) was enormously convenient. Some authorities like Dugald Clerk emphasized the need to conserve our coal stocks,[17] and the increasing relative cost of coal was obvious to all. In the years after the First War, accordingly, the utilization of hydroelectric power in Britain, and especially in Scotland, was discussed more than once in this room; in fact what we would now call the 'energy problem' of those days was effectively canvassed by the Society. In those days too the *Journal* reprinted or summarized papers from other sources, and thus Fellows were certainly well informed about the general march of events, if not given the technical detail that appeared in other contexts.

Coming now to that related and important British invention, the *steam* turbine, I confess to surprise that the *Journal* contains no record of it before 1900, when there is a brief report of a lecture given by Charles Parsons at the Royal Institution.[18] His invention was then sixteen years old and had been devoted to the generation of electricity for twelve. The famous intervention of *Turbinia* at the Spithead naval review had occurred in 1897. Thus the Society of Arts had felt no sense of urgency about imparting to its readers a great and very obvious technical innovation. Some amends was made, however, by the devotion of three Cantor lectures to the Parsons turbine in 1909.

Marconi was more swiftly honoured by the Society in that he was invited to speak on his method of wireless telegraphy by tuned circuits within five years of his first arrival in London;[19] but Marconi was conscious of the value of publicity and his work had captured the imagination.

Yet one could read about Marconi and his black boxes in the *Strand Magazine* long before 1901! All these instances prove, I think, that the evaluation of technological innovation in engineering terms, or the communication of information about technical developments, has not been a principal rôle of this Society. Let me turn finally to a more elaborate and long-continued debate, where I think the Society is displayed acting in a far more useful capacity.

The Channel Tunnel, of which I shall speak, has a curious history. The project to create a direct railway link between Britain and France seems to have been regarded in a very matter-of-fact way in the 1860s, when our railway system was already nearing completion. The Channel crossing was simply 'a nuisance or a terror to the Continental traveller' and therefore it must be removed. Even as late as 1882, when Sir E. W. Watkin, the engineer of a large part of what is now Southern region, was addressing this Society on the subject he could declare: 'in this room, I take it for granted that increase in the means of intercourse or journeyings to and fro between nations means the augmentation of wealth and the expansion of civilisation'.[20] By this time, however, non-technical objections to the project were already strong, as we shall see. It must be remembered that when the Tunnel was first mentioned before the Society, in 1866, there was no cross-channel train-ferry service (not to say a night sleeper);[21] the packet boats were lively and the harbour facilities poor, though even then the journey between London and Paris took only two or three times as long as the fastest journey to-day – between 10 and 13 hours – but this was regarded as almost intolerable. The recent development of train and car-ferries and of course the air service has done much to reduce the need for the Tunnel in relation to passenger travel.

Little was made of the technical difficulties involved. It was generally believed in the 1860s, and subsequent borings confirmed the belief, that a continous line through unbroken impermeable chalk could be traced between the two shores; Watkin explained to the Society how, using a machine driven by compressed air, he had cut a trial bore 7 ft in diameter 100 yards in a single week. He estimated that the work could be accomplished in five years, implying an average progress of well over 150 yards per week. A much earlier speaker, Zerah Colburn, had instanced the successful driving of a water-supply tunnel for the city of Chicago two miles beneath Lake Michigan, and later the completion of the first Alpine railway tunnel, the Mont Cenis, strengthened the opinion that almost nothing was impossible in tunnelling if only the rock were sound, as geology indicated it to be.[22] At this time John Hawkshaw was already engaged in trial bores. Moreover, the alternative engineering possibility of sinking some half million tons of cast-iron tube upon the bed of the Channel for trains to run through, or to be blown through by atmospheric pressure, had its active supporters of whom Mr. Colburn was one. No one really had good suggestions for fabricating and assembling such a gigantic pipe, however.

For a moment, as this Society was duly informed,[23] the tunnel seemed as good as constructed. In the late 1860s the Emperor Napoleon was keenly interested, commissions of engineers were appointed on either side and the granting of a concession was imminent. Bismarck stopped it at Sedan; and the Franco-Prussian war seems to have heightened British xenophobia. In the 1880s some people at least felt that a good deal more than travellers' convenience was at stake, and whether or not the engineers *could* do it, there might be powerful arguments that they should *not* do it. It is interesting that while in the last few years we have grown used to the environmental criticism of attainable technological objectives, the first criticisms limiting the attainment of engineering proposals related to national security. Sir Edward Watkin expressed the fantasies of the Colonel Blimps of 1882: 'at some sudden moment, without any warning, any notice, any declaration of war, the Channel will be unguarded, the commanders at Dover and other strongholds will be traitors, the soldiers will be either asleep or parties to the treachery and 2,000 Frenchmen, soldiers in disguise, will steal through the Tunnel and effect a lodgment on English soil. They will hold the Tunnel mouth and train after train of the enemy will follow, march irresistible to London, and England in 24 hours will capitulate. Then, amidst other discomfiture, the robbing of the Bank of England, the seizure of Buckingham Palace and the Zoological Gardens – the Tunnel itself will be sequestrated as part our temporary ransom, so as to enable the French to invade England again and again.'[24] I particularly like the touch about the Zoological Gardens.

Later communications to this Society on the subject of the Channel Tunnel in 1889 and 1913 make it clear that the Blimps, in the person of Lord Wolseley, really did prevail despite the mockery of Watkin and others, and assurances from the Admiralty that there was no cause for alarm. The earlier favour of such statesmen as Disraeli, Lord Derby and Cobden, the tremendous promises offered by the Tunnel for the restoration of Britain's failing rôle as an entrepôt and general commercial metropolis upon which Colonel H. M. Hozier insisted in 1889,[25] the general enthusiasm of financiers and railway directors made plain by Mr. Fell in 1913,[26] and the continuing suffering of Continental travellers, could not prevail against military timidity. In 1882 it had been necessary to devote a second evening of discussion (28th April) to the warmth of feeling over the national security issue; then in the following year a Select Committee of both Houses of Parliament considered the whole project and reported adversely, on security grounds; by 1913 it was recognized, indeed, that the Anglo-French entente and the success of the aeroplane had outmoded such timidity but the steam seemed to have subsided even though the cost was still reckoned at no more than £16 million. A last paper on the subject was presented by the younger Mr. Hawkshaw in May 1914,[27] and after that no more is heard of the Tunnel in the Society's annals until last year, when as you may recall there was a splendid discussion of the merits of the project from every point of view, and a resolution was passed 'that there are still serious doubts . . . about the value of the project to Great Britain as a whole which need further investigation'.[28]

Clearly this long-debated engineering scheme, touching every aspect of life in these islands very nearly, has been very fully discussed before this Society, and hence before a wide public. Every strong thrust has been reflected here, except that of 1928-30 which led to the appointment of a government committee whose majority report was favourable to the Channel Tunnel scheme,[29] and it may well have been felt then that action by the Society was needless. It is hard to know what more the Society could have done either to inform the public, or to apprise the financial and engineering interests pushing the Tunnel of the public's concern for broader issues than technical practicality and transport convenience. The latter rôle seems to me highly significant; it has never been the Society's intention to promote arts and commerce *at all costs*. Perhaps Lord Wolseley's fears of foreign invasion seem absurd now; but there was in his day a genuine (and not easily solved) problem of making sure that the British end of the Tunnel would be impregnable. The doubts voiced about the Channel Tunnel to-day are far more varied, and I fear substantial. Note that the Society took no definite stand on the question until the passing of the resolution I have just mentioned in this room on 5th June last year; previously the Society had provided a forum for debate—generally highly favourable to the Tunnel interests—without exercising pressure one way or the other. But it has constantly been, in this matter as in others, a guardian of morality in relation to technology and commerce in the manner I indicated in the beginning.

I have said little about science in this talk for the good reason that neither the advancement of pure science nor its dissemination to the public have been the direct concerns of this Society, which has rightly left these tasks to other bodies. The *applications* of science have of course often been discussed in this room, and I have touched on a few such instances; in the *Journal* of the earlier part of this century you will find much technical material of this kind treated in a professional way, mathematics and all. But of course the application of science to technology and engineering is the business of other bodies too, the engineering institutions in London, for example, which can deal with these matters in a specialized way impossible here, and I do not myself believe that the Society of Arts ever did, could or should find its principal rôle in this area. I will go further and say that in my opinion, based admittedly on a limited knowledge of the Society's history, the straightforward presentation of technical information and the professional discussion of technical problems have not been and are not activities for which this

Society provides a proper forum. I do not mean that the Society has not a duty to keep its members informed of major developments in the technological activities of this country, and that is the justification for such lectures as Marconi's on wireless telegraphy and others of the same kind since. But no one, I believe, would fail to agree that the proper places for the professional discussion of these new technologies were the Institutions of Electrical or Mechanical Engineers. The role of the Society of Arts in this context has been, in my view, to relate technology to the public interest, to transcend technology in the narrow professional sense by introducing moral and social considerations. It is for the engineers to debate whether something is best done in a particular way, and sometimes to advise on whether or not money can be made by doing the thing this way; it is and has been very much the concern of this Society to enlighten us on whether the thing is worth doing at all. And this is no negative position, for the Society has rightly insisted many times, from chimney sweeps and drains down to the present day, that many things *must* be done for moral and social reasons that blind financial necessity would have left undone. Of course the manner of the Society's concern for the public interest has changed often; once it was most directed to the cities, and recently it has been much concerned with the countryside. But the principle seems to me to have been substantially constant; the principle that society advances through technological progress mediated by moral values. Long may it remain so!

NOTES

1. *Gentleman's Magazine*, 26, for 1756, pp. 61-2. For information on the Society's history I have relied heavily on Sir Henry Trueman Wood, *A History of the Royal Society of Arts* (London, 1913) and Derek Hudson and K. W. Luckhurst, *The Royal Society of Arts, 1745-1954* (London, 1954). I am also most grateful to Mr. D. G. C Allan for his assistance.
2. Wood, 240ff. His observation that people are generally hostile to industrial innovation whether true or false, is irrelevant to the historical question of whether the Society could or did attain its declared objectives.
3. Wood, 259-62, also *Journal*, LX, 1911, 73-86; CX, 1962, 529-31.
4. The one well-known textile inventor linked with the Society of Arts was the Rev. Edmund Cartwright, who patented a power loom (1786) and a wool-combing machine (1792); he received medals from the Society for agricultural improvements and was a candidate for the Secretaryship in 1799.
5. On the relations of the Bramahs and the Wedgwoods with the Society see *Journal*, CVIII, 1959, 62-5 and CXIX, 1971, 327-31, 407-10.
6. See *Journal*, CXI, 1963, 577-81.
7. See A. and N. Clow, *The Chemical Revolution* (London, 1952), 99ff.
8. See *Journal*, LVIII, 1910, 354-68.
9. *Transactions*, I, 1783, vi.
10. Wood, 443.
11. *Journal*, XVI, 1867-8, 627-42.
12. For a detailed picture of the state of hygiene in British towns in 1876, see *Journal*, XXIV, 1876, 547-64.
13. E.g., the lecture by my colleague, Mr. F. E. Bruce, *Journal*, CII, 1953-4, 470.
14. See *Journal*, XXIV, 1876, 737-8.
15. Ibid., LXVI, 1918, 205.
16. Ibid., LXVII, 1919, 355ff. LXVIII, 1920, 492.
17. Ibid., LXVII, 1919, 337ff.
18. Ibid., XLVIII, 1900, 268-9.
19. Ibid., XLIX, 1901, 506-15.
20. Ibid., XXX, 560-72; 21st April 1882.
21. Ibid., XIV, 1866, 563.
22. Ibid., XVIII, 44-53, 3rd December 1869.
23. Ibid., XIX, 811.
24. Ibid., XXX, 565, 21st April 1882.
25. Ibid., XXXVII, 124-35, 18th January 1889.
26. Ibid., LXII, 88, 19th December 1913.
27. Ibid., LXII, 1914, 592-600.
28. *Journal*, Supplement: 'The Channel Tunnel, A Public Discussion', 5th June 1973, p. 57.
29. Cmd. 3513, 1930.

VIII

Guido's *Texaurus*, 1335

In 1328 died Charles the Fair, last Capetian in the direct line. His Valois successor was Philip VI. For a still very small France - even Dauphiné was yet to be acquired from the Empire - it was a fleeting moment of euphoria before the outbreak of the Hundred Years' War. Philip, who had earlier sought to join Alfonso IV of Aragon in reconquering Granada, was persuaded by the Avignon pope John XXII to take the cross only four years after his succession, a pledge which he renewed in 1336 before Benedict XII. How many knew that such a pledge was an empty dream or mere histrionics? Jerusalem had been in Christian hands only for a few intermittent years since Saladin retook it in 1187; the Templars and Hospitallers were no nearer to the Holy Land than Cyprus; Michael VIII Palaelogus had re-established the Greek empire in Constantinople. Yet officially, even after the debacles of St. Louis, the crusade was still incumbent on Christian rulers.

Did Guido da Vigevano suppose his king might really take ship for Acre, Damietta or Tunis, or did he merely perceive a peg on which to hand a learned tribute to his patron? One cannot guess, for he wrote with absolute seriousness. At the time when his first work was composed (1335) Guido had been in France an uncertain number of years, and was supposedly aged about fifty-five. Possibly qualified in medicine at Bologna, certainly a practitioner in his native city of Pavia, Guido had attached himself to the imperial cause, becoming personal physician to Henry VII during the latter's Italian adventure (1310-13). Since Guido appears in France as physician to the Emperor's daughter, Mary of Luxembourg, who became Queen to Charles the Fair, it seems likely that he followed Imperial patronage though he is said to have returned to practice at Pavia after 1313. Upon Mary's death he went into the service of Jeanne of Burgundy, wife of Philip VI.[1]

The occasion of his writing a "Treasury of the King of France for the recovery of the Holy Land beyond the sea, and of the health of his body, and of the prolongation of his life, together with a safeguard against poisons" is best expressed in his own stiff, legalistic phraseology:[2]

"Whereas in this present year 1335 an expedition beyond sea was ordered for reconquering the Holy Land on which account because I happened to know and understand almost all the sciences as befits a chief physician, I, Guido da Vigevano of Pavia, formerly physician to the Emperor Henry who am now by God's grace physician to the reverend and most holy lady Jeanne of Burgundy by God's grace Queen of France, thinking night and day how and in what manner the Holy Land can be more swiftly and easily gained, considering that the Saracens are furnished with many bodies of water both fresh and salt, and infinite number of towns and castles fortified with walls and moats; and because all that is good and beautiful comes from on high, as may be gathered from the chapter on the eternity of the world [in the book] concerning *The Heaven and the Earth*,[3] therefore God has given us a method of making an easy conquest of the Holy Land beyond sea which I write out for his highness the Lord Philip, by God's grace King of France because [he is] the Lord of the expedition."

But before describing this method, Guido does not neglect his duty as a physician; as he says, nothing in this world can be accomplished without good health and (here the profession speaks) good health cannot be maintained without medical advice; and perhaps this was especially neccssary for a middle-aged man (in his forties!) verging on elderly decrepitude. The treatise on the conservation of health, divided into eight chapters (including that on safe-guarding oneself against poison and bad food), need not detain us though it occupies some nine folios of the manuscript as compared with fourteen devoted to military technology; since the latter are well illustrated the number of words in each part of the *Texaurus* is probably about the same.[4]

Two treatises of the Middle Ages merit some comparison with the *Texaurus*. In Book III, Part IV, Chapters 18 and 19 of the *De regimine principium* of *Aegidius Columna* (c. 1285), there are descriptions of various forms of trebuchet and of escalading towers for use in attacking fortified places; Aegidius describes the same method of finding the height of a tower as Guido. More closely contemporary and nearer in purpose was the *Liber secretorum fidelium crucis super Terrae Sanctae recuperatione et conservatione* (c. 1306-13) of Marino Sanudo (Torsellus the Elder) who was born about 1260 and died in 1337; Marino devoted his whole life to the recovery of the Near East for Christianity and made no fewer than five journeys to the Orient. In Book II, Part IV, Chapter 8 there are some remarks on the construction of balistae and other war-machines, their transport by sea, and the quantities of stores required for crusading expeditions. And in Book III, Part XV, Chapter 5 Marino discusses the siting of castles. Neither of these writers entered into detailed engineering descriptions, nor did either offer original proposals in the manner of Guido.[5]

Thus the military chapters of the *Texaurus* stand out as unique in medieval literature between the Anonymous' *De rebus bellicis* and the engineers of the second half of the fifteenth century - Roberto Valturio, Francesco di Giorgio Martini, Leonardo - who seem to have pillaged Guido da Vigevano, or some intermediate source, as well as each other, pretty freely.[6] This is not to deny that other writers, Roger Bacon among them, had airily projected the possibility of ships without oars or sails, and carts without draught animals, but they certainly gave no technical description of the manner in which such objects may be attained, as Guido does; nor can we learn from them anything of contemporary practice and terminology, as we may from a study of the *Texaurus*. The meaning of all the technical words can be established with fair probability and they are listed at the end of this paper; the fact that virtually all are (like Guido himself) Italian indicates, I believe, the prevalent influence of the vernacular in technical contexts. Latin was fine for learned men but lacked the large vocabulary required by smiths and carpenters.

It would be too optimistic to claim that all Guido's devices can be clearly understood, however, and he himself frequently relies on the discretion of the skilled executant because he cannot express everything as he conceives it. The lack of punctuation in the text, the lack of exact definitions of terms, Guido's own frequent imprecision, prevent one's gaining a sharp image in every case of the spatial arrangements he intended. Moreover, though the writer appeals to his illustrations for further enlightenment, as we have them the figures do not really assist much. In many respects in fact they contradict the text; and whereas the reader longs for a detailed drawing - of the manner of joining sections in that post which Guido describes as all-important, for example - he gets what amounts to only a general view, not at all precise as to detail. In reading this text over many years I have failed to gain a clear idea of how the various pieces of the boat in Chapter VIII fit together, or why the paddle-shaft has a length exceeding twelve feet; nor can I clearly understand how Guido means to arrange the steerable axle of the waggons described in Chapters XI and XII, since although the pivoting of a fore-train is clearly implied, no detailed instructions are given for bringing this about while at the same time mounting the crank-and-gears drive in the manner described. Possibly Guido really had clear notions about *every* technical point in the *Texaurus*, but if so he failed to communicate them to twentieth century readers. Bertrand Gille, on the evidence of *The Renaissance Engineers*, seems to have formed only a very general (not always quite accurate) picture of Guido's technology, based largely on the manuscript's illustrations.

In such respects as these, it need hardly be added, the superiority of the late-fifteenth century manuscripts (though not of Valturio's 1472 woodcuts) is very marked; even the drawings of Villard de Honnecourt, about a century older, may compare advantageously with those of Guido. The Paris manuscript (see below) of the *Texaurus* is indeed a pretty thing to examine; but the fact is that in Villard's machines one can see what relations of the parts were intended, while in Guido's one cannot. Gille, indeed, has remarked that Villard was far more of a 'practical' man than Guido; while I think that this comment is inappropriate (consider Villard's perpetual motion machine, or his rotating angel!) his drawings are certainly more workmanlike.

Perhaps the original manuscript would have made some points clearer. There are three extant MSS of the *Texaurus*: (A) Paris, Bibliothèque Nationale MS. 11015, ff. 32-55; (B) presumably in private possession, ascribed to Martin of Aachen, written in Cyprus in 1375; (C) another later copy, said bv Gille (p. 28) to be in Turin, of which nothing more is known to me.

A) was not always in the *Bibliothèque du Roi*; it was acquired from Jean-Baptiste Colbert. Nor is it an archetypal manuscript but itself a copy from the late fourteenth century. This is evident from the unfinished initials, the omission of some essential phrases, some scribal errors, and the failure to supply all but one of the small textual figures (for which blank spaces are left). This MS was consulted in the preparation of the modern edition of Ducange (Niort, 1883) where many passages are quoted as examples of 'Latin' (really, Italian) words not employed elsewhere.

(B) was once the property of Conte Gugliemo Libri-Carruci (1809-69), and it would be interesting to know how he acquired it. At the Libri Sale in 1862 it was bought by Sir Thomas Phillips (£47) from whose vast collections it was sold by Messrs. Robinson (when I had the privilege of examining it) and was last known to me in the hands of Mr. Lawrence Witten of New Haven, Connecticut, who courteously corresponded with me about it. This MS. contains the whole *Texaurus* except the introductory matter, and has the phrases and text-drawings missing from (A). It was, therefore, not copied from (A) which may even be of later writing - nor even, perhaps, from the same progenitor from which (A) was derived.

The translation below is based upon (A), but I have added the few missing phrases from (B).

Guido's text contains the following technical peculiarities or innovations. (1) Because of the shortage of timber in the Holy Lands, and the difficulty of transporting it thither or there in large bulk, he proposes a sectional, pre-fabricated construction for his war-machines, each section being so compact that one or more of them may be carried by a horse. Sometimes sectionalised pieces are simply made to fold up for easy transport (e.g. the portable boat, the waggon-wheels), sometimes the sections are made to be broken apart, and are then interchangeable (e.g. the 'trunks' of the post, the boxes or pontoons for the floating bridge). Certain ideas of Guido's for the sectional assembly are explained in some detail (Chapter XIII, for example) but in other cases (e.g. the fighting-waggons) Guido really does not explain at all how the sectionalisation is to be effected. To a modest extent he envisaged *multiple* uses for the sectionally assembled components: the rope-ladders, for example (Chapters IV, V) can be used in more than one way; the parts of the pontoon bridge make a boat, and the sectional posts or poles reappear several times. Thus, up to a point, Guido conceived of a principle of military engineering which has only been realized in practice (and in steel) in the twentieth century.

    To permit such sectional assembly, two kinds of joints are much employed by Guido. The first is the hinge, particularly to make a screen-like alternate fold (figure 1). Door-bolts - bars passing through iron hoops - are used behind the hinges to make the joint rigid on assembly, or to join discrete sub-assemblies together. Thus a continuous surface or cladding could be made, as wide as the sections are long, and theoretically of infinite length. This cladding could be a protective fence, a parapet, or an approach road; of course more than one size would be required. The portable boat is sectionalized in much the same way, but I cannot understand precisely how this was supposed to be done to make a reasonably tight vessel of the right shape. Leather, caulking and pitch were, however, proposed as aids to sealing the joints.

    Secondly, Guido utilises a butt joint to assemble timber sections (or 'trunks') of varying diameter into posts and poles (these sections seem to be of unnecessarily short length). More than one joint is described (Chapter II) and neither is completely clear; one procedure may be interpreted thus (figure 2). Each sections is bored out longitudinally at either end to receive a stout iron pin, which thus joins them. The pin has four holes pierced in it, through which spikes are driven to fasten it to the sections; before assembly of the post, each section is carried with an empty hole at one end, and a pin (permanently fixed with two spikes) protuding from the other. This male-and-female joint is further strengthened either by welding protusions to the iron pin fitting into slots in the female joint, or by hooks and eyes, and in any case by iron strips or strained ropes running the length of the whole beam. It appears, too, that the fastening-spikes are slightly bent, in order to force the joints tight home.

(2) Guido employs for flotation casks, leather-covered frames, and pontoons. Presumably the basic idea is not at all original, but the sectional structure of the frames is.

(3) For the propulsion of his folding, portable boat Guido employs paddle-wheels (last heard of in *De rebus bellicis*) rotated by a double-compound crank, for which there is no previous evidence; a hint is given in the text that a crank-handle was then known in hand-mills, such as are illustrated by Taccola a century later. One must assume that the crank-handle was already widespread in association with the grindstone and the carpenter's brace, at least, though neither is well attested.

(4) It is worth noting Guido's rather frequent use of iron, for shafts here and in the waggons, and even apparently for structural purposes in the waggon (if so, the first structural use of wrought-iron save as a masonry cramp!)

(5) Chapter VI contains perhaps the first description of some sort of mechanical lift - perfectly practicable - and incidentally of a most primitive railway-like arrangement. I doubt whether this can have any relation to mining practice.

(6) Guido gives the first description of a self-propelled fighting-waggon, itself exploiting the crank as well as geared drive. Chapter XI clearly implies the basic model of a four-wheeled waggon having a steerable fore-carriage, that is, a solid and independent front axle turning about a vertical pivot attaching it to the waggon-bed, the shafts of course being linked to this front axle. This pivot is crudely shown in the drawings.

(7) Guido gives the first description of a tower-windmill upon whose mechanism (as he explicitly states) the waggon of Chapter XII is based.[7] Now it used to be supposed that the tower-mill was known not much before the later fourteenth century, the earliest illustrations of them dating from c. 1400, but a Pipe-roll of Edward I clearly refers to the construction of a *masonry* windmill ("uno molendino ventrico de petra") at Dover Castle in 1294-5 which must surely have been a tower-mill.[8] There is thus no improbability in my assertion that Guido shows us the working of a contemporary tower-mill, which is of course considerably more sophisticated than a post-mill. As he described it,

VIII

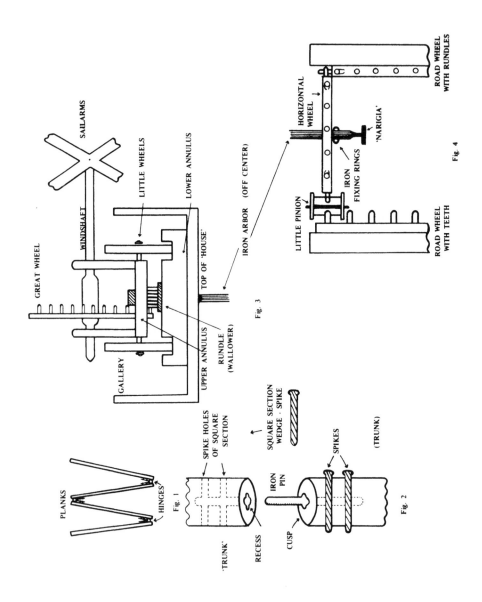

SAILARMS

GREAT WHEEL

WINDSHAFT

LITTLE WHEELS

LOWER ANNULUS

TOP OF 'HOUSE'

IRON ARBOR (OFF CENTER)

GALLERY

UPPER ANNULUS

RUNDLE (WALLOWER)

Fig. 3

HORIZONTAL WHEEL

ROAD WHEEL WITH RUNDLES

'NARIGIA'

IRON FIXING RINGS

LITTLE PINION

ROAD WHEEL WITH TEETH

Fig. 4

PLANKS

HINGES

SPIKE HOLES OF SQUARE SECTION

SQUARE SECTION WEDGE - SPIKE

'TRUNK'

RECESS

Fig. 1

IRON PIN

SPIKES

(TRUNK)

CUSP

Fig. 2

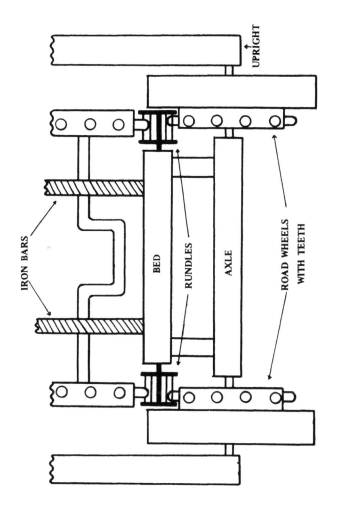

Fig. 5

omitting any reference to the upper covering, bonnet or cap of the mill, the windshaft is mounted on an upper annulus around whose circumference are a number of small wheels riding in a hollow track cut in a lower annulus resting on the tower of the mill (or "house" of the waggon). The windshaft carries a face-wheel engaging a pinion on the upper end of a vertical iron shaft, the upper bearing for this shaft being provided in a cross-timber fastened to the upper annulus, but in such a way (as Guido says) that it does not interfere with the face-wheel (figure 3).⁹ Guido also realises clearly that the axis of the drive-shaft (with its pinion) and of the annular bearings must be the same, and explains how the windshaft-annullus is levered round to bring the sails into the wind. At the lower end of the drive-shaft (figure 4) the chief point of interest is the provision of an idle-pinion between the drive-cog and one road-wheel; only thus, of course, could both roadwheels be made to rotate in the same direction, a point Leonardo forgot in a famous sketch.¹⁰

Capstan-power is provided lest the wind fail, and the front axle (or rather, in Chapter XII, the rear axle) may also be driven like the axles of the waggon in Chapter XI, by crank and gears (figure 5). There are obscurities in the description at this point, but I suppose that the iron bars ("ferramenta") mentioned are placed vertically, and that the holes in them are bearings for the crankshaft. But perhaps another reconstruction may be attempted.

Certainly Guido da Vigevano was quite well acquainted with the craftsmanship and language of artisans. He had some notion of what smith's work was possibie, knew the distinction between wheels with cogs on their faces, and those with cogs on their edges, and distinguished both these from rundle-wheels; he knew that the matching ("adaequandum") of gears and determination of ratios was the business of miilwrights; he is aware of a distinction between the specialists of wind and water-power. One gets a glimpse of that impressive mechanical technology, whose consequences rather than whose technical methods are best displayed in other medieval documents, and whose importance Lynn White has clearly displayed in *Medieval Technology and Social Change*, and other writings.

Translation of the XIII Chapters
De Rebus Bellicis

The text is that of MS. (A). Words enclosed within brackets marked with an asterisk are taken from MS. (B). All other words in brackets are the translator's additions.

The text figures after the words "... made thus" are all lacking from the MS. except that of the windmill-sails, figure 19.

In translating measures, brachium *has been rendered loosely as "half a yard" (it should be rather more), semmissis (generally) as "half a foot". Works like "span" are retained.*

[41r] Now I revert to the knowledge of the conquest of the Holy Land beyond the sea. The way of conquering the Holy Land, in the name of Christ and with dedication to the King of France.

Towns, cities and castles are to be conquered in war, whether by sea or by land, and they can be won by applying ingenuity and possessing a just cause, with the aid of devices constructed in many ways, by a divine miracle. And because it would be a mighty business to transport beyond sea so many heavy [and] ponderous devices for the conquest of the Holy Land, that method cannot be adopted. And since it will be necessary to break suddenly into towns, cities and castles both by night and by day, and to cross with horse and foot rivers both great and small, without delays, I Guido with God's help (as it is a just cause to wrest the Holy Lands beyond the sea from the hands of [our] enemies) will so compose all the devices for the conquest of the Holy Land both by sea and by land that they can be easily borne by horses, and disposed without delay for the actual work of conquest. This I will declare step by step in the chapters below, and after each chapter I always give a clear example [illustration].

Chapter I. How those attacking cities towns and castles ought to protect themselves from the arrows of the Saracens and how to make protective structures and transport them on horses.

Chapter II. On the way of making posts and platforms when there is need to capture towns, cities and castles, and of carrying them folded up on horses.

Chapter III. On the way of making bridges on dry land for placing on the walls of cities and castles, and of carrying them on horses.

Chapter IV. On the way of taking towers of any height.

Chapter V. On the way of making assault ladders and carrying them folded up on horses.

Chapter VI. On the way of making an assault-castle which is placed near the wall of a city, and higher than the wall, and of carrying it in small pieces on horseback.

Chapter VII. On the way of making bridges for rivers, and of carrying them in small pieces on horseback and placing them in the water in one hour.

Chapter VIII. On the way of making boats [capable of] sailing on any stretch of water and on the sea, and of carrying them folded up on horses.

Chapter IX. On the way of riding horses through water.

Chapter X. How infantry may cross large stretches of water.

Chapter XI. On the way of making an assault-car which is propelled without wind or animals, in order to throw large forces into confusion; and it may be carried in sections on horseback.

Chapter XII. For the building of another assault-car which is propelled by the wind without animals, and runs about the [battle-] field with a tremendous violence, [enabling] a small troop to throw a great army into confusion; and it can be carried in sections on horseback.

Chapter XIII. On the way of constructing "panthers" to contain large forces.

CHAPTER I. HOW ATTACKERS SHOULD PROTECT THEMSELVES FROM THE ARROWS OF THE SARACENS

Let them take cotton quilts, and on top of the cotton before the quilt let the powder of iron filings be lightly scattered; it is to be made in this wise. Let filings fair of waters [washed clean] be taken and placed in a copper or iron cup, and this cup with the filings held over live coals, always stirring the filings with an iron spatula until the filings are seen to be red, and then let them be removed from the fire and thrown hot into very sharp vinegar and mixed with the vinegar; and when the filings are cold let them be taken out and again placed in the cup upon the fire, and when it is red let them be thrown into the vinegar, and so on doing this many times the powder will be made exceedingly fine; and that powder is to be well mixed with powder of mastic and scattered upon the cotton and smoothly and firmly rolled up between two cloths; and the quilts are a bit thicker than those we use, and they are to be as long as the makers see fit. And these quilts are to be stretched out along poles with rings, as curtains are, and the quilts are held upright and erect with other poles in face of the arrows; and they make so many of these in number as it seems proper to send out, and in these quilts they make many loopholes.

And because of large crossbows they should make another device viz.: let them take light thin planks two and a half yards long and half a yard broad or thereabouts and six of these are to be joined together with hinges placed in opposite senses so that they can be gathered together [i.e. folded like a screen]; and at the ends of each board a thin cord is fixed with a nail so that when needful two thin rods may be tied into postion so that they may hold all the boards stretched out, save the two at the ends. A loophole is to be made in each board and a stick is joined on about the middle of each board to prop it up [like] an upright door and with these sticks the "door" may be moved about. And because of doubt about the crossbows of [illegible] the backs of these planks may be covered with thin iron; let as many of these "doors" be made as seems proper to the maker and stouter and longer; and the structures described can be made effective by [using] these "doors" and the soldiers safeguarded from harm.

(42r) CHAPTER II. ON THE WAY OF MAKING POSTS (Fig. 6).

A suitable post is necessary for many tasks. For that reason one must begin with this alone; and that may be done in this way. Let "trunks" [sections] be prepared from a single good log, let them be about two yards long each "trunk", and as thick as a man's leg; and every "trunk" has two holes (namely one at one end and one at the other, [bored] lengthwise) and the hole is to be nine inches deep from one end and the same at the other. And further let one hole only at one end have two hollows, that is one on one side and one on the other, as large and deep as the point of a knife, and let those holes be as wide as a large thumb so that one section may be joined to another with an iron pin in this way; namely: let there be an iron pin made half a yard long and as thick as the hole of that section, and on one side of the pin, nine inches along are two projections one on one side and one on the other as high and thick as the point of a knife so made that the pin with its projections may be placed in that hole which has two hollows, and that pin may have four holes (two at one end and two at the other) large enough to take a stout spike. And those two holes which are towards the projections are a little bit oblong and the other two (at the other end) are to be round; and when the pin is thus prepared that part of the pin which is without projections is placed in one end of one "trunk", and so the pin is firmly nailed in so that it will never shift. And when it is necessary to make the post the remaining [part of] the iron pin with the points is placed in the other "trunk" and fixed with two other spikes so arranged that those two spikes can be taken out and removed when it is necessary to make the beam. And those two holes [in the protruding end of the pin] are so matched with the holes in the wood that the two "trunks" are forced tight together. And near the two ends the "trunk" is bound with iron bands. And four iron strips as wide as a thick finger and as thick as the thick point of a knife are placed lengthwise from one ring to the other down the length of the section and these are firmly nailed to the "trunk"; or the "trunks" or rings may be held tightly together by cords lengthwise so that the "trunk" post?] may be strong and light, as shall seem fitting to the maker of it.

And in order that they may be pierced correctly, let them be pierced with a lathe [drill?] and then of necessity [the post] will always be straight so long as the iron pins are straight. And so you will have a complete post, strong, and as long as the maker wishes. And the post thus fabricated is the basis for the whole recovery of those lands, whence the greatest degree of attention should be paid to this post. And next let a round platform be made, which is [to be] placed on the head of that post, in this way: namely, have two [circular] base-pieces made of larch-planks an inch thick, and let them be three-quarters of a yard in diameter or so much that one or two men with crossbows can wind themselves up inside it. And let thin planks be taken nine inches wide and one and a quarter quarter yards long, and let them be joined together with hinges placed inside and outside in such a way that the planks can be rolled up into a package. And when it is necessary they are laid out flat, and then curved round those [two circular] base-pieces and nailed to them, one base-piece being placed at least nine inches distant [vertically] from the other; and let the planks be firmly nailed upon [both]. And let two pulleys be placed on top of the post by means of which the platform is hauled up by [pulling on] two ropes tied to the base-pieces of the platform. And let two struts be provided each four yards long, made in sections in the same manner as the post. And let there be four hoops on the post at about the sixth section with which the struts are knotted or tied on. And when it is necessary to erect the platform against some tower or the wall of a city or castle it is done thus. First have the post of the platform laid out a good way from the tower or wall of the castle, and place the platform on the post as far as the struts; and have the four struts tied on to their hoops. And take

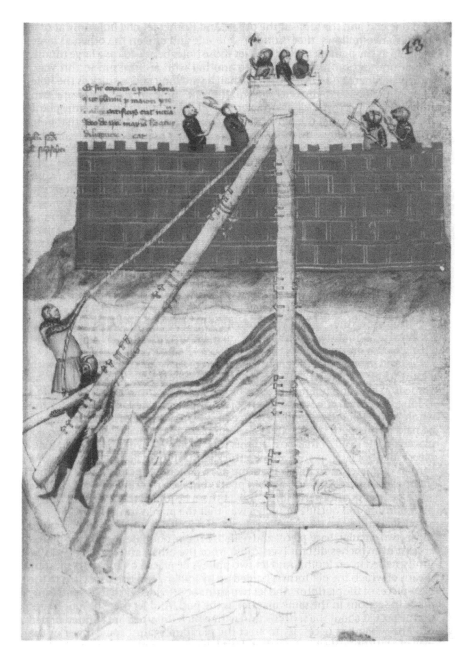

Fig. 6

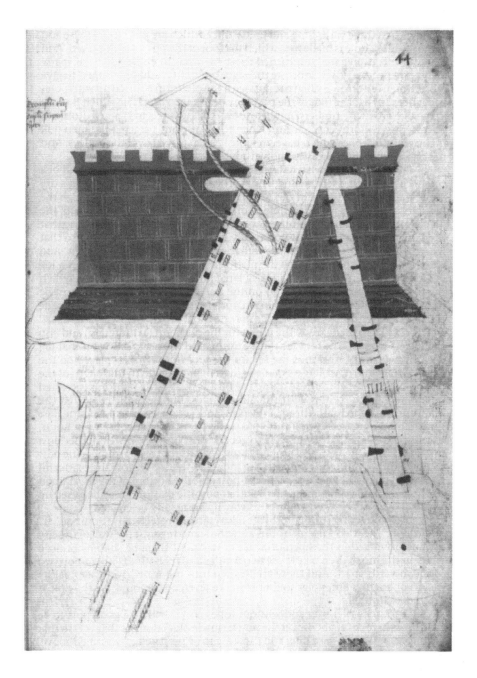

Fig. 7

two logs each three yards long and pretty thick which are placed crosswise and in the middle a hole is made into which the foot of the post may be placed. And when all this is prepared it is carried to where it has to be erected by as many men as are necessary, while before them go the "quilts" and the "doors" to give protection from arrows. Then the post with the platform is raised up with poles and lances so that the foot of the post may be put into that hole which is made in the two crosswise timbers. But before everything is erected those who are to go into the platform enter it. And when the platform is erected the men pull [it] up with the ropes which have previously been placed over the pulleys. And the platform is erected easily and without risk.

But if we should wish to lean this platform or post against the wall of a city or castle [so that it is inclined, not vertical], and to place the same in a dry moat, then it may be done in this way. Let the platform be pierced through the [double] base-pieces, with the hole in one base made close to one edge [of the platform] and the hole in the other [above or below it] close to the other [so that the post now passes through not axially but aslant] and this is done so that when the post is inclined against a wall the platform will remain upright. And when this has been done place the platform upon the post, and obtain a log of wood three yards long, and make a large hole in the middle of it in which the foot of the post may rest; and on the post are two struts, which must be fastened to this log so as to keep the post upright. Then obtain two trestles made thus which are placed in the bottom of the moat crosswise, two yards apart. And with all this made ready the post with the platform (as aforesaid) and also that log of wood as the foot for the platform, and the post resting on that log with its two struts. And thus the log for the foot of the post is placed on the trestles in the moat, which are made as high in proportion to the length of the post as seems proper to those doing the business, and the platform is erected upon these same trestles with the aid of poles. But before it is erected the platform is fastened at the head of the post, and a man gets into it, and the post is restrained by ropes so that it may be rested gently against the wall. But if, because of its length, it is difficult to get the post upright, have another post made as long as the first, and properly built with its foot and struts like the post of the platform, and the two posts are joined at the top (below the platform) by an iron pin; and when the platform-post is at the bottom of the moat upon the trestles, then with hands and the other post the platform may be raised up to rest it against the wall, and in this way it can readily be erected.

Those erecting the platform are always to be protected by those "quilts" and "doors". And as many platforms are to be leaned against the wall of the city as those involved in the operation think fit. And if there be shallow or deep water in the moat, then take those two trestles and fix both together upon two logs two yards apart. And these are thrown into the water over the moat by means of engines and poles so that they are level, and upon these the pole is erected as aforesaid.

Another easier way and without ropes and it will be a good post to be held up with ropes as above. Let the post be made of sections or "trunks" so that each section is half a yard long and these are to be prepared as described above; and the sections are joined together with an iron pin as above. But the pins are each half a yard long (all but two inches), and the pin has four holes and the section another four into which four iron wedges are inserted, and with these two sections are made fast together upon the pin, and so with the remainder; and at the end of one section let there be an iron band four inches broad with three spikes made thus. And in another section let there be another hand with three grooves into which the three spikes enter, made thus; and each spike is joined up with an iron wedge. But when the post is made in this way the hole in the section is round and so is the iron pin. And I will draw this pole in the illustration because it is easier to sketch, with an iron band four inches wide placed in the middle of each section where they are to be joined together. And thus a good post is completed, which will be very necessary for the most part of the other devices. And so the greatest of care is to be taken with it.

CHAPTER III. ON THE WAY OF MAKING A BRIDGE AND RESTING IT UPON THE WALL OF A CITY (Fig. 7).

First, let thin, light planks be taken, nearly two yards long, and half or a quarter of a yard broad in the middle, in number proportionate to the length we wish to give to the bridge, and let them be joined together by iron hinges placed cross-wise and on either side so that the planks may be folded up into a bundle and conveyed on horses. And let eight holes be made in each plank, that is to say [four at one end of the plank and another] four at the other end, nine inches from the end, with the second two [holes] two inches away and three inches from the edge of the plank. And these planks are threaded through the four holes with two ropes [on one side and two] on the other side of the plank, that is to say in opposite senses; and these ropes are as thick as a thick finger. And when this has been assembled let the planks, thus threaded, be stretched out and the ropes stretched tight with wooden levers as big as a lance, long enough so that one lever may be linked with another so that they can be lashed together. And the levers of one side of the planks are lashed to the levers of the other side, and holes are made as those doing the job see fit. And when this is done the bridge will be stiff, strong and light. And at the end of this bridge, that is half a yard from the top of it or as much as the thickness of the wall, an axle is placed below it which sticks out nine inches on either side. And upon this axle are fastened two posts rather stouter than the post of the platform, as long as the distance from the bottom of the moat to the top of the wall where the bridge is to rest. And at their feet those two posts are three yards apart. But because it would be a tricky business to take the measures for the length of these posts, take two logs four and a half yards long and pretty thick and upon these make trestles, upon which the foot of the posts is placed over a log four yards long. Upon these let the foot of the posts be placed as has been said concerning the platform. Upon these the bridge is raised and lowered so that it falls on the wall correctly, [that is] the portion beyond the axle. And when this has been made ready another bridge is to be made by the same method as before, and about three times as long as the estimated height of the wall within the town, and this is joined to the head of the other bridge with two hinges, that is to say to the part of the bridge beyond the axle. And with all this made ready let this drawbridge be held up a little way, suspended by two ropes fastened to the [main section of the] bridge, so arranged that when the bridge is leaned against the wall and those two ropes are taken away the drawbridge will fall down inside the town. And the first [i.e. the main section] of the bridge should be so far from the wall across the moat that the bridge is level enough for horse and foot to pass over it.

And many lances may be thrust against the bridge beyond the poles or support of the bridge so that with their aid it may be flung up upon the wall. And thus the bridge may be transported with its drawbridge [held] above it with ropes as has been said. And when the bridge is transported across the moat, let trestles be placed in position first and then the foot of the bridge upon them, and it is immediately heaved up upon the wall by means of lances and hands and immediately the cords are raised allowing the drawbridge to fall into the town. But before the bridge is made ready it should first be known whether the drawbridge will fall on the ground or upon houses, and if it fall on the ground well and good. And if it fall on a house let them have at hand many bridges both long and short and one of these is laid down from the drawbridge to the ground-level; it is to be lashed on to the end of the draw-bridge and flung cross-wise on to the ground. And to the end that the bridge may be placed in its position in security first of all two platforms are placed on the wall and then the bridge between them, and no one will dare to attack the bridge. And to enable the foot of the bridge or post to be transported on horseback they are made in advance in sections, as has been said of the post for the platform; and this bridge may be provided with a parapet as is said below of the river bridge, if those in charge think it necessary. And thus the bridge for gaining entry to cities and towns is completed.

CHAPTER IV. ON THE WAY OF ASSAULTING TOWERS (Fig. 8).

Let the height of the tower first be taken in this wise: let a man take a stick so long that it reaches from the ground to his eyes, and let this stick be held upright while the man lies on his back on the ground with his feet against the stick and pointing them towards the tower; and let the man lying on his back on the ground look towards the tower, carefully sighting the top of the tower against the tip of the stick, and when he has made this correct sighting let him measure how many stick-lengths there are from his feet to the foot of the tower, and it will be so many stick-lengths tall plus one more. And let four posts be made according to the length of this measure or rather more than the height of the tower, using larch-wood in sections in the manner employed for the post of the platform, and from these let them make two ladders, [two yards]* wide at the base and one yard wide [at the top]*. One ladder, namely that which is placed nearer to the army, is made rather longer than the other; and the rungs of that ladder or ladders are made in this way, namely as long and short as the ladder is at top and bottom, and of good wood and thick in order to be strong, and at the ends these timbers are thinly covered with iron and bored through. Then they are threaded upon two ropes as big as a thick thumb and as long as the ladders are, and tightly lashed on at such a distance apart as the operatives see fit. And these ladders thus made from ropes can be packaged up and extended when necessary and transported upon a horse. And when it is necessary to bring them into use they can be stretched out and strongly lashed to the posts and thus made into good and sufficient ladders. And afterwards these two ladders are linked together tightly at the top by means of an axle in such a way that the ladders can pivot upon the axle. Next let them make a bridge of thin boards one yard wide and more, made in the way described for the "dry bridge" described before, of such a length as the people doing it see fit, and that bridge is to be joined by means of [iron] rings or hoops to the linked top of the inner ladders, that is to say on the side towards the tower; and this bridge is held up above the ladders, having two ropes joined to the foot of the bridge which pass over two pulleys on the top of the ladders, so placed that those remaining on the ground can, with the ropes, raise and lower the bridge. And this bridge may be fitted with a parapet as is said below of the river bridge. And thus with everything made ready the ladders along with the bridge are brought up, the ladders being held manually with the bridge towards the tower, and the foot of the ladders is set in the moat (if there is a moat), whether it be dry or wet, or if there be no moat, and the ladders are set up with other ladders and the bridge is held up above the ladders, and as soon as the ladders are set up let the bridge fall down upon the tower or near it; and all these things are to be arranged by those in charge as they judge best, always adjusting things according to the arrangement of the tower and the ladders. And when the ladders have been set up thus four [men-at-arms] climb up them and cross the bridge to assail the tower. And the ladders [on the side] towards the tower are covered with the above-described quilts, and they ascend the other ladders and will be safeguarded from the crossbows of the tower; and let those carrying and erecting the ladders not forget to have the defences placed in the first chapter. And thus the structures for overcoming towers however high they may be are completed, against which no tower can be defended.

CHAPTER V. ON THE WAY OF MAKING FORTIFIED LADDERS AND CARRYING THEM ROLLED UP ON HORSEBACK (Fig. 9).

Let ladders be prepared secretly in the following way: have them make rungs of logs half a yard long and more and as thick as a large thumb, and in good quantity, and of stout wood, and at the end[s] of each of them make a small hole no bigger than a slender straw. And all the rungs are strung along two long ropes as thick as a thumb and firmly lashed with coarse thread, as the makers shall see fit, as far apart as the rungs of a ladder ought to be, and so made they are to be rolled up into a bundle so as to be easily carried. After this let them obtain larch-poles as thick as a big lance, and made in sections joined together in the manner described for the post of the platform. And when many sections have been joined together the ropes with the rungs are to be lashed to them; and a log is obtained, some three feet long, upon which the foot of that ladder is placed. And next let them make platforms in this way, to be placed at the top of the ladders: take a thin and light plank one yard wide and half a yard broad and something more, which is set down as the base, and take four planks to be the sides to this base which are to be joined to the base; being so joined with hinges that one side falls [inwards] upon another, so that the platform can be carried the more easily, and opened up by four door-bolts connected to the tops of the side-pieces, as the makers shall see fit. And let them take as many of these poles so put together as we wish to make long ladders, and on one side of the

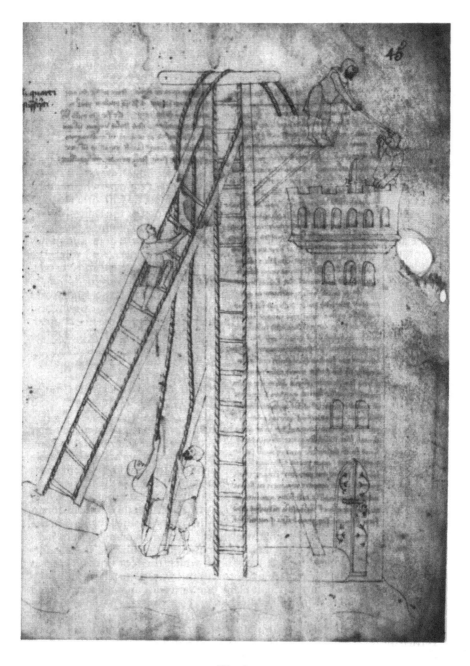

Fig. 8

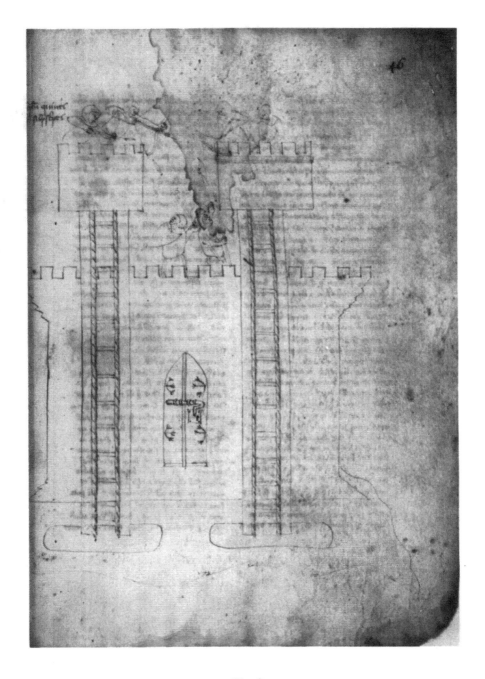

Fig. 9

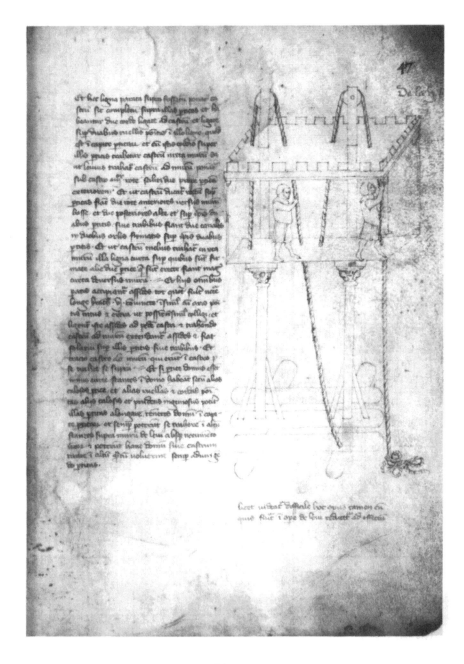

Fig. 10

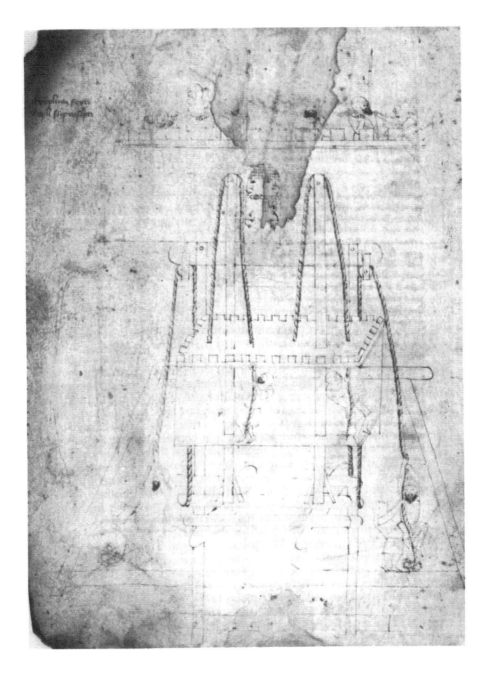

Fig. 11

base of the platform let rings of iron be placed, so wide that they can be placed over the tops of the poles and fixed with nails and irons driven in the top section of the pole so that they can be fixed and removed when necessary, as the makers shall think best. And on the other side the platform is supported by two iron handles [?] fixed on the poles of the ladders as the makers shall see fit, for it is not possible to set out every detail. But when someone undertakes this work he will himself make provision for undertaking this platform. And so in this way a single horse will carry fifteen or twenty of these ladders with platforms and thirty or forty without platforms, each three and a half, five or ten yards long, and thus these secret fortified ladders are finished.

CHAPTER VI. ON THE WAY OF MAKING A FORTIFIED CASTLE WHICH IS POSITIONED AGAINST THE WALL OF A CITY AND HIGHER THAN THE WALL, AND IS CARRIED ON HORSEBACK (Fig. 10 and 11).

Take first a larch plank three inches thick, half a yard wide and four yards long, and this plank is jointed in the middle with four strong iron hinges so that it can be doubled up on one side and carried on a horse; and two door bolts as thick as one's arm with four thick rings for each are fixed to the other side of the plank, at three inches from the edge. And let two holes as wide as a man's leg be made in that plank, one about thirty inches from one end and the other the same distance from the other end. And next let them take two posts as thick as a man's leg made in sections and joined by the method described for the post of the platform, and as long as the height we wish the castle to attain, that is to say six, eight or ten yards. And after this let them take eight light squared timbers, one and a half yards long and as thick as one's arm and these eight timbers are placed crosswise above that plank half a yard apart and firmly nailed to the plank; and upon the ends of these timbers are fixed two [further] timbers four yards long and as thick as a big arm, and let these two be nailed down upon the ends of those [eight]. And upon the four corners of these timbers four uprights [?] are fitted, as thick as a big arm. And upon these uprights are placed two crosspieces of wood, fore and aft, in which two holes wide enough to admit the posts are to be made. And upon the [first] plank and [other] timbers a fortified house is built, two and a half yards high, with a roof and walls [made] of thin light planks; and in the roof (centred on the two holes in the [first] plank) two holes are to be cut, somewhat wider than those two holes. And with the plank thus made ready with its house, after this let them take three squared timbers the size of a man's thigh and five yards long, and the two shorter beams are fixed cross-wise under the other longer ones, namely [so as to lie under] the centre of the two holes in the [first] plank, being so positioned that the two central crossings line up with those two holes in that plank; and above in the middle of the [two central] crossing-places holes are to be made large enough for the feet of the posts to be fitted inside. And after doing this let four struts be made for each post to be placed below the [first] big plank as the makers shall think best in order to keep the posts upright.

And when the castle with its foot has been thus constructed let two pulleys be fitted at the [head]* of each pole by means of which the house is raised up using ropes fastened to the [first] big plank, and so arranged that those

standing within the house can pull themselves up. And thus we have constructed a castle which can be carried by fifteen men and conveyed in pieces
on horseback. And when this has been made ready, so that the castle may be
moved up against the wall of the city beyond the ditch (whether wet or dry)
proceed thus. Let them take timbers as thick as one's thigh made in sections as
was described above of the post of the platform, and let them be fastened
together as above to make two poles as long as the distance from the lip of the
moat to the wall of the city. And a timber of the same thickness as above two
and a half yards long is placed crosswise under these two poles, about the
middle of them more or less as the makers think fit. And two feet or props are
placed under that [cross] timber of such a length as the makers think fit. And
these feet are lashed to hoops underneath that big [cross] timber where the
makers think right, and the [two] poles or beams with [the crossbeam and its]
two feet are thrust [across the moat] up to the wall, and another wooden
crosspiece is fastened to the end of the two beams, like the one in the middle.
And when these timbers have been made ready across the moat the castle thus
completed is placed on those poles [or beams]; and two ropes are tied to the
castle passing over pulleys which are mounted on the [cross] timber at the end
of the poles, and the castle is pulled upon the poles up to the wall by means of
these ropes. But in order to draw the castle more easily up to the wall, four
wheels are placed under it, namely two close to [one] outside pole [and the
other two near the other outside pole]*. And to make the castle go straight upon
the beams let the two fore-wheels towards the wall be small and the two rear
wheels large, and let two channels be made on the two beams by fastening two
[pairs of] laths upon them. And to draw the castle closer to the wall let the short
timbers upon which the two upright poles are mounted be made shorter at the
front towards the wall. And with all thus made ready let them take as many
planks as are necessary, three yards long and joined together by means of
hinges placed on this side and that so that the planks may be folded together,
and these are tied to the feet of the castle so that as this is pulled towards the
wall the planks are extended to form a walk upon the poles or beams. And when
the castle has been hauled up to the wall those who are within it draw
themselves up. And if the posts of the "house" should be too short those
standing within it should have with them spare sections of the posts and spare
pulleys and rope fixed to the spare sections, and with these a skilful engineer
can lengthen the poles, keeping the house at the top of the posts; and they will
always be able to heave themselves up above the level of those standing upon
the wall, easily and without danger to themselves; and they can raise this house
or castle as high as they wish by always adding more [sections to the] posts.

CHAPIER VII. ON THE WAY OF MAKING BRIDGES OVER WATER, AND
CARRYING THEM IN PIECES UPON HORSEBACK AND GETTING THEM INTO
THE WATER WITHIN AN HOUR (Fig. 12).

Let them first take thin larch-planks about two yards long and half a yard broad;
taking two of these, let them place one over the other with a space three inches
deep between, and in between laths of wood are placed, one inch wide, so that
the planks cannot be shaken loose. Then round the sides and ends of the two

Fig. 12

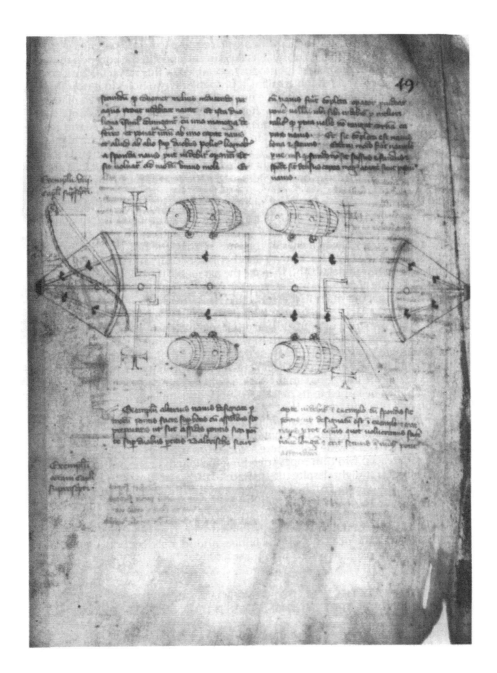

Fig. 13

planks strips of wood nine inches wide are placed, and so firmly fastened with nails and glue that neither air nor water can gain entrance [to the pontoon thus formed]. And from the planks [pontoons] assembled in this way a bridge is to be made as follows: each doubled plank [pontoon] has two holes in it near the end of each side; these holes are to be nine inches from the end of each plank and three inches from the side. The hole is to be [rectangular], three inches across and one inch wide, so that a rope three inches wide can be placed within it. And the holes are [to be cut so as] to incline more to one edge of the plank than towards the other. And after this one plank is to be joined to another by means of four pintle hinges, strongly nailed on; that is to say the planks are joined together with hinges on both sides so that they can be folded up together into a package.

And thus all the planks, so "doubled" and linked, may be threaded together evenly upon ropes of the width stated passing through the four holes [in each], to the number of ten or twelve together as much as a horse can carry, and the said ropes have buckles at the ends so that one horse-load may be immediately joined and buckled [to another horse-load]* and quickly placed in the water without delay. And to make the parapet of the bridge let them fix at the end of the plank nine inches from the edge two iron rods with twin ends made thus and about thirty inches long; these [hinged] iron rods lie lengthwise on the plank so that each plank has four of these rods lying upon it; and when needful they are stood upright as is made clear and obvious in the illustration because I cannot so well set it forth in words as I see it in my mind's eye. But the picture will show it. And from these the parapet to the bridge is made so that it can be put in place or removed again quickly and without fuss. And at its end each plank of the bridge has an inch-wide hole three inches from the side so that each may be lashed to its neighbour, and so each plank has eight holes, four on one side and four on the other.

Also each plank has four iron rings on its upper surface a span distant from the ends and three inches from the sides - the upper surface of the plank is that which has the wide holes more inclined towards the edge - and in these rings are threaded two thin ropes stretching the length of the bridge. And to keep the bridge straight despite the strength of the water [flow of the river] let an inch-thick rope be fastened to an anchor and thrown into the water a bowshot from the bridge, and under this rope let there be a couple of dinghies to hold it above the water; and if necessary several ropes may be put out to keep it straight. And these dinghies will be described in the chapter on boats, because they will always be needed with the bridge. And so a perfect bridge may be obtained for crossing all waters with horse and foot.

CHAPTER VIII. ON THE WAY OF MAKING BOATS SAILING UPON ALL SORTS OF WATERS AND CARRYING THEM FOLDED UP ON HORSEBACK (Fig. 13).

Let them first take three planks (or more, or fewer) such that their combined breadth is nearly two yards and their length three yards. And from these let the bottom of the boat be made, thoroughly hammered up and made staunch so that no water can enter it. And this bottom is to be somewhat pointed towards the ends thus; in order to make the bow and stern of the boat, which are not very sharp. And this bottom is split lengthwise down the middle and on its upper surface near the split two thin planks are fastened, one on each side of the split, each a span broad and as long as the division; these are nailed to the bottom and the [two halves of the] bottom are joined together by pintle hinges placed beneath the bottom about half a yard apart and so joined that the boat-bottom can be folded back on itself [lengthwise]. And the hinges and the slit are covered at the back with well tarred leather nailed on to the bottom; and the leather itself is covered as far as the hoops with thin planks nailed to the bottom; and the edges of the planks along the division [of the bottom] are chamfered a little inside because of the curvature of the boat so that the boat may be given a curved shape of a span; and on the outer sides of the base inside at three inches from the edge two other thin planks are fastened, one span wide, and strongly nailed on to the bottom; and these are to be a little chamfered towards the edge as needed because of the curvature so that the sides of the boat may come straight up but leaning outwards a little as they should. And many thin cross-pieces are fastened on top of the bottom between those two thin planks in order to strengthen the bottom.

And after this let them make the sides of the boat, each of them two and a half feet broad. And to enable them to fit the bottom at the ends they must slope inwards on one side so that towards the ends they rise up and may be joined to the base, thus. And the planks of these sides are firmly hammered up and made staunch. And thin planks a span wide are attached to the exterior of the sides near the edges where they are joined to the bottom. And the sides are joined to the bottom inside by means of hinges so fixed that the sides can pivot and fall [inwards] on the bottom. And these hinges are half a yard apart. And so the bottom shall be folded back on itself and the sides [shall be] inside upon the bottom; and at the end is a [rubbing] strip outside such as other boats have. And thus the boat can be carried folded up on horseback. And after this let them prepare seven ribs, namely two complete ones made thus which are positioned at the ends of the boat and on which the planks of the boat are strongly nailed, and another five ribs made thus positioned in the middle on which the planks are firmly nailed when the need arises. But before the boat is fastened to its ribs let them have tow ready, pitched and waxed, which is inserted into the three divisions [between the halves of the bottom, and the bottom and the sides] and then the boat is strongly nailed to the ribs and afterwards caulked with tow using chisels and hammers. A skilled man will easily understand this because I cannot write it more clearly. And so that the boat may be made unsinkable let them attach to its sides four wooden casks each a yard and a half long and a span wide. And thus is completed a good and safe boat for four horses, which can be got ready in an hour and put into the water either by night or by day; a single horse will carry one (or three horses, two) of them folded up separately. And so that the boat may be strongly propelled through the water by fewer men, let them make paddles in this way, namely let them take two logs as thick as a man's arm and two yards long each, according to the judgement of the makers, and on their ends let them fit two paddles crosswise, a foot broad and a yard long more or less, as the boatmen shall judge best for propelling it through the water. And those two timbers are joined together by an iron [crank] handle and placed one at each end of the boat upon two polished pieces of wood as high above the sides as the makers think fit. And so they are turned like

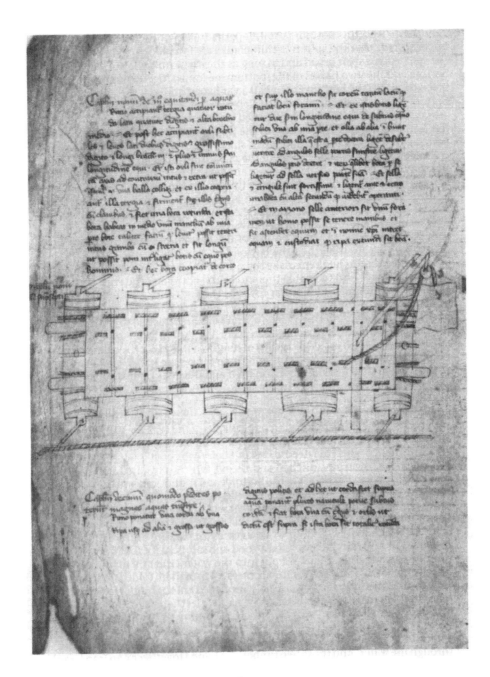

Fig. 14

Fig. 15

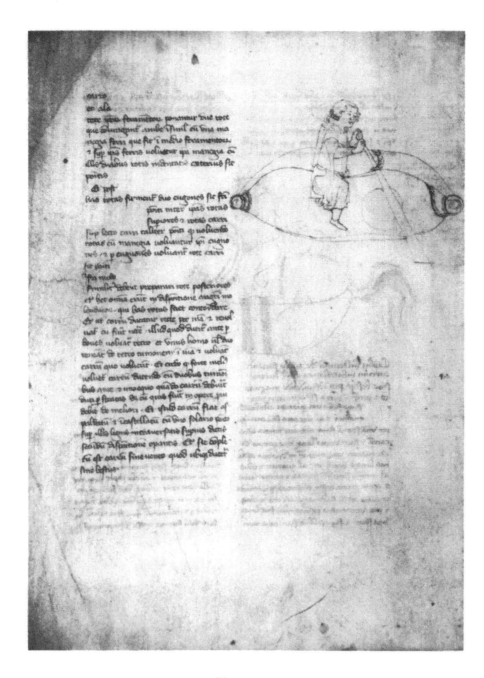

Fig. 16

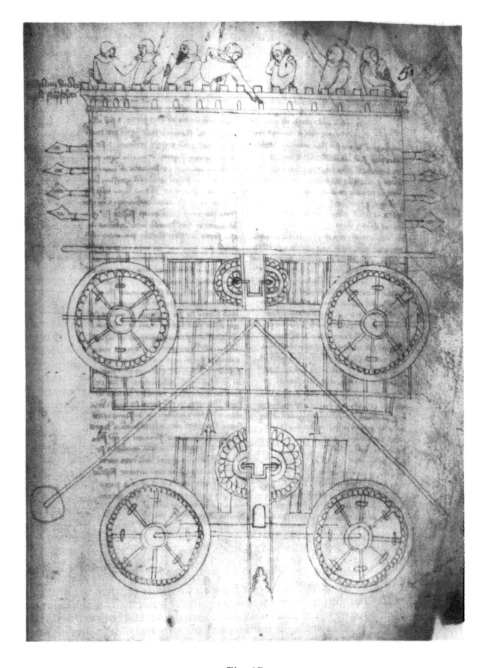

Fig. 17

a [hand-]mill. And when the boat is finished the maker may take care to place a sail where he thinks best provided that the mast does not touch the rib at the front of the boat. And thus a good and safe boat is completed. Dinghies are made in the same way except that the bottom is not divided and the bottom and the sides are made more pointed towards the ends as in the prow [?] of a boat.

(Fig. 14). An illustration of another boat designed like a floating bridge made on casks, with planks prepared in the manner of the [floating] bridge placed on two platform-poles, as is readily seen in the picture, with side-walls attached as shown in the picture; and the boat will take as many horses as we wish by making it longer and will be safe so that it can never sink.

CHAPTER IX. ON THE WAY OF HORSE-RIDING THROUGH WATER (Fig. 15).

Let them first take four round shields four inches thick and nine inches wide, and after this take thin light laths two inches wide and [one thick]* and a yard and a half long or more or less according to the horse's length. And these laths are joined with hinges placed this way and that so that they may be rolled up into a package; and the shields are covered with these laths and hoops are fixed upon them with nails so as to make a round cask; and this cask has in the centre a piece cut out towards one side, so made that the [rider's] leg and stirrup can easily be placed within it and long enough so that the man's foot can easily be placed between the lashings of the cask to the horse. And this cask is covered with leather and over the cut-out place there is a [flap of] leather wide enough to take the place of the hole. And two of these casks are lashed lengthwise to the horse, beneath the horse so that one is on one side and the other on the other, and the lashing is done thus, viz. the cask on the right side is lashed underneath the horse's belly to the saddle-girth on the left side [and the other cask on the left side] is lashed to the saddle-girth on the right; and also each cask is lashed to the saddle on its own side. And the saddle and its girths are made very strong, and the casks may also be lashed fore and aft as those doing it see fit. And in the saddle bow at the front a hole is made so that the man can hold on with his hands; then mounting his horse and commending himself to Christ he enters the water, but let him have a care that the far bank be good for climbing out again.

CHAPTER X. HOW INFANTRY MAY CROSS GREAT STRETCHES OF WATER (Fig. 16).

Let them first stretch a rope from one bank to the other as thick as a thick thumb, and so that the rope may not sink beneath the water many dinghies are placed beneath the rope. Let them make a cask with shields and laths as was said above. But this cask is to be quite round, one yard long and nine inches high, and in the middle it has towards one side a small hollow place so that a man may sit in it, and may have two stirrups for his feet; and a man may mount on this cask commending himself to Christ and lash himself fore and aft to his belt and so enter the water always holding on to the rope with his hands and pulling himself to the [far] bank. And if he shall have three casks, viz. one in the middle and one on either side lashed together for and aft as those doing it see fit, then he can safely cross all stretches of water without a rope, holding either one or two paddles in his hands as seems best to himself.

CHAPTER XI. ON THE WAY OF MAKING A FIGHTING-WAGGON, WHICH MAY BE PROPELLED WITHOUT DRAUGHT-ANIMALS IN ORDER TO DISOR-DER LARGE FORCES, AND IT CAN BE CARRIED IN PIECES ON HORSEBACK (Fig. 17).

Let them first make four wheels two yards high, and these wheels are fitted with teeth on the inside towards the waggon, running all the way round a span in from the rim, the teeth being four inches long. And these four wheels are fastened upon two axles a yard and a half or two yards long (more or less, according to the [width of] the streets or wide open fields which the wheels may traverse without any obstacle) - the wider the waggon is made the better it will be. And the front wheels are one span distant from the back wheels. And on the four axles they are to make the bed of the waggon in the proper way as other waggons are made. And to the ends of the axles beyond the wheels four uprights are attached, two yards high or more and as thick as an arm, and upon their [upper] ends are placed two transverse timbers lining up with the axles [beneath]; and when this has been done they are fastened upon the bed of the waggon over the first axle; and this I mean [to be done] with the one axle as well as the other, namely in the fore part above the bed of the waggon, in the middle of the axle, two [vertical] iron bars are fixed, of such a length that they can be fastened upon the bed of the waggon at one end, and at the other end fixed in that timber which was fastened [above] ;* and the two iron bars are two spans apart, an inch thick and three inches wide; and at about their centre the two iron bars [each] have a large hole, and so they are fastened above the bed of the waggon, and at the side of the iron bars are mounted two [toothed] wheels which are linked together by an iron [crank-] handle which is in between the iron bars; and the handle with its two wheels turns upon the iron bars, the wheels being toothed and placed outside thus.

And after these [toothed] wheels two rundle-wheels made thus are placed between those upper wheels and the [toothed] wheels of the waggon, above the bed of the waggon, and so positioned that by turning the upper [toothed] wheels with the [crank-] handle the rundle-wheels are revolved, and the wheels of the waggon (placed thus) are revolved by the rundle-wheels. The rear wheels should be similarly prepared in this way, and all these things shall be arranged by the master millwright, who knows how to match these wheels together. And in order that the waggon may be propelled in a straight line along a road and steered when necessary, the part which is pulled by oxen in front is turned back, and one man (or two) hold the reversed shafts in their hands and steer the waggon where they wish. And I believe the waggon will be much better steered by two shafts in front and a horse, when the waggon is to pass through streets. But whoever is to do this will provide for the best. And this waggon is made with a parapet and castellated with an open platform built on those

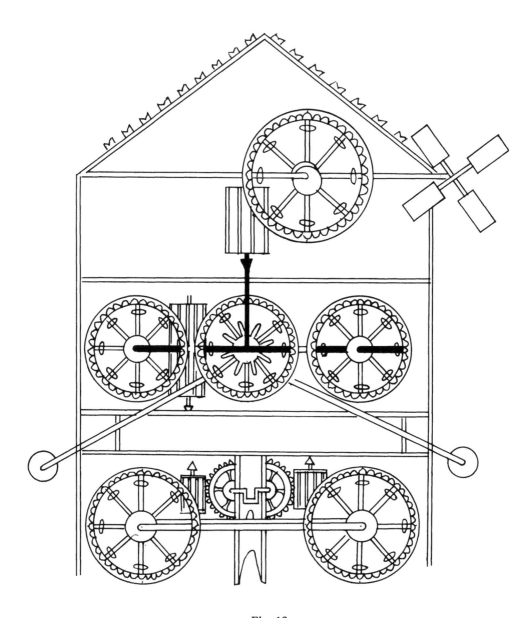

Fig. 18

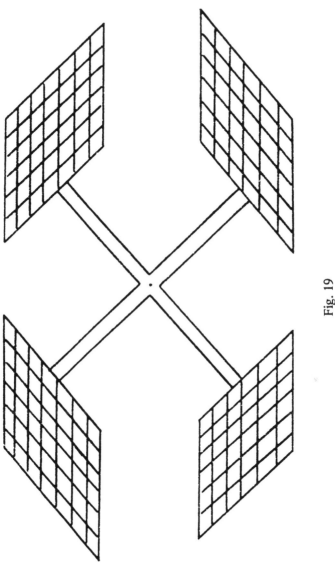

Fig. 19

transverse timbers spoken of above, according to the maker's plans. And thus the waggon to go everywhere without draught-animals is completed.

CHAPTER XII. ON THE WAY OF MAKING A SECOND WAGGON WHICH IS PROPELLED BY THE WIND WITHOUT DRAUGHT-ANIMALS, AND WHICH DASHES VIOLENTLY OVER OPEN COUNTRY TO THE CONFUSION OF ALL TROOPS (Fig. 18).

For propelling another waggon by the wind proceed thus: let the two fore-wheels of the waggon be made two yards high as aforesaid - I call those the fore-wheels that do not have the shafts because these ought to precede - and one of these two wheels of the waggon is to be fitted with rundles and the other is to be toothed on the side towards the waggon at a distance of one span from the outer rim of the wheel; and transversely above the bed of the waggon five timbers are fastened each as thick as an arm, that is to say before and behind those wheels, being so placed and so long that a round "house" may be built upon them enclosing these two wheels; so that these two [fore-] wheels remain within the "house" and the other two wheels with the shafts remain outside the "house"; and these two rear wheels [with the shafts] are as far removed from the [fore-] wheels as the maker thinks fit. And this "house" is made of planks, three yards high and three round [that is, in diameter], more or less, as the maker judges; and either inside or outside the "house" a palisaded gallery is to be made so that a man may pass round the "house" from one side to the other. And on the top of the "house" is a round wheel lying flat which is hollowed out inwardly and in the hollow are many little wheels [fixed] with iron pins at a distance of nine inches one from another, and upon those wheels is placed another round wheel so that it may turn about upon the little wheels in the proper way, just as they do in windmills; and upon that [upper] wheel is fitted an axle [windshaft] which protrudes beyond the wheel for half a yard at one end; on this extension a sail made thus is fitted crosswise, on the arm that sticks out beyond the wheel or the "house" (see Fig. 19). And at the other end the windshaft protrudes only so far beyond he "house" that a man standing on the gallery of the "house" can move the windshaft round (with the wheel upon which it rests) using a lever, as is done in windmills. About the middle of the windshaft there is fixed to it a great wheel toothed on one side; and below the windshaft is fixed a single crosswise timber (so that it does not touch the windshaft), and this is mounted upon the [upper bearing-] wheel. And in the middle of that timber is mounted one end of [another] arbor [of iron], within a thick square [plate] of iron, the arbor being long enough to reach from the single [crosswise] timber down to the bed of the waggon, which is about a yard and a half. And all around that iron [arbor] rundles are fixed so that they may engage with the teeth of the [great] wheel on the windshaft. And below the rundles four rings are fixed [welded?] at right-angles to the iron [arbor] so that the horizontal wheel mentioned below may be properly held firm. And at the bottom of that iron [arbor] an iron *nariga* [?bearing] is placed upon the bed of the waggon, made thus, upon which rests the foot of that pointed iron [arbor] so that the iron [arbor] revolves upon that bearing; and that iron [arbor] is so positioned that it is closer to one wheel of the waggon than to the other by the breadth of the little pinion [see below]. And so the round wheel of the "house" must be broader [i.e. be placed off-centre] towards that wheel of the waggon which is fitted with rundles than towards the waggon-wheel fited with cogs by the breadth of this little pinion, so that the iron [arbor] may hang down properly inside the "house". And just above the bed of the waggon a wheel is fastened to that iron arbor which is cogged all the way round outside; this lies [horizontally] over the waggon-bed, two inches above it, and is firmly attached to that thick iron [arbor] by those rings. This toothed wheel engages on one side with the waggon-wheel in such a way that its teeth enter the rundles of one single wheel of the waggon only; on the other side between the horizontal wheel and the other waggon-wheel a little pinion made thus is placed in the middle. Both the teeth of the horizontal wheel and the teeth of the [second] waggon-wheel - for this must be fitted with cogs - revolve in [engagement with] this little pinion. Thus one waggon-wheel has [rundles and the other]* teeth. And when all is done and the wind blows, the windshaft is driven round with its [great] wheel, and that wheel drives the long pinion and the long iron [arbor] which in turn revolves the horizontal wheel and that wheel drives a waggon-wheel [directly] on one side, and on the other side drives the other, toothed wheel of the waggon through that little pinion indirectly. All these matters are the concern of the master millwright and especially the master windmill-wright, for I have based this structure on a likeness to windmills. And the "house" itself may be fortified and castellated as the maker sees fit. And to prevent the waggon toppling over two timbers as thick as a big arm are attached to the "house", [extending] on either side of the waggon beyond the wheels as the maker sees fit from the height [of the "house"] towards the ground, but [terminating] half a yard above the ground. And at the ends of those timbers are two small wheels made thus [to run on the ground] . And because the wind does not always blow let them make many large holes four inches from the outer rim of the horizontal wheel [at the lower end of the iron arbor] and let strong wooden capstan-bars half a yard or more long be fixed in these holes, so that many men standing all around that wheel can heave it around by hand. And, if the maker shall see fit, the other devices described above for the other waggon can be employed outside the "house" and above the bed of the waggon between the other two wheels which pivot with the shafts [these are now the rear wheels]; that is to say these wheels may be revolved by [crank-] handles and the waggon may be propelled either by the wind or without the wind. And in order to drive it straight and steer it [the fore-carriage] with the shafts is to be reversed, as was said of the other waggon. And in order that the wheels of the waggon may be portable in pieces let the felloes be linked by hinges placed this way and that so that they may be folded up into a package and on the opposite side are fastened locking-bolts; and each felloe has an iron pin [at its end]; and when necessary these are strongly and firmly nailed together into a wheel as the maker judges best. And thus waggons are finished which are propelled without draught-animals and without wind, of which every part can be carried in pieces on horseback and then assembled in a single day or night, from which may ensue much of benefit to the recovery of the Holy Land.

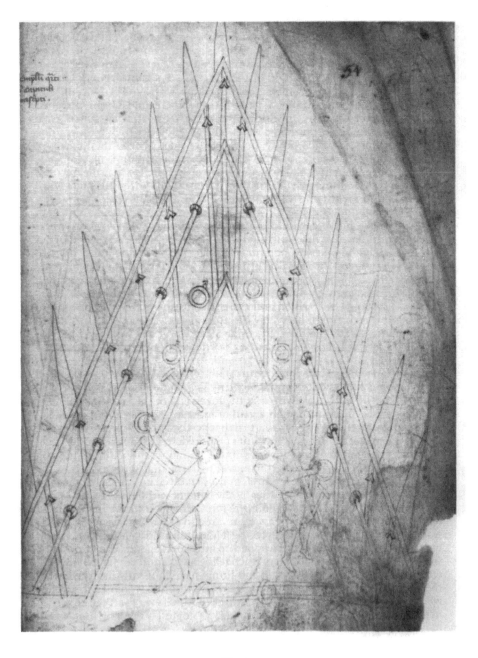

Fig. 20

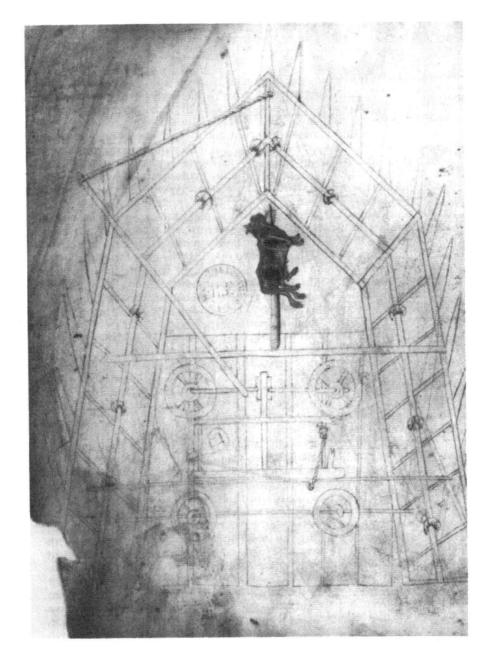

Fig. 21

CHAPTER XIII. USEFUL AND NECESSARY: ON THE WAY OF MAKING DE-
VICES OR "PANTHERS" FOR HOLDING OFF GREAT FORCES WITH A MOD-
ERATE NUMBER IN ONE'S ARMY (Fig. 20).

A "Panther" is made thus: let them take as many planks as are needed in
proportion to the length we wish to give the "panther", that is, twenty five, fifty
or a hundred yards; each plank being two and a half yards long they are joined
together by iron hinges placed inside and out [along the long edges] so as to
combine a number into single horse-loads; and then when needful one horse-
load may be linked up with another as the maker thinks best. And when
needful the planks, placed upright, are joined together in such a way that at one
end [the fence so formed] comes together in a point [two sides of a triangle
meeting] while the other end [third side] is left open as the maker thinks best.
And they are so joined at the apex that this open end may be widened at the
wish of those within the "panther". And each plank has a loophole and a post
acting as a strut for the "panther", and also a ring; and holding the rings and
the posts those within the "panther" can transport it anywhere and open it out.
Inside [the fence] two poles are fastened down the length of the "panther",
namely one on each side so that the planks may be lashed to them and the
"panther" thus made rigid and firm. And if it should be feared - because of the
height of some house or other - that those within the "panther" might be
wounded [from above] let the "panther" be covered over with the quilts
described above but let them not obstruct the opening and closing of the
"panther" as the maker sees fit for I could not sketch all this in the figure.
However, let each plank have another little ring fastened to it at the top for
attaching quilts or boards necessary for protection against [sling-] stones as the
maker thinks necessary. And at the back of the "panther" [forming the third
side] two poles are so placed that they hold the "panther" upright, and they can
be shortened or extended with the "panther" itself by means of [sliding] rings
placed towards the middle [of their overlap] as will be seen in the figure. And
all around the outside of the "panther" are many sharp slender iron spikes each
a yard long or more so that no one can approach close to the "panther"; and
of these let them make as many as seems appropriate for these "panthers" are
easily made. And in order to attach the iron spikes rapidly and without fuss to
the "panther" let them be made thus. Take beams made in sections each a yard
long, the timber being four inches broad and three thick; these sections are to
be linked together by iron hinges so that they can be folded into a package, and
against the broad side of the timber these iron spikes are nailed on a foot apart,
and they are so nailed to the beam that they can be folded into a package with
the beam itself, and put away in boxes. And when needful they are stretched
out and lashed all around the "panther" as the maker sees fit. And three or four
of the beams thus spiked are fastened all around the "panther" outside. And if
anyone will but ponder well upon this device many a good result will flow from
it because a great army may be thrown into confusion by a small force with a
few of these devices. And a skilful and ingenious man may encircle a waggon
and oxen by such a "panther" in such a way that the oxen are safely fenced in
by the "panther" as is made obvious in the next figure (Fig. 21), and the men-
at-arms may stand on the bed of the waggon inside the "panther". And upon
that waggon a mangonel may be built which will throw rocks and stones in all

directions as may be seen from the figure of the "panther" with a waggon. And if one should choose to surround a vast space of ground (say a league or more) and hold it with these devices so that [enemy] forces cannot pass over it, let them fasten the posts of platforms (as described above) between one waggon (as described above) and another, of which the longer ones can be fixed upon the waggons themselves, so that the poles may be placed in position or removed as necessary; and these poles are to be armoured with long iron spikes facing towards the [enemy] army, lying in this way between one waggon and the next, and in this way a small body of troops can confuse and hold off a great army.[11]

NOTES

[1]For general information on Guido da Vigevano see George Sarton, *Introduction to the History of Science*, III (i), Baltimore 1947, 846-7; Elie Berger, 'Guy de Vigevano et Philippe de Valois', *Journal des Savants*, 12, Paris 1914, 5-14; Ernest Wickersheimer, 'L'Anatomia de Guido de Vigevano', *Archiv fur Geschichte der Medizin*, 7, 1913, 1-25.

[2]Texaurus Regis Francie acquisicionis terre sancte de ultra mare nec non sanitatis corporis eius & Vite ipsius prolongactionis ac etiam cum custodia propter venenum.

Cum Anno currente millesimo Trecentesimo Trigesimoquinto passagium ultra mare fuerit ordinatum occasione terre sancte conquirende, quapropter quia intelligere & scire contingit circa omnes scientias ut habetur primo physicorum Ego Guido da Vigevano de Pavia olim medicus Imperatoris henrici Et qui nunc per dei gratiam medicus reverende ac sanctissime domine Johanne de Burgundia per dei gratiam Regine francie.

Cogitans die noctuque qualiter & quomodo terra sancta levius et citius haberi possit eo quod Saraceni muniti sunt multis aquis tum dulcibus & salsis, & infinitis villis & castris munitis muris & fossatis. Et cum omnem bonum et pulchrum sit de sursum ut habetur de celo et mundo capitulo de eternitate mundi. Igitur a deo datus est nobis modus leviter conquirendi terra sancta de ultra mare Quem rescribo serenissimo principi domino philippo per dei gratiam francorum Regi ex eo quod dominus passagii.

[3]Perhaps the allusion here is, in fact, to the end of Chapter 9 of *De caelo* (279a. 29-30).

[4]The chapters of the medical section are as follows: Capitulum primum de cibo et potu; secundum de sompno & vigilia; tertium de motu & quiete; quartum de repletione & evacuatione; quintum de aere; sextum de accidentibus animae; septimum de visu, auditu, dentibus & memoria [et] de rauso; [not numbered] de veneno et malo cibo. On Fol. 32r of MS. fonds Latin 1015 in the list of chapters the topics of chapters 5 and 6 are reversed from the actual order of their treatment, as given above, and the last chapter is numbered eight.

VIII

[5]I referred to this text in Jacques Bongars, *Gesta Dei per Francos* (Hanover, 1611), Vol. II.

[6]This section of the *Texaurus* has been discussed by Carlo Promis, 'Gl'ingegnari militari', *Miscellanea di storia italiana*, XII, Turin 1871, 418-9; by myself in *Actes du VIIe Congress Int. d'Histoire des Sciences* (Florence, 1958), 966-69, and in *Technology and Culture*, II, 1961, 17-21; and by Bertrand Gille in *The Renaissance Engineers*, 28-31. The last-named undervalues the technical content of Guido's verbal descriptions.

[7]In *History of Technology*, II, Oxford, 1956, 625, Rex Wailes puts the invention of the tower-mill in the early fourteenth century.

[8]John Salmon in *Journ. of the British Archaeological Assn.*, 3rd Ser., Vol. 29, 1966, p. 75. The construction occupied two building seasons and seven masons, so it is hardly likely that only the base of a post-mill was built of, or enclosed in, masonry. The site of this mill is indicated on the plan of Dover Castle in H. M. Colvin (ed.), *The History of the King's Works. II. The Middle Ages* (1963).

[9]This is my interpretation; the text states that the cross-timber is fastened to "the wheel" [annulus]. But unless it is attached to the upper annulus it will foul the face-wheel when the 'cap' turns into the wind.

[10]It is curious that Guido in the text figure shows windmill sails of a highly improbable form, repeated by Taccola and Valturio (cf. Gille, p. 88), though the main illustration shows sails of the right shape, if far too small. Distortion of the former kind is quite common, however (see *History of Technology*, II, Oxford, 1956, figs. 563-5).

[11]The author wishes to thank Michael C. Mahaney for the drawing of Figs. 1-5 and 19.

ILLUSTRATIONS

Numbered in the preceding translation, these have been previously reproduced as follows:
Figs. 6, 10, 13, 14, 16-18, 21 in Gille, pp. 28-31, 88
Figs. 11, 12, 13 in *History of Technology*, II Oxford 1956, Figs. 658, 594
Figs. 17, 18 in *Technology and Culture*, II 1961, Figs. 1. 2.
I have omitted the first figure reproduced in the first printing of this paper as being barely legible, and unnecessary.

GLOSSARY

Archera, *It.* arcuera = loophole

Arzonus, *It.* arcione = saddle-bow

Aspaldare, *It.* aspaldo = parapet

Aspus, *It.* aspo = handle

Assale, *It.* assale = axle

Assis, *L.* axis, *It.* asse = (1) plank, board (2) axis, spindle (3) hinge

Axis, *L.* axis, *It.* asse = hinge

Balla, *It.* balla = bale, package

Baltrisca, *It.* baltresca, bertesca = platform or brattice

Bota, *It.* botte = cask

Briga, *It.* briga = fuss, worry

Calchare, *It.* Calcare = to press, stuff (calcastoppa = caulking-tool)

Calosa, or perhaps better 'talosa' if related to *It.* tagliare, = section

Casula, dim. of *It.* cassa = chest, box

Catenatia, *It.* catenaccio = door-bolt

Cavigia, *It.* cavicchia, caviglia = pin, peg, bolt

Coraterium, ? gallery

Coroginellus [? *It.* correggere = ? upright]

Corrigia, *It.* correggia = leather strap

Corva, *It.* corba = rib (of a boat)

Cossa, *L.* coxa = thigh

Cugnonis, *It.* conocchia = rundle-wheel, pinion

Cultra, *L.* culcita, *OF.* coulte = quilt

Dubiones, *It.* doppino = hoop

Fibla, *L.* fibula, *It.* fibbia = buckle

Gavilia, *It.* gavello = felloe

Impesato, *It.* impeciato = tarred

Infuselatus, (*L.* fusus, *It.* fusello = a small spindle), fitted with rundles but 'fusellato' = tapering

Laresi, *L.* larix, *It.* larice = larch

Macius, *OF.* masse = club, hammer

Manegia, *It.* maneggiare = to handle

Orlus, *It.* orlo = edge, rim, lath

Pala, *It.* pala = shovel, paddle

Panthera, *L.* panthera = protective enclosure

Rampo, *It.* rampone = nail, spike, hook

Redrizare, *It.* raddrizzare = to erect

Sponda, *It.* sponda = side-piece (of a cart, etc.)

Spranga, *It.* spranga = cross-piece, bar

Stopinus, *It.* stoppone, stoppa = tow

Tergia, ? *It.* targa = shield

Tumo, *It.* timone = rudder, shaft of cart

Vera, *It.* vera = band, ring

IX

SCIENCE, TECHNOLOGY, AND

WARFARE, 1400-1700

No historian ought to have the impudence to speak in public on the topic I have chosen for today; certainly no historian can do justice to it without a polymathic ability which I myself, as a historian of science having only a nodding acquaintance with the evolution of technology, do not possess. As for the history of war, I have followed it only in desultory fashion. But at least I have examined one aspect of this multifarious problem in some detail—and not everyone who has had ideas about the historical relations of science and war has done even that.

It has been and I suppose still is commonly assumed that there is a casual if not a causal connection between the growth of science and the development of military techniques: weapons use explosives, and these are products of chemical science; navies navigate, and navigation is a special kind of applied astronomy; weapons have been improved through metallurgical skills, which themselves arise from science; and the engineer who employs a lot of difficult formulae printed in a little book is also a sort of mathematician. These crude correlations enable people to argue either that scientific research has been pursued for the sake of military advantage, or that soldiers have always been successful in harnessing scientific philosophy to their chariots. Without war we would have had a more primitive science, without science a more primitive art of war. Now I would not deny that waging war and preparing for war have been most potent agencies in the formation of our civilization, nor that rewards dazzling the eye of the inventor (though ever so rarely won) have caused virtually every piece of human knowledge to be exploited in war, at least in imagination. Nevertheless, the attitude I have described—which we may think of as extremely ancient, since it compelled the Romans to misunderstand

---

*Professor Hall was unable to attend the Symposium because of illness. In his absence, the paper was read by Professor Lynn White, Jr.

4

the true character of Archimedes—is very largely a piece of wishful thinking. It happens to have a measure of justification for our own age, but no earlier one. Rather over a century ago the Reverend Francis Bashforth, who ought, one might suppose, to have had his mind on even higher and swifter things, devised an electric chronograph for measuring the speed of projectiles, whereby ballistics ceased to be a speculative branch of applied mathematics and became an applied science. With that event, to my mind, the science of war begins to replace the art of war. The era in which a few men in white coats can, with a year or two's work, determine the fate of battalions, ships, or airplanes had begun. But not before then.

Throughout the period with which I am concerned today military affairs were determined by three things: organising ability, including what we today call logistics; basic craft skill; and courage, which is perhaps some sort of social attribute. They no more depended on scientific knowledge than they did on the relative size of the combatants' populations or their real economic strength—think of the wars of France and Holland in the seventeenth century. If you make any study of the competent military commanders of the sixteenth and seventeenth centuries, you will find that they insisted on proper equipment and supply (so far as was possible in those days), that they trained their men in use of weapons and military evolutions, and that they sought to inspire them with courage; you will find no evidence that commanders believed in science or mathematics, or expected any other skill in their troops than that of managing their pieces dextrously and carrying out orders on the field. Perhaps a critic may object: but what of gunnery and ballistics, what of fortification and geometry? I will deal with these points later.

It seems to me that the notion of war's indebtedness to science, and indeed of the pursuit of science for utilitarian ends, is a very unhistorical projection backward from our own age, in which warfare has indeed been transformed by science, countenanced by reference to special pleading on the part of some early propagandists of modern science. I do not by any means deny that Francis Bacon wrote of the "Empire of Man over Nature" and so forth, or that a few other apologists, in the search for public acclaim and finance, boasted of the practical good that science would do, especially in its applications to agriculture and medicine. I do deny that these propaganda claims affected what the natural philosophers, mathematicians, and scientific societies actually concerned themselves with. The argument that inventive craftsmen, with their special concern for promoting the "Empire of Man over Nature" by rational and experimental investigations, formed the spearhead of the scientific revolution of the seventeenth century has in my opinion rested on a very dubious use of the historical evidence and indeed downright errors from the time of Edgar Zilsel to that of Christopher Hill.

Before I can proceed to the limited section of this historiographical debate that is relevant now, I must propose some distinctions and definitions. For without some rigour in this respect anything is demonstrable. I shall define science as a rational and theoretical inquiry into nature, having as its objects description, analysis, and explanation. Technology is the practical knowledge and skill by which material ends are accomplished. Science and technology are often associated. A man can be a chemist and a skilled glass-blower, a mathematician and an airplane pilot. Science may be required for the design of a certain instrument, technology for its actual fabrication. But we should not confuse things, when the use of terms is critical, by speaking of astronomy as a craft, or surveying as a science. Thirdly, we should distinguish between the case where the technologist applies, for his own purposes, existing scientific knowledge and the case where a scientist tries to solve a problem in order to present its solution to technology. As examples of the former one might cite the development of scientific navigation and cartography by the Portuguese in the fifteenth century, and the introduction of the gunner's quadrant in the sixteenth. This quadrant in its various forms was an inclinometer. No doubt its ultimate parent was the astrolabe which brought forth such a large progeny, and it appears in the *Trattati* of Francesco di Giorgio Martini (1439–1501), written in the late fifteenth century. In my own view this famous gunner's quadrant was never much more than an elaborate piece of mystification intended to enhance craft prestige, but let that pass. As examples of the scientific search for the technically useful solution to a problem, it is commonplace to indicate the interest of seventeenth century astronomers in the determination of longitude, associated with the foundation of the observatories of Paris (1666) and Greenwich (1675). When he has made such a distinction clear the historian may also appreciate the significance of a second one arising from it: the fact that he builds a technique upon science does not make the craftsman or technologist a scientist. The navigator with his backstaff, the gunner with his quadrant and range table, is not an astronomer nor a mathematician. Conversely the scientist who dreams (shall we say) of smelting iron with coal instead of charcoal does not thereby acquire the practical skill of an ironmaster. Even taking out a patent will not make him a craftsman. As for the inventor, he is only too often rightly to be dismissed as a "projector," a dreamer lacking both philosophic reason and craft experience.

Now let me try to isolate more particularly the possible areas of interaction between science and war in the renaissance age. First, two exclusions. I shall not consider navigation as it affected naval operations, for the simple reason that though indeed navies generally took more trouble over navigation and cartography than the merchant marine, government service has never furnished the sole opportunity for the exercise of navigational skills. For instance, when Galileo offered his new-made (not invented!) telescope to the Signoria of Venice, he was clearly thinking of Venetian commerce, not her defence. For similar reasons and lack of time I say nothing of medical science and war. It is a commonly held view that pressures of war have stimulated advances in medicine, and in the sixteenth century one could point to the career of Ambroise Paré as exemplifying this; but equally civil life presented enormous problems of which all were aware, for example in this period syphilis and plague, to which great medical endeavours were devoted. These matters set aside, and disregarding such obvious trivialities as the use of arithmetic in ordering troops into formation and the computation of logistic needs, we are left with these possible areas of interest:

TABLE 1.—MILITARY TECHNOLOGY

	Edge Weapons	Cannon and Mortars	Fortification and Siege Warfare	Small Arms
Ballistics (ext.)	—	X	X	—[1]
Engineering[2]	—	—	X	—
Explosives	—	X	X	X
Mechanics	—	X[3]	X[4]	X
Metallurgy	X	X	—	X

[1] No one at this time thought of using calibrated sights or other complex aids to long range shooting on small arms.

[2] I use the word here in the older, more restricted sense, including everything from the design of fortifications and siege works to the moving of earth and construction of masonry, but excluding mechanical engineering.

[3] This includes cranes and hoisting devices, boring machines, carriages and limbers, and much paraphernalia needed for the manipulation of heavy guns.

[4] This includes both mechanical devices used in engineering and those employed by besiegers against a fortification.

I shall deal with weapons first, remembering always that close-range weapons are the decisive ones in this period: sword and pike, pistol and musket, fieldpiece or small naval gun. It must strike us at once that the all-important metallurgical knowledge and craft skill by which weapons were fabricated throughout this period, and long after, were quite unrelieved by any light from science. The armourer forged, formed, finished, and ornamented steels for helmet and cuirass; the bladesmith carefully compounded soft and hard steels to produce cutting, flexible swords; the locksmith worked with file and chisel on half a dozen curly little levers and so on. All this was totally innocent of chemical science and crystallography. Locks, barrels, blades were made by specialised developments of the common blacksmith's skill. The gunfounder, again, excelled in his art because he prepared larger and often more beautiful castings in bronze or iron than any other metal-founder; but he too knew nothing of metallurgy, not so much as the nature of the difference between cast iron, wrought iron, and steel. Of course, I do not mean to say there was no metallurgical invention; invention of bellows and furnaces, fresh alloys of bronze, and even treatment of iron castings went on all the time. But none of it was invention based on science.

Consider now the mechanics of weapons. As my diagram showed, this question is relevant to small arms and siege warfare. I think it would be pedantic to go into such details as the mechanical design of gun carriages, based on conventional waggon practice, or such a will-of-the-wisp (first found in the fourteenth century) as the "armoured car"; it was a favourite of the fifteenth century engineers, and Leonardo's drawing, so perfunctory as to be impossible, is familiar. Siege engines were always a speciality of war. They too existed, or I should rather say were illustrated, in Leonardo's notebooks and those of Francesco di Giorgio Martini, as well as the printed work of Roberto Valturio (1472). I have no evidence that such cumbersome, complex machines

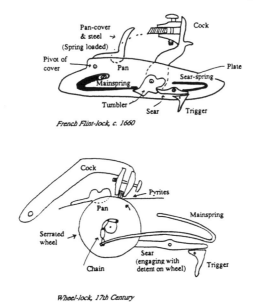

French Flint-lock, c. 1660

Wheel-lock, 17th Century

Fig. 1. Lock Mechanisms.

were actually employed in war at this time; again, I think you will agree
that their construction presented no unusual problems, save that of scale
perhaps.

Far more interesting, and really important, is the history of the
small arms lock mechanism. You will recall that three main devices
appear in overlapping succession: the matchlock, the wheel lock, and the
flintlock, the last showing many variant forms. The objects constantly in
view through the evolution of the weapon were reliability, especially
with respect to damp, and convenience of recocking. The matchlock was
cheap, simple, unreliable, dangerous, and tedious. The wheel lock was a
beautiful, expensive, and reliable device, slow to recock. Therefore it
did not wholly replace the matchlock. The flintlock was good in every
way, also fairly cheap and robust; therefore it remained standard for a
century and a half. So far as I am aware we know nothing of the first
inventors of these devices, all in the sixteenth century, still less of the
origin of the local variants. But I have never seen it suggested that the
inventors were not working gunsmiths, presumably masters of the
locksmith's craft. I find it extraordinarily difficult to see how science or
mathematics could have had any part in this evolution of small arms.

Casting around a little more in the same area of technology, we
discover that hunting weapons were rifled from the early sixteenth
century, that breech-loading mechanisms were introduced in such
luxurious weapons,[1] and that the multi-shot idea was applied in a
variety of ways (one of the simplest, of course, being the revolving
cylinder).[2] All these inventions, and there are many of them, are related
by experts on firearms history to the craft, and it is worth noting that
this adventurous technology, never (except for rifling) really successful,
was devoted to expensive, personal weapons that are beautifully
ornamented, not to the common soldier's musket. The reasons are
obvious. And with respect to rifling, observe that it was the rich sporting
marksman who was interested in accuracy; the tactical practice was such
that accurate individual fire at aimed targets had no place on the field
of battle, at any rate before the War of Independence.

The only gift of science to small arms technology in this period was

---

[1] As one example among many, the Marquis of Worcester's patent of 15 November 1661 refers to an "invention to make certain guns or pistols, which in the tenth
part of one minute of an hour may . . . be recharged, the fourth part of one turn
of the barrel (which remains still fixed) fastening it as forceably and effectually
as a dozen threads of any screw . . . ." Clearly this is a multi-thread, quick-closing
device for a breech block.

[2] Again, one example: Pepys refers on 3 July 1662 to "a gun to discharge seven
times, the best of all devices that ever I saw, and very serviceable and not a bauble";
similarly on 4 March 1664.

the air gun. Scientific interest in pneumatics through the mid-seventeenth century revealed the exceptional force not only of atmospheric pressure (von Guericke's "Magdeburg experiments," 1654) but of compressed air also, and provided the pump technology. The rest was easy. But the "philosophic" gun was never more than a curiosity, or an assassin's tool (as in Conan Doyle's story) except when it was adopted in the Austrian army in 1780. This 30-shot rifled sharpshooter's weapon was no toy, since its highest ball velocity approached 1,000 ft/sec, and the calculated energy was 400 ft/lbs, but it was far too complex and costly for ordinary use.[3]

We can almost as briefly dismiss explosives. There was a literature of gunpowder-making at least from 1420, when the tradition of the *Feuerwerkbuchen* began; this was first printed in 1529, and Biringuccio also dealt with powder in 1540.[4] There was a constant search for the means *"Wye man noch eyn besser und stercker Pulver machen soll,"* resulting in scores of recipes, whereby not only were the proportions of the normal ingredients varied, or the ways of preparing them (especially the charcoal) modified, but other irrelevant ones added, like camphor, sal ammoniac, sublimate of mercury, vitriol, or brandy. An important step was introduced towards the end of the sixteenth century when powder was "corned" or granulated, instead of leaving it in a fine, floury state. Early in the seventeenth century a new, very touchy explosive was discovered: fulminating gold. This remained a chemical curiosity of no destructive value. By Louis XIV's reign a considerable industry was devoted to recovering saltpetre from the efflorescence on walls and nitrogenous animal wastes, to obtaining sulphur from volcanic and other deposits, and to the milling of these ingredients, just as there was a considerable art devoted to preparing various grades of powder as needed for blasting, cannon, small arms, or fireworks. But there was no chemical science in this. Philosophy speculated about volumetric changes manifest in the passage from fluid to "air" (steam) or from solid to "air" (powder fumes). Philosophers realized that the force of powder sprang from the expansion produced by heat, and even endeavoured (from 1675) to capture this force in a piston-and-cylinder arrangement. Some philosophers—not all—accepted it as a fact that gunpowder could burn in the absence of air, and argued from this that saltpetre and air possessed in common a "something" necessary to fire; a few imagined fire to be a reaction between "sulphureous," that is combustible, particles and "nitroaerial" particles. However, all this fascinating speculation was of no value to the gunpowder-men.

---

[3] F. H. Baer, "The Air Rifle That Went to War," *American Rifleman* 115 (Dec. 1967): 32–35.

[4] Wilhelm Hassenstein, *Das Feuerwerkbuch von 1420* (Munich, 1941).

It would be foolish to deny that the medieval invention of gunpowder furnished the philosophers with a fascinating subject for speculation, but I know of nothing that would indicate gunpowder's making any major contribution to the evolution of chemical science, or vice versa. Perhaps I should mention that in both the early Royal Society and the Académie des Sciences there was some interest in little devices by which the quality of samples of powder might be tested; but this, and one or two like points that cropped up, seems awfully trivial.

I propose to leave gunnery and mathematical science to the end of my talk, turning now to the technology of static defence and attack. First mechanical engineering. There were two ways in which machines could be of great assistance: they were useful in earth-moving for defence, and in providing cover for attack. Leonardo shows us a sketch of a great ditch-digging machine, for use in swampy ground particularly; Ramelli pictures a soil-conveyor replacing the more usual ramp-and-barrow. Obviously these machines were as applicable to civil engineering as to military; recall that there was very considerable activity in canal building during the sixteenth and seventeenth centuries. However, authors and inventors seem as regularly to have associated earth-moving with fortification, as they associated dredging and pile-driving with peaceful commerce. And there is this curious difference too. We know that the dredges and pile-drivers really existed, for there are accounts of such machines in action. But I am not aware that excavating machines or conveyors were used at all in practice (unless we call the tram-road and truck, as in Agricola, a conveyor). I believe we will not go far wrong in supposing that pick, shovel, and barrow were the appliances really in use—the ones that built our canals and railways.

Earth and stone for defence; timber obviously for protection of the offensive. The permutations of a few simple technical ideas extending back to remote antiquity were endless, from Leonardo's promises of portable bridges and armoured cars, scaling ladders, mantlets, "and other instruments" in his lette. to Ludovico Sforza, to the Marquis of Worcester's invention of a "transmittible Gallery over any Ditch or Breach in a Town-wall, with a Blinde and Parapit Cannon-proof." Some devices of this sort were quite complex, as may be seen in the late-fifteenth century sketches of Francesco di Giorgio Martini. Were these ever really used? That they were in modern times seems unlikely, though they certainly had been in antiquity. In any event, there is nothing very involved about their construction.

With fortification proper we enter a very different world, and one that could devour the revenues of even a great prince. The site had to be carefully surveyed, and then the fortifying walls and towers properly sited in order to protect the desired area; sometimes areas of housing had to be knocked down if they stood in the way of the defences, or impeded their usefulness. Expert engineers might differ as to the principles to be followed, as happened with regard to Berwick-on-Tweed in Elizabeth's reign.[5] But whatever the difficulties of application to a particular site, and these were grave, the military engineer throughout this period followed fairly clear principles—the ones that evolved first in Italy, in reaction to the introduction of cannon and the failure of medieval methods of static defence. The medieval castle repelled besiegers by the height of its stout wall, strengthened by round towers at intervals, and further protected by a moat which also impeded attempts to batter the walls or mine beneath them. Defensive missiles were hurled directly down from above. After the introduction of firearms a breach could be opened by cannon placed at a relatively safe distance from the wall, and the defenders could be annoyed by explosive bombs flung from mortars (this type of bombardment was familiar to both Valturio and Leonardo in the fifteenth century). If an attack was made on a breach, or the besiegers attempted a direct assault by scaling the wall, it was difficult for the defenders to concentrate their fire against them, harassed in turn by the counter fire of the besiegers. The first step the Italian engineers took in strengthening the defence was to break up the straight or curtain

[5] Lynn White, Jr., "Jacopo Aconcio as an Engineer," *American Historical Review* 72 (Jan. 1967) : 425–44.

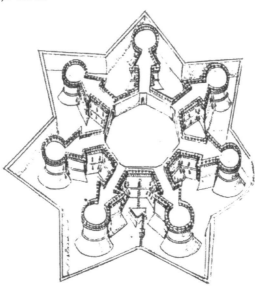

Fig. 2. Fortification Sketch by Francesco di Giorgio Martini.

wall between the towers, the weak spot, into a series of strong points or bastions, so that the whole circumference of the city bristled like a hedgehog. Here is a rather elaborate plan by Francesco di Giorgio Martini, also near the end of the fifteenth century. His bastions are imperfect, in that the bases of the round towers are not well covered by enfilading fire from other points. Enemy troops trying to get between the bastions would, however, be terribly exposed to a crossfire.

Defending troops using small arms were protected by parapets. Defensive cannon, both for counter-battery work and the immediate protection of the walls, were installed in strong casemates such as are still visible in the Elizabethan fortifications of Berwick.

In Bonaiuto Lorini's work on fortification (first published in 1592, but I have used the edition of 1609), the new principles fully elaborated in Italy by the middle of the sixteenth century, and derived from elementary geometry, are elaborately treated. Geometry, says Lorini, is not merely useful but essential, the very foundation of all our procedures. Without it and proper rules of proportion it would be impossible to construct a symmetrical, polygonal fortress. The engineer had to learn to set out the angles, slopes, and sides exactly to obtain the required protection.

To reduce the vulnerability of the walls to round shot, they were decreased in height and increased in thickness, earth replacing masonry at the upper level. To prevent escalading, the ditch was made much more formidable. Lorini's sectional illustration gives the general picture: note the sharpshooter at the end of the mined gallery.

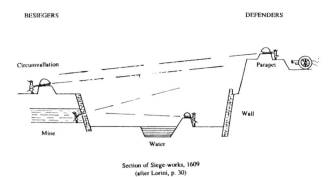

BESIEGERS   DEFENDERS

Circumvallation   Parapet

                                              Wall

Mine

Water

Section of Siege-works, 1609
(after Lorini, p. 30)

Fig. 3. Section of Fortification, 1609.

A great advantage of the new system, as regards the defender's firepower, was that it permitted flank fire by which not only the short curtain walls but the bastions themselves were protected against assault.

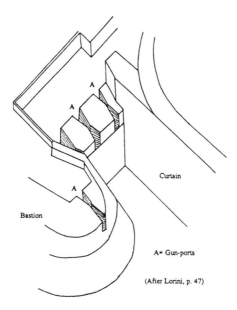

Curtain

Bastion

A = Gun-ports

(After Lorini, p. 47)

Fig. 4. Bastion and Curtain.

Lorini's drawing shows how the embrasures for cannon might be designed in order to provide this enfilading fire. Note that the guns at the root of the bastion, by being set back, are protected from round shot by the whole thickness of the bastion. It became the chief concern of the later military engineers, among whom Vauban was outstanding on the French side and Coehoorn on the Dutch, so to plot the angles of the perimeters of wall and bastion, together with the various supplementary ravelins, horn-works and so on, whose names were so loved by Uncle Toby, and the position of the defensive firepower on each, that the maximum amount of crossfire was directed at any point of attack. However, as may be seen from this rather detailed military scene published in a book of fortification in 1696, neither the attack nor the defence underwent further radical changes in the seventeenth century, though of course what is shown here is crude and out of date for the time.[6]

The question may now be asked: did the development of fortification in this period increase the pressure on the engineer's mathematical skill? I think to some extent it did. The problem of working out the

---

[6] Sebastien Fernandez de Medrano, *Ingenieur pratique ou Architecture militaire et moderne* (Brussels, 1696). General Medrano was "Directeur de l'Académie royale et militaire des Pays-Bas."

Fig. 5. Late Seventeenth Century Fortification.

form of a late seventeenth-century fortified city, with its elaborate rings
of defence in considerable depth, multitude of planes, and countless
firepoints was far more complex than making a plan for a medieval
castle or a renaissance fortress. But not incomparably more difficult.
Medieval and renaissance architects, we know, prepared designs of their
buildings and employed geometry; we can follow the development of
their tradition from Villard de Honnecourt (c.1250) onward. All
renaissance architects and engineers claimed that geometry was the
foundation of their work, that without geometry no building could be
completed. Both Filarete and Francesco di Giorgio Martini were
familiar with the Vitruvian notion that perfect architectural proportion
was geometrically derived from the proportions of the human body.
According to the former the mathematics can be learned from Euclid
and Campano da Novaro (13th century) .[7] Or as the poet Lydgate wrote:
"by craft of Euclid mason doth his cure." Geometry was indeed the
original secret of the freemasons.[8] It was very elementary; the skilled

[7] John R. Spencer, ed., *Filarete's Treatise on Architecture* (New Haven and
London, 1965) , 1:9.

[8] When this secret became devalued, the lodges in the seventeenth century
became cult-centres.

military engineer of the seventeenth century needed considerably more mathematical sophistication than that cherished in the masons' lodges. Even so, no competent mathematician of the time regarded what was needed for engineering as other than trivial geometry and arithmetic, and it is almost ludicrous to see here an application of science as though something abstruse were going on. One might as well say that the literate engineer "applied literature" to his work. What we should see is a change, resulting ultimately from the introduction of artillery, in the attitudes and training of the architect-engineer, pushing him further along the road he had long before commenced to tread. I do not think the mathematicians had much to do with spurring him on; it was necessity that did so.

My final conclusion is this: the profession of the architect-engineer embraced the most highly sophisticated technology existing in the fourteenth, fifteenth, and sixteenth centuries; it was the one technical profession making large demands on organising and planning ability, drawing-office skill, taste, craft knowledge, and mathematical learning. We know from the lives and writings of Alberti, Filarete, Francesco di Giorgio Martini, Leonardo da Vinci, and others that the architect-engineer practised the arts of war as well as those of peace. We know that his had long been a proud, independent profession, only rarely willing to admit in some abstract speculations the superior ability of the academic mathematician, and in my view this situation changed little through the Renaissance. It was the architect-engineer who saw what cannon did to the old style of fortification and it was he who devised a new one—turning as always to his ancient sources of inspiration and strength, geometry and the laws of proportion.

The gunner, villain of the last few minutes, becomes hero of the next. He was socially, intellectually, and educationally the inferior of the architect-engineer. The books written by real gunners, like Robert Norton, Master-gunner of the Tower of London, are poor, dull, derivative books. The best books on artillery were written by gentlemen and generals, though perhaps they are not the truest. Perhaps I should make exception for one who saw service as a gunner (at Tilbury and Gravesend) in Elizabeth's reign, William Bourne, and who wrote about artillery (*The Arte of Shooting in Great Ordnaunce*, London, 1587) as well as other usual topics of the mathematical practitioner. But even he is by no means one of the great exponents of the art.

What is surprising about gunnery is that the mathematicians took it up—not of course the really dangerous and dirty business but the definable and fascinating problem of the flight of bodies through the air. It is not strange to find trajectories of roundshot and mortar shells

IX

16

drawn by such an architect-engineer as Leonardo, trajectories that are in fact a good deal more realistic than his abstract ideas, but why did it become a mathematical commonplace? The answer is that, in its most general form, projectile motion had been since the days of Aristotle one of the unsolved problems of natural philosophy. Aristotle and common sense suggested that nothing moves without a mover, there can be no motion without cause. What is the cause of the free flight of a projectile, separated from the first cause of its motion? Again, according to Aristotle, unsupported heavy bodies fall down freely towards the centre of the universe, while only violent effort can force them to move away from the centre; the former motion being natural is accelerated, the second decelerated. What kind of a path does the projectile describe, then, to satisfy these conditions? Why, for example, when projected horizontally does a heavy body not fall straightaway toward the centre, but follow a sensibly horizontal line for a while, called by gunners the point-blank range?

One can discover attempted solutions to all these problems in the scattered notes of Leonardo da Vinci, who was an eager, self-taught, and naive philosopher as well as architect-engineer (and anatomist). None of his notes on motion represents an original idea; they are wholly unsystematic, and often contradictory. I use them as an example only. First we might observe that Leonardo confutes the still common belief that interest in projectiles came as a consequence of the invention of gunpowder. This belief is logically unsound and historically inde-fensible; it arises, I think, from a confusion about the history of firearms. It took only about half a century (that is, to about 1370) for cannon to replace mechanical siege engines—though the latter survive in books thereafter. It took well over two centuries for small arms to replace the crossbow in war.[9] Further, there was nothing very interesting about the behaviour of early bombards, whether built-up wrought iron cannon or the monsters used by the Turks against Constantinople; these were short range pieces. More interesting problems arose with the introduction of light bronze cannon firing cast-iron shot to longer ranges toward the end of the fifteenth century, and contemporaneously of mortar-bombs, with their indirect trajectory. However, these improved weapons did not create an intellectual difficulty that had been recognised for centuries, and which the mathematicians rather than the practical gunners attempted to solve. For experimental or imaginative purposes the crossbow was often the more convenient device; so Leonardo asks

whether if a bolt is shot from a crossbow four hundred *braccia* a crossbow made

---

[9] The crossbow was commonly employed for sport in the seventeenth century; as the Lancashire prodd it survived among poachers and others into the nineteenth; and it is still manufactured for sporting purposes at the present day.

in the same proportions but four times the force and size will not send the bolt four times as far.

And again,

> I ask if a crossbow sends a bolt weighing two ounces a distance of four hundred *braccia*, how many *braccia* will it send one of four ounces? [10]

Here incidentally we may note Leonardo's propensity—the regular resort of medieval technology—to suppose that simple linear proportionality is applicable to every problem. One of the first practical successes of the new science of mechanics was to prove that this is not the case.

We also find common misapprehensions of fact stated by Leonardo, for example:

> In the centre of the direct path taken by heavy bodies which traverse the air with violent movement, there is greater power and greater striking force when an obstacle is met than in any other part of its line. [11]

Leaving aside the physical interpretation of projectile motion to concentrate on kinematics, [12] we note next that Leonardo embraced an analysis of the trajectory originating with the scholastic philosopher Albert of Saxony in the latter part of the fourteenth century, which analysis in turn derived from the Oxford and Parisian schools of the previous generation. The trajectory was divided into three portions, the first and last rectilinear, the middle curved. If a crossbow bolt is shot upward at an angle to the horizon, the violent motion it receives overcomes both gravity and the natural resistance to motion so that it flies straight. As the violent motion weakens, the trajectory becomes curved; when gravity and the resistance overcome the motion of projection, the projectile falls straight down to the ground. Many writers (including Tartaglia) depict such a trajectory. If the assumption is made, as it is by Daniel Santbech for example, that the first straight-line segment is always proportional to the force of projection, then the range becomes proportional directly to the charge and the cosine of the angle of elevation. The regular decrease of the cosine from angle zero to the right angle puts this rule at variance with experience, and accordingly

---

[10] *Codex Atlanticus* 314v b; *The Notebooks of Leonardo da Vinci; Arranged, Rendered into English, and Introduced* by Edward MacCurdy (London, 1938), 1:531.

[11] Paris, Institut de France, MS A, 43v; MacCurdy, *Notebooks*, 1:540.

[12] As is now very well known, the impetus theory expounded with variations by all the most important writers on philosophy of motion in the sixteenth century (Tartaglia, Cardan, Benedetti, Buonamico, and the young Galileo) was of medieval origin, going back (probably through Islamic philosophers) to the Byzantines. See Marshall Clagett, *The Science of Mechanics in the Middle Ages* (Madison and London, 1959) and Stillman Drake and I. E. Drabkin, *Mechanics in Sixteenth Century Italy* (Madison and London, 1969).

more practical writers on gunnery of the late sixteenth century proposed arbitrary mathematical schema relating increased range to increasing angles of elevation from zero upward.

The Italian mathematician Niccolò Tartaglia, who was born about 1500 and died at Venice in 1557, was the founder of ballistics since he devoted a whole book to it (*Nova Scientia*, 1537) and much of a second (*Quesiti, et inventioni diverse*, 1546). The former of these opens in the dedication to the Duke of Urbino with the following highly circumstantial piece of autobiography:

> When I dwelt at Verona in 1531 I had a very close and cordial friend, an expert bombardier at Castel Vecchio, . . . [who] asked me about the manner of aiming a given artillery piece for its furthest shot. Now I had had no actual practice in that art (for truly, Excellent Duke, I have never fired artillery . . . or musket); nevertheless, desiring to serve my friend, I promised to give him shortly a definite answer.[13]

This has always been taken *au pied de la lettre*, though I fail to see why Tartaglia's venerable bombardier should not rather be put with the Ancient Mariner, Shelley's traveller from distant lands, and countless Masters and Scholars of didactic dialogue. However, we can hardly suppose Tartaglia would have renounced experience had he possessed it.

The solution, resting on no very clear argument, is that 45° of elevation gives the extreme range. I have no doubt but that this was an intuitive result based on proportional symmetry; Tartaglia claims it was verified by trial. He also knew that complementary angles should give equal ranges, and claimed further that the extreme range is always ten times the point-blank. Hence he did not argue that the initial rectilinear segments are always equal at any angle. The curved segment he took to be an arc of circle to which the linear segments were tangential.

Tartaglia's theory is not much more than a dressed up version of Albert's or Leonardo's, and the mathematical garnish is really quite arbitrary. His most original contention was that no part of the trajectory—not even the point blank—is truly rectilinear; yet in geometry he always treated it in the way I have described. His conceptions are Aristotelian ones, modified by the impetus theory. As Koyré has remarked, it was exceedingly difficult in these matters to step outside the tradition, and in so far as Tartaglia departed from it—especially in abandoning the idea of strictly rectilinear segments—this did not help him to solve the geometrical problem.

The recondite and sometimes absurd philosophical arguments that

---

[13] The translation is by Stillman Drake. Drake and Drabkin, *Mechanics in Sixteenth Century Italy*, pp. 63-64.

constitute the greater part of Tartaglia's writings on ballistics were of no value to the compilers of practical manuals on gunnery, and though these writings of Tartaglia were plagiarised in other vernacular texts, it was necessary if range-tables were to be given—for Tartaglia, having promised them, did not give them—to derive them in some arbitrary fashion. Thus Diego Uffano, "captain of the castle of Antwerp," proposed a simple arithmetic series increasing the range steadily from zero to 45 degrees; if the point-blank range was 200 paces, he said, then at each degree of elevation the range would be 244, 287, 329, etc., to a maximum of 1,190 paces. The only real interest in these arbitrary tables is that they prove how great an authority mathematics possessed. There is no reason to believe that they were ever used, or were usable. But they made excellent propaganda. In some of the treatises on gunnery, and many other books on applied mathematics, great emphasis was laid on the importance of the arts of measuring distance, heights, depths, of preparing plans and maps, and of familiarity with simple arithmetic and geometric rules, to the scholarly or gentlemanly soldier.[14] The writers of these books have in mind a figure who is not by any means the engineer-architect of the earlier Renaissance, who was not a leader of men in battle or tactician, but a noble, scholarly soldier who shall be master of the established mathematical arts, and also of the mathematical art of gunnery, as well as of all the practical aspects of warfare. So Thomas Digges writes (*Pantometria*, 1571) : "for science in great ordinance especially to shoote exactly at Randons (a qualitie not unmeete for a Gentleman) without rules Geometrical, and perfect skill in these mensurations, he shall never know anything." [15] Such a gentleman-artillerist was perhaps too ideal a figure, but other writers insist that the gunner have skill in surveying and so forth to raise him above the ordinary level of under-officers. Even so, some study of accounts of battles on sea and land suggest that the average good gunner was a man who knew how to load and fire his piece efficiently and safely, and while aiming by line of sight make such allowance as experience and trial suggested for long range and other factors. We have to remember that seventeenth century cannon were very idiosyncratic and irregular in their shooting, each gun being made from a unique mould, that the charge and quality of the propellant was highly variable, and the projectile occupied only 90 per cent of the area of the bore. Consistent

[14] See Peter Whitehorne, *Certain Waies for the Orderyng of Souldiers in Battelray and Setting of Battailes* (London, 1562) ; Cyprian Lucar, *Three Bookes of Colloquies* (London, 1588) ; and Walter Ryff, *Der Furnembsten . . . Architectur . . .* (Nurnberg, 1547).

[15] Sig. A iii. "At randons" means, elevated above the point-blank (hence, tellingly enough, the more usual *random*).

IX

20

practice was impossible, and rules likely to be less effective than the good gunner's experience and correction of his aim.

Some historians (Edgar Zilsel and recently Christopher Hill) have made much of the existence of a whole group of superior artisans at this period, from the architect-engineers through painters and musicians to gunners, surveyors, navigators, cartographers, and instrument makers, apothecaries, opticians, clock makers, and so on. Obviously, levels of craft skill did exist; some crafts employed simple mathematics, others chemical knowledge. Clearly too these superior craftsmen contributed to the refinement of technology. But one should be cautious of taking "mathematics" in too grandiose a sense; one should remember that gunners and sailors were simple men, and that, for sure, they were consumers, not creators, of mathematics.

However, we can now see how well the world was prepared for at least one feature of the kinematical discoveries of Galileo; the writers on practical mathematics and artillery had long been confident that their art must follow some rational mathematical scheme, and they were prepared to believe that Galileo had discovered it. The writers did not test his theory by experiment nor enquire about its application in the field; it was enough that the new theory looked right, even if sometimes the explanation of its curved trajectory harked back to Tartaglia, rather than Galileo himself.

As everyone knows, Galileo rediscovered and applied to actual bodies falling at the surface of the earth the square-law of acceleration; he understood perfectly the vectorial combination of motions, and this gave him the parabola as the path of a projectile—neglecting the curvature of the earth itself. Accordingly, at the end of his *Discourses on Two New Sciences* (1638) he was able to produce that great desideratum, a theoretical range table, and a number of accurate propositions about projectile motion. This work on ballistics was developed further by Galileo's pupil, Evangelista Torricelli, who generalised and completed the theory, after which it passed into general circulation.

What was Galileo's interest in solving the problem of projectile motion, which as we know occupied him for over thirty years? To my mind the utilitarian aspects of the *Two New Sciences* have been grossly exaggerated. Galileo was above all a mathematical philosopher; most of his life work was devoted to the general theory of mechanics, not to say astronomy and cosmology. But he liked to display his abilities in the most direct and conspicuous fashion. There can be no doubt that the *Two New Sciences* was written to demonstrate the falsity of the simple rules of proportion followed in the old craft tradition, and the superiority of the new, philosophical laws devised by Galileo himself. He was not, so to speak, on the same side as the artisans; he was proving that the philosopher understood things *better* than they did. For a century and more, gunners had fumbled at the mystery of ballistics; Galileo's new treatment of kinematics unlocked it at once. Galileo was quite explicit about this. In 1632, after Bonaventura Cavalieri had first put the parabolic theory in print, he complained to a friend about the loss of "the renown, which I so keenly desired and had promised myself from my long labours" in mechanics, saying that to master the trajectory of a projectile had been their chief objective. It was, after all, the most celebrated of all problems in mechanics, quite apart from any question of the usefulness of its solution.

Did Galileo believe his own solution to be useful? If we suppose Galileo to have been drawn all along, as Tartaglia said he himself was, to a practical problem of artillery, and if Galileo really thought that he had solved this practical problem, then the answer clearly is, Yes. Certainly Galileo talked about his ballistic theory in a very practical way. But he was also quite aware that when movements are very swift they are greatly impeded by the resistance of the air: this resistance was especially strong in the case of musket and cannon balls. "The enormous impetus of these violent shots," he wrote, "may cause some deformation of the trajectory, making the beginning of the parabola flatter and less curved than the end;" but so far as this book is concerned, he went on, "this is a matter of small consequence in practical operations, the main one of which is the preparation of a table of ranges for shots of high elevation . . . and since shots of this kind are fired from mortars using small charges . . . they follow their prescribed paths very exactly." [16] Hence Galileo correctly enough supposed that the parabolic theory could have a limited application. However, he was by no means always scrupulous in making this clear—his tables do include the small angles—while Torricelli was even more realistic in his language, thereby creating the impression that the parabolic theory had completely solved the problem of exterior ballistics, at least in principle. When challenged, Torricelli attributed discrepancies in practice to the imperfections of guns and gunners, being seemingly reluctant, unlike Galileo, to admit that a large physical factor had been omitted from the parabolic theory.

Later writers on the theory of gunnery until well on in the eighteenth century were content to rely on this beautifully idealist conception, which became general from about 1670 onwards. Such influential "practical" treatises as Robert Anderson's *Genuine Use and Effects of the Gunne* and François Blondel's *Art de jeter les bombes*

[16] Galileo, *Dialogues Concerning Two New Sciences*, trans. by H. Crew and A. de Salvio (Evanston and Chicago, 1946), p. 246.

22

appeared in 1674 and 1683 respectively. Blondel's book is very thorough both in its critique of older ideas about the flight of projectiles and in its exposition of the parabolic theory; here he called in some of the mathematicians of the Académie Royale des Sciences to solve its most abstruse proposition. Blondel claims that the theory is most exactly applicable to mortar fire but does not exclude its use for cannon, nor admit plainly that air resistance is a disturbing factor. In 1731 Belidor's *Bombardier François* employs the parabolic theory, limiting his tables strictly to mortars. Benjamin Robins in 1742 was the first to show decisively that the parabolic theory was inadequate for all but very slow projectiles.

The mathematical study of the motion of bodies in resisting fluids was by this time two generations old, since it had begun with investigations by James Gregory, Wallis, Huygens, Newton, and others beginning about 1670. To connect these investigations with artillery practice at that time seems to me wholly unrealistic, though of course I do not deny that the mathematicians were conscious of the fact that artillery projectiles like ships did exemplify resisted motion. As I remarked before, one must remember that military orders commonly forbade gunners to fire at other than point-blank range, especially at sea; commanders were sceptical, to say the least, of any attempt at long range practice, except with mortars.

To sum up, it seems to me that the historian of any branch of technology must be careful not to read the present back into the past, nor to credit the writings of armchair specialists and propagandists without some other check that such authors describe things as they are, and not as they might be. Just as there have always been soldiers, artists, and industrialists of the no-nonsense brigade who have dismissed all attempts at rational theorisation (whether mathematical in form or otherwise) as absurd and needless, so there have always been experts trying to convince the world that they alone hold some particular theoretical key to reality. Time has proved the philosophy of the latter group correct. Everyone today knows what abstruse computations enter into the calculation of trajectories; fundamentally the methods used today go back historically to the theoretical mechanics of Newton. But there was not either in principle or in historical fact any role for theoretical mechanics to play in the warfare of the seventeenth century, and any mathematics used was of a most trivial kind. As in attempts to put physiology and medicine on a chemical basis, or to construct a machine enabling man to fly, the imagination of the seventeenth century ran forward to what was realised only in the twentieth. But we should not overrate the importance of such imaginative foresight, or conclude that experimental research and technological invention have always been exclusively devoted to turning such visions into reality. It is always dangerous to disregard the force of tradition, and the strong conservative elements in even the most original minds. In both science and technology many of the most persistent and ultimately the most fruitful of problems have been traditional ones, tackled in different manifestations by successive generations.

BIBLIOGRAPHY OF WORKS
NOT CITED IN THE NOTES

Biringucci, Vanoccio. *De la Pirotechnia*. Venice, 1540.

Cavalieri, Bonaventura. *Lo Specchio Ustorio*. Bologna, 1632.

Dircks, H. *The Life, Times and Scientific Labours of the Second Marquis of Worcester*. London, 1685.

Ffoulkes, Charles. *The Gunfounders of England*. Cambridge, Eng., 1937.

Francesco di Giorgio Martini. *Trattati di architectura ingegneria e arte militaire*, a cura de Corrado Maltese. Milan, 1967.

Frankl, P. "The Secret of the Medieval Mason." *Art Bulletin* 27 (Mar. 1945) : 46–64.

Galilei, Galileo. *Discorsi e dimostrazioni matematiche intorno a due nuove scienze*. Leiden, 1638.

[Gregory, James]. *Tentamina quaedam geometrica de motu penduli et projectorum*. Glasgow, 1672.

Guerlac, Henry. "John Mayow and the Aerial Nitre." *Actes du 7ᵐᵉ Congrès Internationale d'Histoire des Sciences*, pp. 332–49. Jerusalem, 1953.

Hale, J. R. "The Early Development of the Bastion." In *Europe in the Late Middle Ages*, edited by J. R. Hale and others, pp. 466–94. Evanston, 1965.

Hall, A. Rupert. *Ballistics in the Seventeenth Century*. Cambridge, Eng., 1952.

_____. "The Changing Technical Act." *Technology and Culture* 3 (Fall, 1962) : 501–15.

_____. "Merton Revisited." *History of Science* 2 (1963) : 1–15.

_____. "The Scholar and the Craftsman in the Scientific Revolution." In *Critical Problems in the History of Science*, edited by Marshall Clagett, pp. 3–23. Madison, 1959.

Halley, Edmond. "A Discourse Concerning Gravity . . . and the Motion of Projects." *Philosophical Transactions*, London, vol. 16 (1687) .

Hill, Christopher. *Intellectual Origins of the English Revolution*. Oxford, 1965.

Koyré, Alexandre, "La dynamique de Nicolo Tartaglia." In *La science au seizième siècle*, Colloque de Royaumont, 1957, pp. 93–113. Paris, 1960.

Lorini, Buonaiuto. *Le fortificationi*. Folio edition. Venice, 1609.

Lot, Ferdinand. *L'art militaire et les armées au Moyen Age en Europe et dans le Proche Orient*. Paris, 1946.

MacIvor, Iain. "The Elizabethan Fortification at Berwick-upon-Tweed." *Antiquaries' Journal*, vol. 45 (1965) .

Nef, J. U. "War and Economic Progress, 1540-1640." *Economic History Review*, 1st series, vol. 12 (1942) .

Newton, Sir Isaac. *Philosophiae naturalis principia mathematica*. London, 1687.

Ramelli, Agostino. *Le diverse et artificiose machine*. Paris, 1588.

Rattansi, P. M. "The Intellectual Origins of the Royal Society." *Notes and Records of the Royal Society* 23 (1968) : 129–43.

Robins, Benjamin. *New Principles of Gunnery*. London, 1742.

Santbech, Daniel. *Problematum astronomicorum et geometricorum sectiones septem*. Basel, 1561.

24

Tartaglia, Niccolò. *Nova scientia*. Venice, 1537.

_____. *Quesiti, et inventioni diverse*. Venice, 1546.

Torricelli, Evangelista. *Opera geometrica*. Florence, 1644.

Uffano, Diego. *Artillerie*. Translated by Th. de Brye. Frankfurt, 1614.

Valturio, Roberto. *De re militari*. Verona, 1472.

Walter, E. J. "Warum gab es im Altertum keine Dynamik?" *Archives Internationales d'Histoire des Sciences* 3 (1948) : 365–82.

Webb, Henry J. *Elizabethan Military Science, the Books and the Practice*, Madison, 1965.

Zilsel, Edgar. "The Sociological Roots of Science." *American Journal of Sociology* 47 (1942) : 544–62.

X

GUNNERY, SCIENCE, AND THE
ROYAL SOCIETY

More than twenty-five years ago I published a book called *Ballistics in the Seventeenth Century*.[1] Today I might call it "Early Science and the Art of Shooting." I had found a fascinating thesis topic; it constituted not only my apprenticeship in the history of science but also my first introduction to some of the giants of the Scientific Revolution: Galileo, Mersenne, Huygens, Newton, Leibniz, Johann Bernoulli. I have never subsequently moved very far from them. The subject also drew me into the darker mysteries of the use of guns in war.

It is a curious historical fact that while there is much historical evidence about the making of cannon, their military organization, their weights, numbers, and cost, and so on, there is very little to show how they were actually deployed in war. Until relatively recently, before Michael Lewis cleared it up,[2] there was even a good deal of doubt concerning one of the most celebrated gunslinging episodes in history, the repulse (defeat is too strong a word) of the Spanish Armada. The puzzle of that prolonged Elizabethan Battle of Britain turned precisely on the characteristics and use of naval artillery; and you may be sure that Howard of Effingham, Drake, and others left no narrative to explain it to us. This was true of

Reprinted from *The Uses of Science in the Age of Newton*, ed. John F. Burke, by permission of the publisher. © 1983 University of California Press, Berkeley, California.

nearly all battles and sieges. The technical management of guns in tactical situations was a curiously neglected branch of "baroque" military history.

There was, however, an extensive printed literature of *Buchsenmeisterei*—beginning perhaps with the German *Feuerwerkbuch* of 1420,[3] which gave rise to printed progeny branching off, on the one hand, to deal with the Brock's Benefit of what we still call fireworks, and on the other, to the great artillery handbooks of such men as Luys Collado,[4] of which Robert Norton's *The Gunner* (1628) is a late, imitative, and rather feeble descendant. These books, some of them in the form of a dialogue between "the Ancient Gunner" and "a Novice," contain a great deal of technical information on subjects ranging from the making and testing of both ordnance and gunpowder, to carriage's, hoists, and horses. They instruct one in distinguishing good powder from bad, in making allowance for the decreasing thickness of metal in a piece between breech and muzzle, in determining the caliber of the shot and the charge of powder, in mastering a proper gun drill, and in handling the piece safely. Yet they say rather little about the use of the artillery in active service; they are very discreet about the range at which the various types might be employed, their placing with respect to troops of other arms, their normal rate of fire, and their effect. As for the gunners' own narratives of their experiences in actual battles or sieges, they simply do not exist.

Moreover, these early artillery manuals are highly misleading and untrustworthy in at least one important respect—and I think they share this defect with all early technical treatises. They are overelaborate and overrational. They describe the craftsman or artificer—in this case the gunner—as far too clever a fellow, weighing and measuring and calculating in a highly improbable manner.[5] Consider those many pretty mathematical books (Digges's *Pantometria* [1571], for example) in which the gunner is shown measuring the range to a castle or a ship at sea by careful triangulation or by another geometrical process involving instruments. Can one imagine that this was often done? Or that a precise measure of the range would have been very helpful in laying the gun to hit a given target? There were indeed instruments called gunner's quadrants which were supposed to be helpful in setting cannon to the right angle of elevation, but there still seems something fanciful about pictures of cannon tilted up like howitzers with their trails dug into the ground so that no recoil would be possible. Many of these laborious studies smell of the lamp; they surely contain many purely literary effects, in the manner of the navigational manuals that promised an accuracy of determination that never was attained on sea or land. And it is easy to see why. The early technological books were written not only to instruct but also to impress; and the higher their coffee-table quality, the less their didactic utility. Most were dedicated to a grandee—a monarch or successful noble commander; hence there was a patron whom the author must impress with his skill and learning. In addition, since the real gunner in the field was presumably often illiterate and very unlikely to be learning his trade from books, these handsome volumes must have been intended for the whiter hands of the gentry for whom even in the mid-seventeenth century the profession of arms was still a normal social expectation. The young gentleman had to be properly impressed by the master gunner's excellence, just as he was to be impressed in other books by the skill of the riding master, the fencing master, or the dancing master.[6] Who would wish to buy a book on a technical subject without a touch of mystery to redeem its prosaic bread and butter—the touch of mystery so blithely promised by Sir Hugh Platt in *The Jewell-House of Art and Nature* (1594) or by John Bate in his *Mysteries of Nature and Art* (1634).

The gunnery writers flattered the sophistication of their readers, who probably knew no more of triangulation, trigonometry, or even the rule of three than most real gunners; they blinded them with science. Of course no writer would ever admit openly to a bit of coney-catching, but we are told about it in satire—as in Ben Jonson's *Alchemist*, where Surley complains

That Alchemie is a pretty kind of game,
Somewhat like tricks o' the cards, to cheat a man with charming
What else are all your termes,
Whereon no one o' your writers agrees with other?
Of your *elixir*, your *lac virginis*,

Your *stone*, your *med'cine*, and your *chryososperme*,
Your *sal*, your *sulphur* and your *mercurie*,
Your *oyle of heighte*, your *tree of life*, your *bloud*,
Your *marchasite*, your *tutie*, your *magnesia*,
Your *toade*, your *crow*, your *dragon* and your *panthar*,
Your *sunne*, your *moone*, your *firmament*, your *adrop*,
Your *lato, azoch, zernich, chibrit, heautarit*,
And then your *red man* and your *white woman*,
With all your broths, your *menstrues*, and *materialls* . . .
And worlds of other strange ingredients,
Would burst a man to name?

All their talk, Subtle, the alchemist, rejoins, was used "to obscure the art" and, as his gull, Mammon, chimes in, so that "the simple idiot should not learn it, and make it vulgar."[7] But real gunnery, like other real crafts, was in fact learned by "simple idiots," ploughboys and sailor boys without benefit of letters, and we are entitled to wonder whether the alternate "literary" approach to these mysteries, which remains our only source of information, was not highly deceptive and unrepresentative of what actually happened.

The reality of the use of guns in seventeenth-century warfare is confirmed by the rich anecdotage and the pictorial evidence of a later, but still technically similar, epoch of warfare. We are all familiar with the order "Don't fire till you see the whites of their eyes," with the compliments exchanged before the battle of Minden, and with the yardarm-to-yardarm cannonades of Nelson's ships of the line. There is much to suggest the incredible resolution of combatants, little to indicate ballistic finesse.

But what of the relationship of gunnery to science in the wider sense? It has long been supposed that the various technical branches of the art of war, especially ballistics (or the science of shooting) and fortification, held the attention of men of science and mathematics, who sought for rules to define the proper practices. And indeed an interest in the flight of projectiles figures in the writings of Tartaglia, Thomas Harriot, Cavalieri, Mersenne, Galileo, Huygens, Wallis, James Gregory, Halley, Newton, Leibniz, and many more, though not necessarily in the works published in their lifetimes. In other words, the motion of projectiles, and especially the motion of projectiles in air, was studied by a considerable number of men, and by men of the highest abilities. Why? The common supposition has been—and perhaps among the less critical still is—that these men were naively eager to solve "useful" (or at least utilitarian) problems, not so much with the object of doing good (for even in the seventeenth century to increase the accuracy of shooting could hardly be regarded as clearly an act of kindly benevolence, and we have evidence that warlike inventions were then already regarded with horror) but rather in order to prove the general utility of the scientific approach to practical problems, and to demonstrate the particular powers of the individual writer.

In a classic study more than forty years ago, Robert K. Merton wrote briefly of the presence of both pure and applied aspects in ballistics research as suggesting that "to some extent at least," the interest of those who carried it out was directed to it "because of the practical utility derivable therefrom."

The effort to attain mathematical precision in artillery fire was a model for the industrial arts and a link with the current science. In any event, military needs, as well as the other technologic needs previously described, tended to direct scientific interest into certain fields.[8]

Merton's position is more subtle and more carefully qualified than this excerpt alone might indicate; for instance, he was well aware of the purely intrinsic interest of ballistic problems to mathematicians and of the force of active tradition in guiding problem choice. He was certainly far less extreme than B. Hessen, for whom science was no more than industry and commerce conducted by other means:

If we compare this basic series of themes [in seventeenth-century physical science] with the physical problems which we found when analysing the technical demands of transport, means of communication, industry and war, it becomes quite clear that these problems of physics were fundamentally determined by these demands.[9]

And indeed the military realism of seventeenth-century mechanics had been emphasized long before by William Whewell, who opined that "practical applications of the doctrine of projectiles no doubt had a share in establishing the truth of Galileo's views."[10]

In my 1952 book I argued that such views about problem choice as determined by technological need were not sup-

ported by the evidence. Existing military art was incapable of
adopting a mathematical theory of projectile flight and applying it to practice, nor (so far as one can tell) did it ever attempt
to do so, at least before the death of Newton. Therefore, if the
learned men developed the theory out of a desire to solve
useful problems, they were mistaken. It was my view, however,
that they did not tackle ballistic problems for this reason but
because these problems were intellectually fascinating and to
some extent open to solution. I pointed out that the tradition
of interest in projectile motion goes back to Aristotle; that it
became a crucial issue in the medieval philosophical development of mechanics; and that it is really rather absurd to suppose that this tradition can have been immediately affected by
the invention of gunpowder and the introduction of artillery.
If Leonardo da Vinci was the first to bring together the post-
Aristotelian tradition of academic mechanics and the brute
craft of artillery, it was the mathematician Niccolo Tartaglia
who first published the marriage (in 1537) and tried, without
success, to make it mathematically fruitful.[11] Tartaglia's discussion was founded, on the one hand, on perfectly conventional academic ideas and, on the other, on improbable empirical facts that he unwisely supposed were certified by the
experience of practical gunners. The one significant and correct idea he had about the shape of the trajectory was rejected
by almost all his successors, before Galileo.

Tartaglia's recipe for practical success—to try to graft craft
experience onto academic learning—failed, as it must always
fail;[12] chemistry was not born from a marriage of the Aristotelian theory of elements with the experience of goldsmiths,
nor was geology the product of combining miners' lore with
the Book of Genesis. Galileo, for whom the theory of projectiles was an obvious extension of the most general principles of
motion, relied entirely on a mathematical argument. Given
that horizontal motion is uniform and vertical motion is accelerated under gravity, the parabolic trajectory necessarily follows, and from that follows an endless family of mathematical
propositions which Galileo happily explored. Whether or not
Galileo believed the real cannonballs or mortar bombs that
combatants fired in anger actually followed parabolic trajectories is not, perhaps, very important; surely he must have had

the common sense to recognize that his elaborate family of propositions was only of interest to mathematicians.

Of course, to intellectuals it was a very considerable discovery that a geometric curve was realized in physical nature, even if only under impossible conditions. Apart from the circle, only one other curve, the ellipse, was known to be so realized; and Galileo had rejected Kepler's demonstrations that it was. Galileo saw the parabola in other places too—in perfectly economical beams and in ropes suspended by their ends. And so a new issue was raised: To what extent do ordinary, actual events correspond to idealized events? Or, expressed another way: To what extent do ordinary, actual events obey idealized laws of nature? Archimedes gave one definition of a perfect fluid—How closely do real fluids approach it? Boyle's Law describes an aspect of the behavior of an ideal gas—How similar to this ideal is air? And so forth. It is one thing to declare that nature is geometrical, but (as Galileo also knew) when one comes to deal with real solids, liquids, and gases the geometry may become very perplexed.

This issue was new in the seventeenth century—or perhaps I should say it appeared new in the post-Copernican world.[13] It had not existed for the Greeks or in the medieval world—or, in fact, as long as the qualitative belonged to philosophy and the quantitative to mathematics. Light as an experience belonged to philosophy; a light ray, an imaginary abstraction, belonged to mathematics. Within such a tradition no one expected to find mathematical relationships exemplified in the real world. One recalls Simplicio's complaint: "After all, Salviati, these mathematical subtleties do very well in the abstract, but they do not work out when applied to sensible and physical matters."[14]

While the Aristotelianism of the Simplicios crumbled, the new metaphysic of mathematizing nature, sometimes called Platonism, combined with experimentally derived, quantified laws such as Snell's or Boyle's laws to reshape the problem of ideal and reality. Consider Kepler's ellipses, for example: Might not the elliptical orbit be a very close ideal approximation but the reality a complex, awkward curve barely distinguishable from the geometric ellipse? Before Newton no one could be certain, and one had to be a confirmed Platonist to assert with assurance that the geometrical ellipse is the real, as well as the ideal, path of every planet. It was for this reason that many astronomers about the middle of the seventeenth century experimented with alternatives to Kepler's version of the elliptical orbit, which Newtonian dynamics alone was later to justify and explain.

Or consider another case, that of the timekeeping pendulum. Galileo's rather informal, Platonic confidence that the ordinary circular pendulum is perfectly isochronous was proved false by Huygens in 1657; the true isochronal curve is not geometric at all but mechanical, the cycloid. What is more, Huygens proved that the center of oscillation of a mass such as a round pendulum bob is not the same as its geometric center; for all practical purposes one can measure the length of the pendulum from the center of the bob, but to be exact one must take second-order effects into account. This recognition of second-order effects as modifying, though possibly only slightly, more-or-less obvious first-order Platonic laws has been one of the ways by which science has developed.

As soon as we consider—and in due course of development carry out experiments upon—fast-moving projectiles propelled by exploding gunpowder, we hit upon an interesting example of this state of affairs. Although the first-order trajectory is clearly a parabola, the most ordinary experience suggests a whole variety of effects that must complicate this simplicity. First, as Galileo himself was well aware, a swiftly moving projectile must be slowed down by the resistance of the air beating upon it in a violent gale. Second, as tennis players know, the motion of a projectile in air may, because of inevitable asymmetries, produce changes in the trajectory. And third, the motion of the earth itself, or variations in the composition of the atmosphere, may further modify the apparent trajectory. So complex is the interplay of these various factors—all of which except the last, atmospheric variation, were recognized before the end of the seventeenth century—that there can be no completely general solution to the problem of the motion of a projectile from A to B, unless A and B are very close together.

Air-resistance, a second-order effect, attracted a great deal of attention because it was so obvious that the first-order parabola was just too naive a solution. Still, once the parabola had been proposed, it was up to the military bombardiers to make what they could of it, which was little enough, though it was to provide the official theory for mortar shooting for some time.

Rather than follow much further the contorted evolution of fluid mechanics in relation to projectiles, let me just point out that the great Scottish mathematician James Gregory, an older contemporary of Newton, solved the rather simple formulation in which the retardation is supposed to be proportional to the flight time and to act along the original direction of flight (therefore the vertical or gravitational component is supposed not to be resisted). In this case the first part of the trajectory is not much altered, but the end is squeezed in so that the range is shortened. Gregory proved that this new trajectory is itself a different parabola with its axis inclined to the horizontal to produce the asymmetric curve. Actually he was not the first to do so: Thomas Harriot did the same at the beginning of the century, but Gregory knew nothing of it as it was discovered only recently in Harriot's notes.[15]

It will be obvious that this hypothesis produces less plausible results as the parabola becomes steeper; therefore, only where the trajectory is flat can it be considered even a reasonable first approximation. In practical terms it tells us only what common sense would—that for the same range, air-resistance requires the aim to be a little higher than it would be according to Galileo's parabolic theory.

The only realistic way to analyze the problem, *pace* Gregory, was to consider the resistance of the air as acting continuously along the projectile's line of flight, directly opposing it at every point. Huygens in 1668 and Newton in 1684 succeeded in solving this mathematical problem for the case in which the resistance is supposed to vary directly as the velocity of the projectile, but both of these great mathematicians were defeated by the case in which the resistance is proportional to the *square* of the velocity, which seemed to them (rightly) more physically plausible. Newton in the *Principia* produced an approximate answer.

Thus, in a couple of generations, applied mathematics had run from the easy success of Galileo's parabola, straight into a blank wall—in the case of air-resistance only, not to mention the other complexities. No general solution of the problem of exterior ballistics is possible. Newton was fond of remarking that "Nature is always simple and consonant to herself"; and if we examine the metaphysical foundations of Newtonian science we find his confidence justified. It is in *detail* that Nature reveals the thorny problems of her complexity—whether it be in the three-body problem of gravitational force, the theory of the tides, the precession of the equinoxes, or fluid mechanics. For all his assurance of the "simplicity" of nature, there were many such problems that even Newton could not completely analyze. The hoped-for logical chain linking theory (ideal) with practice (reality) could not be completely forged. Even the near successors of Newton could not predict precisely the fall of a cannonball, any more than they could predict with perfect accuracy the exact track on the earth of the next solar eclipse, or predict to the day the return of Halley's comet. There was still an unaccountable margin of error, even if by now it was, perhaps, below one percent in most instances.

The motion of projectiles presented itself as a compelling and obvious instance of the problem of matching mathematical ideal and experimental reality. Other instances may be recalled, not least among them the related problem of fall, studied by Alexandre Koyré,[16] and the problem of longitude at sea—a valuable example because it shows so transparently the immense gulf between in-principle solutions and solutions that will give the required degree of accuracy in practice; that is, within margins of error very much less than one percent.

The problem of longitude at sea is another example of the role of practical desiderata in problem choice. No one would deny the usefulness of a workable method of finding longitude at sea; nor would anyone deny that Flamsteed's choosing to devote himself largely to the problem of lunar motion was conditioned by considerations of utility; apart from everything else, it was required of him as Astronomer-Royal at Greenwich. But suppose Flamsteed had looked less at the moon, would he have paid more attention to, say, variable stars or nebulae, or looked for new planets? Two things may be said. First, I doubt whether Flamsteed would have given as much time to the moon if the problem had not been one that

interested him, and one that was capable of conferring immense prestige. The problem was clearly a scientific one.
Second, it was an absorbing problem and therefore worthy of
his effort. No one at this stage in the history of astronomy
could find in variable stars or nebulae, or even the search for
new planets, subjects likely to give an astronomer work for a
lifetime. The moon was such a subject. The theory of projectiles, by contrast, offered only a limited interest. And though
to some degree its problems arose from experience with projectiles and were made more dramatic by the cannon and
mortars of sixteenth- and seventeenth-century warfare (firearms were high technology and remained so until Hitler built
Peenamunde), insofar as projectiles won some little attention
from mathematicians after Galileo, it was because projectile
theory presented an intellectual challenge. I find it impossible
to believe that mathematicians such as Gregory, Huygens, and
Newton thought their complex curves would change the art of
war.

Allow me to explore Newton's involvement a little further.
Historians have demonstrated that Newton devoted very little
effort to mechanics between the two plague years, 1665 and
1666, and the commencement of what was to be the *Principia*
in late 1684. We know that he did some interesting and swiftly
progressing work on central force during those early Cambridge years, when he was also most active in mathematics, and
that he touched on some other topics in mechanics also; but
nothing more happened until Robert Hooke opened a correspondence with Newton in 1679. There is one incident we
should attend to, however. In 1674 Robert Anderson published a book on gunnery. Although he was apparently not
without mathematical learning—his was the first gunners'
manual to make use of logarithms—Anderson clung in an
already naive way to the simple parabolic theory. He was quite
prepared to contemplate ranges of ten miles or more (then
wholly imaginary) and still suppose the projectile to be unaffected by its passage through so much air. He was criticized for
this by James Gregory, who produced in rebuttal the inclined
parabola theory. Gregory's London friend John Collins, the
correspondent of so many mathematicians at home and
abroad, sought to secure support for Gregory's poor opinion

X

of Anderson's book by soliciting the opinions of mathematicians such as John Wallis, Savilian Professor at Oxford, and Isaac Newton, Lucasian Professor at Cambridge. It seems now rather like taking several steam hammers to a small nut. Both Wallis and Newton condemned Anderson. Fortunately we have Newton's letter of reply, pointing out precisely the probable error on which Gregory had insisted. Remarking that according to the parabolic theory the projectile must actually accelerate in the downward half of the trajectory, despite air-resistance, he gave it as his view that such an acceleration was very unlikely and proposed (arbitrarily) an alternative construction representing the projectile as continually slowing down throughout its flight, which he thought it must do in practice. However, he told Collins not to mention his name if he passed this opinion on to Anderson, "because I have no mind to concern myself further about it."[17]

These words proved a poor prophecy, however, for when his thoughts were returned to mechanics by Halley's visit to Cambridge in August 1684—and the famous question "What curve is described by a planet under the action of an inverse-square central force?"—Newton spontaneously tackled the problem of resisted motions in the *Propositions on motion* he drew up that autumn.[18] In the *Propositions* Newton introduces the matter as an aspect of a quasi-Aristotelian distinction between celestial and terrestrial motions. Only in the heavens, he points out, can motion accord with the abstract mathematical propositions outlined in the first part of his paper, "For I think that the resistance of pure aether is either non-existent or extremely small." On earth all real motions take place in fluids, and their representation by geometry becomes more complex. Newton then goes on to show how this may be done by hyperbolic constructions on the assumption that the resistance is directly proportional to the velocity (these are equivalent to the results Huygens had obtained in 1669, using the related logarithmic curve). And, he continues, the same method of analysis may be applied to movement in water.

It seems obvious that in these latter propositions, the foundation of the second book of the *Principia*, Newton went far beyond Halley's original request for a dynamic derivation of the planetary orbit. In the *Propositions*, and far more elaborately in the *Principia*, Newton embarked on a general account of rational mechanics appropriate to both earth and heaven, air and ether. Among the "cases" of motion in air that he chose to examine were the oscillation of pendulums, free-fall, and projectiles, cases chosen because they were so well known as to be notorious, and because they offered possibilities of comparing theoretically computed and experimentally measured results. There is little or no practical value in discovering how the motion of a clock pendulum is affected by air-resistance or how much more slowly a stone falls in water than in a vacuum. With Newton, then, the problem of the trajectory of a projectile entered the highly mathematical realm of rational mechanics, where it rested for a very long time. I shall not disturb it further.

If no one was about to put Newton's mechanics to practical test by firing ordnance, it was far different with Galileo's parabolic theory. Here the issue was to find out how far, if at all, theory was faulted by experience. Some of the more practical men in the Royal Society seem to have been upset by James Gregory's assertion, against Anderson, that the parabola theory could not hold in actual shooting, while Anderson himself seems to have accused the king's gunners of the Tower of ignorance, or some other negligence now obscure. At any rate, in September 1674 Collins, in London, was able (with an undertone of amusement) to report to his friend Gregory in Edinburgh that the book had "caused much bustle which hath caused the Master of the Ordnance, Sir Thomas Chichele, Sir Jonas Moore, The Lord Brouncker and the gunners [of the Tower] to go many times a shooting with mortar pieces at Blackheath." Another person involved was Robert Hooke, who had bought Anderson's book on 20 May and who went with those just mentioned to Blackheath on 11 September for the shooting, on which date he notes, "Dined on Lord Brouncker treat and coach." On 17 September they went again: "Tried experiment of bullet with good [results] found it very near a parabola."[19] How the tests were made is not stated; it was presumably done by keeping the powder and shot as constant as possible and seeing if the ranges at different angles of elevation came out as predicted by the tables of parabolas.

Sir Jonas Moore, a latecomer to the Royal Society (he was to be elected only in December of the same year, 1674), was surveyor-general of the ordnance and therefore professionally interested in the Blackheath trials; similarly Lord Brouncker, president of the Royal Society, was a member of the Navy Board, an associate of Samuel Pepys in fact, and was therefore also involved in the defense of the realm. Moreover, Brouncker had long before made some experiments on explosive recoil which had been published in the Royal Society's *History*.[20] We may presume that in the case of Brouncker a mathematical interest in ballistics—he was a mathematician of some note—combined with professional position. The presence of Hooke at these Blackheath trials and also that of Henry Oldenburg, secretary of the Royal Society, is a little strange unless we presume that Brouncker wished to lend scientific color to the trials by bringing along two of the society's officers. Equally likely is that he thought it would be an agreeable day's excursion for his two friends, for the society was in recess, not meeting again until 12 November, and no report of these affairs was ever made to it.

The trials at Blackheath were by no means unprecedented in the Society's history. Nearly ten years previously an entry in the Society's Journal Book indicates that a distinguished officer, Colonel George Legge (1648–1691)—in later years to be lieutenant general of the ordnance and commander-in-chief of the Navy as Baron Dartmouth—was to be thanked by Sir Robert Moray, vice-president, for having obtained from the king a gun for the society to make experiments with; and Robert Hooke, the society's curator of experiments, was instructed, rather late in the day one might suppose, to draw up a series of experiments for the improving of artillery. Why the use of the gun was sought, and whether the experiments were ever attempted, we do not know, since nothing more on the subject appears in the Journal Book. Nor do we know whether the list of artillery experiments proposed by Sir Robert Moray and published in the *Philosophical Transactions* in 1667 had any connection with these events.[21] Moray proposed to investigate three principal "practical" points: (1) the point-blank range; (2) the optimum quantity of powder in the charge for full range; and (3) the length and bore for maximum range. Internal evidence makes it clear that the writer had little or no real experience of artillery and its use.

Though at this time the society's attention was instantaneously diverted from cannon to onions and the vegetation of plants, it was not always so capricious. At its very beginnings, during 1660 and 1661, it had repeatedly urged Lord Brouncker to make those experiments on the recoil of firearms previously mentioned, until at last they were carried out in the courtyard of Gresham College.[22] Brouncker seemed intent to prove that if recoil caused the barrel to deviate from its "laid" line, as in those days was commonly the case, the shot would be forced out of its true course by the pressure of the barrel on one side or the other. It was perhaps an obvious point, but he proved it rather neatly by the experiments—as also that the velocity of the shot could alter the manner of its being diverted by the recoil.

A few years later, there was a sustained period of interest in the propellant of firearms, ordinary black powder, and its comparison with other explosives, especially *aurum fulminans*, fulminating gold. The problem immediately raised was, How can one explosive be experimentally compared with another? especially since an "explosion" was assumed to be by no means a simple, repeatable phenomenon. Part of the effort, therefore, was expended on improving "testers" of gunpowder, devices in which a few grains of explosive might be fired to blow up a heavy hinged lid or a plunger to a measured height. Hooke devised an "improved" form that did not fully justify itself since it all too frequently shattered. It did emerge that a black powder of the ordinary kind but made with extraordinary care and refinement, such as that Prince Rupert provided the society with a specimen of, could be as much as ten times more powerful than commercially made powder.[23]

The strange behavior of explosives when shut up in solid steel was also explored, though for what purpose is far from clear unless it was in order to direct an irresistible force against an unbreakable object.[24] It is not without interest that in connection with these experiments Prince Rupert sent in a report on the efficacy of gunpowder in blasting rock—an early reference to explosive quarrying—and Hooke alluded more than once to his ambition to use gunpowder to bend a spring, which, as he said, would enable a man to command force without limit.[25] These trials did not succeed; but we may perhaps connect them with Huygens's later interest in the use of explosive force to do mechanical work, brought back to England by Denys Papin. Of course Huygens invoked the indirect principle by making atmospheric pressure do the work, which leads to the steam engine tradition.

The interest in gunpowder and explosions was certainly related to the contemporary preoccupation with nitre or saltpetre, which makes up about three-quarters of black powder. Was not nitre, or rather a nitrous component in the air, an essential agent in combustion and respiration and, very likely, in the nutrition of plants? So, at least, Hooke and others were prepared to assert and demonstrate by experiment. To follow this clue would lead us far afield, but if we pursued it we would get a full picture of why on 15 February 1665 Hooke made an experiment to show that gunpowder would burn in vacuo, as Robert Boyle had previously asserted it would, and why a fortnight later other trials were made on the addition of sulphur—the other "fatty" principle of combustion—to hot nitre.[26] The fact is that this villainous fellow gunpowder was thought to partake of the magical quality of nitre, which at this time was cast in the role of the principle of life.[27]

Discussion of this question recurred time and again in the Royal Society's debates,[28] as did that other hardy perennial, the effect of air-resistance on the movement of projectiles. On the whole, the society tended to support Galileo's position. It showed no inclination to doubt his assumption or axiom that a curved motion can be considered as compounded from two rectilinear elements, or (to put it more simply) that the horizontal, uniform component in the motion of a projectile is undisturbed by the vertical, accelerated component. This axiom had been rather crudely verified by the Accademia del Cimento in Florence and adopted by Borelli in *De vi percussionis* (1667).[29] Nevertheless, its strong a priori character had been challenged by Honoré Fabri in *Dialogi Physici* (1669), and the challenge was reported to the society on 24 November 1670. Though the general presupposition was in favor of Galileo and Borelli, the Royal Society thought that an experimental trial should be made, though it did not contemplate field trials with actual artillery such as the Florentines undertook.[30] On 8 December Hooke was "ordered to prepare for an experiment to be made at the next meeting in the assembly room, by having two balls, and projecting the one horizontally from the window over the door, and letting the other fall down perpendicularly from the same height." This was actually done on 26 January 1671 and twice repeated, but there was always doubt about whether the two balls hit the floor simultaneously, and though most of those present thought they did, the question was never decisively settled. As with so many other matters, the general opinion prevailed and became customary because it seemed reasonable rather than because exact experiment falsified the opposite belief.[31]

The effect of air-resistance on the trajectory that, on the basis of this first approximation, was deemed parabolical, was a harder nut to crack, not only in the more refined mathematical analysis already reviewed but also in the experimental or commonsense approach. Hooke raised the question as early as March 1663, in a rather confused series of proposals that came to nothing at that time. When he and Walter Charleton were ordered in the following year to report on the measured velocity of bullets, they came back with the figure 720 feet per second, surely too low.[32] Nothing more happened for several years, until in 1669 at the time when the society was investigating the laws of impact and the "force" of moving bodies, it was suddenly proposed as a possible experiment, "To show the proportion of the resistance of the air to bodies moved through it."[33] It can hardly be a coincidence that, in Paris in the previous year, Christiaan Huygens had begun to explore the same problem, and a link is furnished by the regular correspondence between him and Sir Robert Moray in London.[34] The London experiments were made with a flat pendulum: the larger the arc it swung in, the more it seemed to suffer resistance, "and the impediment to motion decreased in a greater proportion than the decrease of the velocity," though the exact proportion was not discovered, and repetition of the experiments threw no clearer light on the phenomenon.[35]

There followed another interval in which nothing happened. Then, in July 1675, Hooke once again showed an

X

experiment on air-resistance, which may, however (the minute is laconic), have related more to his notions about flying and the action of wings than to projectiles;[36] after this, three more years elapse before the topic crops up again. Anderson's book and the problems it raised were not touched on by the society—perhaps because the "summer" vacation was prolonged that year until 19 November—although John Wallis had written at some length from Oxford of his reasoned rejection of the simple parabolic hypothesis (joining with Newton in this), in letters to John Collins and Henry Oldenburg.[37]

There could be no question but the theoreticians of mechanics took the matter of air-resistance seriously, and now Robert Hooke himself at last spoke out openly to the same effect. Challenged in his belief that a marine depth-sounder, in its descent to the bottom and ascent to the surface, would travel distances (depths) proportional to the elapsed time, he retorted by explaining the notion of "terminal velocity," illustrated by the guinea and feather experiment. "All mediums whatsoever [he said] had some resistance to the motion of bodies through them, and even those which had least, had yet a very considerable opposition to a motion that was proportionably accelerated. Hence it was how birds were able to sustain themselves in the air. . . ." Hooke's friend Sir Jonas Moore (now risen under the new regime to be vice-president of the Society), who had once gone to Blackheath to test the parabolic theory, now recalled that in shooting mortar bombs the greatest range was at somewhat less than forty-five degrees, since the shots at high elevations were more deviated from the exact parabolic line than those at lower elevations; and he expatiated on the atmospheric effect on projectiles.[38]

If the Society's corporate contribution to the study of the flight of projectiles seems uncertain, repetitive, and inconsistant, especially as regards the experimental mode of inquiry, it must be remarked that this is no more than typical of any topic one tries to follow through the Society's Journal Book. Rarely was a topic examined in an exhaustive, or even a systematic, manner; rarely was even a tentative conclusion reported, except in the form of passive recognition of the results of work carried out elsewhere by an active Fellow such as Boyle or

Grew. So often an investigation was begun and dropped after a couple of meetings; or the order was given to "let this experiment be shown" and it was not done; or the apparatus worked indecisively or not at all and the whole matter was abandoned. Out of many instances, the example of the reflecting telescope is notorious, for it is obvious that common sense, physical labor, and enthusiasm could have created a successful large-scale reflector in the 1670s just as certainly as in the 1730s.

The Royal Society was, in fact, a very inefficient agent of experimental research. Baconianism was easier to preach than to practice and led too easily into whimsy and incoherence. In the painfully slow evolution of experimental science we should certainly note, however, that the more serious of the Fellows were conscious of the Society's shortcomings. At the end of 1666 Sir Robert Moray was already urging the Society to take stock, to consider "how the experiments at the public meetings of the Society might be best carried on; whether by a continued series of experiments, taking in collateral ones, as they were offered, or by going on in that promiscuous way; which had hitherto obtained." Just over two years later, no improvement in method having occurred meanwhile, the secretary, Henry Oldenburg, renewed the question, moving that the Council "would think upon an effectual way of carrying on the business of experiments at the meetings of the Society" and that committees be appointed "proper for directing of experiments"—that is, for planning a series of them in advance. Oldenburg's motion was carried and the president said he would see to it, but experimentation remained as promiscuous as before.[39]

In the 1670s, as the Society's financial position worsened (for although the list of Fellows was long, few paid their subscriptions regularly) and the laughter of its critics grew louder (it was notorious that "very many things were begun at the Society, but very few of them prosecuted"), it seemed more than ever necessary to systematize its experiments and discourses.[40] In the ponderous words of Sir William Petty, "The Society has been censured (though without much cause) for spending too much time in matters not directly tending to profit and palpable advantages (as the weighing of air and the like). . . ." If we may judge from Petty's proposals for reform in 1673, even such apparatus as the Society had managed to assemble for making "the experiments of motion, optical, magnetical, electrical, mercurial etc. And such instruments as had formerly been used by the Society" were disorganized, scattered, and broken.[41] Shortly thereafter, with Petty apparently still leading the party of reform and action, a purge of useless, nonsubscribing Fellows was instituted and a committee set up to prepare "a list of considerable experiments to be tried before the Society" and to collect the necessary apparatus.

The outcome, though not wholly negligible, was clearly far from satisfactory. Several months later Petty again lamented the Society's "want of good experimental entertainment at their meetings," which was a principle cause of its sluggish condition. He now proposed that a team of Stakhanovites volunteer to submit an experimental discourse or considerable experiment, in turn, week by week, under pain of a fine for nonperformance. This was done for a time, and with the appointment of a second curator, Nehemiah Grew, things looked up—only to lapse once more into inanity after the Hookian Thermidor following the death of Oldenburg in 1677, when Brouncker and the old guard were ousted. Years later even such young Turks as Edmond Halley or Denys Papin were neither insistent nor successful in causing the Society to investigate in accordance with a rational research program.[42]

There was no institution, no model, no method, for systematic experimental science. Only in observatories was work regularly pursued according to some strategy, clear night after clear night, and even then the program was likely to be interrupted or switched by some chance event such as the appearance of a comet or a big sunspot. Only individuals—Boyle, Newton, Grew—were able, and inspired, to follow a line of investigation with drive and persistence, yet rarely for longer than a few weeks on end. How often does Boyle report the loss of some experiment because of the untoward arrival of visitors, just as Newton himself left his optical experiments incomplete. Set in this context, the launching of the elaborate and sustained experimental investigations that alone would yield permanent results, in the study of fluid motion generally and of projectiles in particular, may be seen as quite hopelessly impossible. What the Royal Society actually achieved was no more than the feeblest gesture of faith in empiricism.

After all was said and done, despite Bacon and *Nullius in Verba*, the Royal Society was as apt as anyone else to believe what it wanted to believe, as the slightest inspection of Birch's *History* will demonstrate. The antischolastic skeptic could too easily become the credulous mechanist, for the mechanical philosophy could as well render up false or misleading answers as could Aristotle's philosophy. And in the case of projectiles, the a priori mechanist could be doubly deceived. He was deceived in the first place if he accepted Galileo's parabolic theory as more than an approximation, imagining air-resistance to be negligible; but he was deceived again, more subtly, by the logically consequent supposition that resistance obeys some simple law—that it must be directly proportional to the velocity, or to the square of the velocity. How seductive the square-law was may be seen from Sir William Petty's treatise on it. On wholly a priori grounds Petty saw the duplicate proportion as everywhere, a fundamental law of nature having far-reaching technological implications. For example, he stated that in ships the speed attained was as the square root of the driving power, which he equated with the sail area; which is to say that the resistance increases as the square of the velocity. By similar reasoning he put the speed of a bullet as the square root of the amount of powder fired. Huygens, Newton, and many others adopted the square-law as virtually self-evident, though Newton admitted it only as an approximation. Why experiment when nature offered so ready and satisfying an answer?

Thus the matter was handed back to the mathematicians, with neither experiment nor experience pronouncing an effective word, and the problem became one of defining the trajectory curve, the reshaped parabola, in terms of a sharply formulated a priori law of resistance. In this form (as we have already seen) the problem could be solved when the resistance was made to vary directly as the velocity; it was insoluble, however—in exact and general terms—for the case of proportionality to the velocity squared. Newton examined an approximation to the unattainable exact solution, but—and this should be emphasized—the mathematical solution (or nonsolution) to the mathematician's problem was only of interest to mathematicians. The a priori character of the assumptions on which the problem of the trajectory curve was based entailed this consequence: because experiment, not to say experience, had been thrust aside, ignored, or found unnecessary in defining the problem, the problem's solution could not relate to experiment or experience, save by coincidence. Moreover, the physically "better defined" trajectory was not only nonexperimental, it was useless too.

This is an example of applied mathematics ascending into a stratosphere of sophistication which is far beyond the sight of ordinary mortals, and from which it is unable to return with any workable proposals. We have to remember that even relatively obvious questions in mechanics, such as that of the height of a jet of water spouting upward in relation to that of its pressure-head, could contain great complexities of theoretical analysis.[43] In this case (the case of *Principia*, book two, proposition 37 of the first edition) Newton was saved from serious errors of reasoning and fact by his editor, Roger Cotes.[44] In the case of projectile motion Newton was less fortunate, and an initial unfortunate slip (there was no error of reasoning) led him into difficulty and embarrassment.[45] No wonder, then, that these problems came to be of purely technical, mathematical interest and of no possible practical import.

Consider also Christiaan Huygens's mathematically elegant but practically clumsy solution to the problem of timekeeping—making the pendulum swing between cycloidal arcs so that it too would trace a cycloid. The mathematical elegance of this design is redundant, since it is enough to reduce the pendulum's swing to a small arc of a few degrees, as was done a few years later with the anchor escapement in clocks. In the small arc, the difference between cycloid and circle is both theoretically and technically negligible.

In ballistics the effect of air-resistance can never be ignored without a corresponding loss of precision. As the more complex rational approach to the problem led only to sterility, however, there was only one alternative to giving up the whole problem as insoluble: to adhere to the original rational solution provided by Galileo in the parabolic projectory. With this simplification accepted, and air-resistance therefore ignored, all that remained for the mathematical ballistician was a simple exercise in geometry—resulting in paper constructions or instrumental procedures developed for the supposed benefit of the practical artilleryman. Since at this time weapons were either guns fired more-or-less at point-blank range or mortars shooting off bombs at high elevations but with little pretense to close accuracy in grouping, parabolic tables were probably as practically useful as any that could have been generated in accordance with more sophisticated hypotheses. The mortar bombardier could, if he took the trouble, at least derive comparative elevation-range correlations from them.

It is well known that the most impressive French study of exterior ballistics at this time, François Blondel's *L'Art de Jetter les Bombes* (written in 1675 but not published until 1683 for national security reasons), was couched wholly in terms of the parabola. It nevertheless contained contributions from some of the best scientific minds working in Paris—Jean-Dominique Cassini, Ole Roemer, Philippe de la Hire, and C. F. Milliet de Challes. Furthermore, Blondel's tone was eminently "practical"; though he was aware of the parabola's status as an ideal approximation to the true trajectory in air, the parabola was, he reported, confirmed by ocular tests. It was therefore worthwhile to handle expeditiously the problem of finding elevations when shooting over sloping as well as horizontal ground. This was the chief business of his book, which the mathematicians dealt with in a variety of ways.

Edmond Halley, following Blondel's track, considered this same problem in two papers in the *Philosophical Transactions* (of 1686 and 1695). He was the first to point out that the greatest range is always obtained when the angle of fire bisects the angle between the ground and the vertical;[46] but his main mathematical interest was in the problem of shooting over rising and falling ground, a problem which Halley mistakenly said had not been treated by Torricelli and whose first solution had been given by Anderson. Actually Torricelli *had* solved this problem and with a construction more elegant and simple than Halley's.[47] Like the French mathematicians, Halley aⁱ

X

gued that "the opposition of the air . . . in large and ponderous shot . . . is found by experience but very small, and may safely be neglected." Thus the parabolic tables could be safely relied on, he thought, "as if this impediment were absolutely removed." He supported this view by a rather vague allusion to some trials made with a small mortar. Halley's early nineteenth-century editors rebuked him for giving this assurance, pointing out that, were the ball not impeded by the air, the range of cannon would be ten or twenty times as great as it actually is.[48] The criticism is perhaps not entirely apropos since Halley himself was well aware that, as applied to "great guns" with a velocity well in excess of 1,000 feet per second, "the theory of Galileo allowing no opposition to the ball from the air is insufficient in great distances."[49]

We must, moreover, credit Halley with the common sense to perceive that tables were not a great desideratum of the field gunner if computed with "an exactness and curiosity beyond what the gross practice of our present cannoneers seems to require, who lose all the geometrical accuracy of their art from the unfitness of the bore to the ball and the uncertain reverse [recoil] of the gun, which is indeed very hard to overcome."[50] This last comment was written in 1701, after two generations of familiarity with the parabolic trajectory, and indicates that the situation then was just as it had been in the beginning: geometrized shooting was an ideal unattainable in practice.

What are we to make of all this curious miscellaneous activity? We see first of all the natural eagerness of the Royal Society's Fellows and of their fellow academicians in Paris— and indeed of numerous private mathematicians—to demonstrate that their science could cope with utilitarian problems. I need hardly recall the parallel work in the geometrization of fortification by the engineers of whom Vauban is the best known.[51] Scientific research had been commended to the state not only because science brings intellectual enlightenment but also because it brings fruit, and some scientists, at least, thought it incumbent on them to justify this promise. As so often happens, propaganda had pronounced a self-fulfilling prophecy.

In addition, we can observe, I think, a consistent distinction between the "practical" men, such as Brouncker, Moray, Petty, and Moore—though Charles II's shipwrights would hardly have called Petty practical—some of whom actually carried out artillery experiments, and the mathematical idealists, such as Gregory, Huygens, and Newton, all of whom regarded the parabolic theory as in principle too simple in its physical assumptions. Halley's work comes between these two positions. The latter group I think we may discharge of any concern for actual shooting; in an age of experiments a man who produces a paper ballistic theory of great mathematical complexity can scarcely be regarded as a utilitarian. For the mathematicians, surely, the interest was in puzzle solving. The former group were more intent on practical accomplishment—this stands out in Moray's 1674 paper—and though we may, with hindsight, doubt whether their efforts could in any circumstances have proved fruitful (for the reasons noted by Halley), the explanation of their failure to push on steadily with the accumulation of performance data lies more in the unsystematic character of experimental science at this time, and its inadequate control of material resources, than in a lack of interest in such an approach. Social reasons, one may surmise, obstructed the social pursuit of science; the mathematical exploration of the problems of gunnery was not similarly limited and was pressed much further forward.

It is curious how the parabolic theory, at its inception so highly scientific and mathematical, rapidly became the standard resource of those who would geometrize shooting. This translation from the scientific frontier to the commonplace world through a process of simplification and codification represents in miniature what was commonly to take place later in the application of science to technical problems and in the education of technical operators. The parabolic theory provided very simple, schematic solutions to all problems of gunnery; it was difficult, if not impossible, to prove that it was seriously in error for all practical situations in the field; and accordingly it was widely, and for long, publicized by the would-be instructors of cannoneers. It is still difficult to determine whether it really influenced practice, however, and I must leave that problem, if it can be solved, to others.

As for the relative significance of these investigations and aspirations within the whole compass of seventeenth-century science, opinions will probably differ forever. I think it is useful to make the sort of distinction between individuals that I have already indicated—and we may at least be sure that in the whole scientific work of a Huygens or Newton the theory of projectiles was a tiny element. I am inclined to think that if we view the work of the Royal Society as a whole we shall not come to a very different conclusion, despite the not uncommon opinion (which Robert Merton is held to have proved) that many of the early interests of the Royal Society revolved around the problems of mining, navigation, and war. In fact, Merton only made a claim that 7.2 percent of his sampling of Birch's *History* related directly to military technology—in which reckoning he may have included all references to air-resistance, the laws of impact, nitre, and so forth.[52] One can only suppose that those who suggest that those three technical issues were constantly, in a technical or utilitarian sense, the object of the Society's attention have spent little time looking through Birch's volumes.

I shall not hesitate to agree with Merton's conclusion that some influence on problem choice was exerted by practical questions such as gunnery; with whose problems a few of the Fellows were plainly well acquainted, though I think Merton's quantitative measure of this influence was high. Once the problem has been chosen, however, may not the historian also ask what contribution the investigation of it may be seen to make to either technique or knowledge—what, in Bacon's words, does it yield of "fruit or light"? In the case of gunnery, the yield seems slight as to knowledge—it contributed to a minor branch of rational fluid mechanics—and nil as to technique. One cannot believe that the Fellows of the early Royal Society—from Ashmole to Wren—would sustain their reputations by such sporadic, slight, and ineffective investigations as those I have recounted here.

The fact is that the investigations that bore fruit were in the quite different areas of astronomy, chemistry, optics, pneumatics, and so forth. And it was necessarily so. The scientific mastery of the complex problems of technology still lay a century or more in the future because it depended on two things: the elucidation of clear scientific models (in chemistry, for example) and the appearance of men, such as Watt and Smeaton (or Böttger and Macquer), who would successfully investigate technical problems in a scientific manner, as was almost never done in the seventeenth century. The notion that technology is simply "applied" or borrowed science has been rightly criticized by historians of technology. We now believe that there is more to the solution of technical, and indeed, scientific, problems than the awakening of interest in their solution. The history (or sociology) of problem choice is a significant study, but it is well to remember that the selection of a problem for study (and those words themselves beg a lot of questions) is only a necessary, not a sufficient, condition for the provision of a usable solution.

NOTES

1. A. R. Hall, *Ballistics in the Seventeenth Century* (Cambridge; At the University Press, 1952; reprinted, New York, 1969). The historical literature has since been enriched by a number of works treating of the history of artillery from the economic, strategic, and political point of view, such as Carlo M. Cipolla, *Guns and Sails in the Early Phase of European Expansion, 1400–1700* (London, Collins, 1965) and John Francis Guilmartin, *Gunpowder and Galleys* (Cambridge, University Press, 1974). Guilmartin especially has provided excellent technical material on sixteenth-century cannon and their use, confirming my contention that long ranges and hence ballistic tables were irrelevant to cannon fire in warfare, save in exceptional circumstances. I should note also H. J. Webb, *Elizabethan Military Science: The Books and the Practice* (Madison, University of Wisconsin Press, 1965).

2. Michael Lewis, *Armada Guns: A Comparative Study of English and Spanish Armaments* (London, 1961).

3. W. Hassenstein, ed., *Das Feuerwerkbuch von 1420: 600 Jahre deutsche Pulverwaffen und Buchsenmeisterei* (Munich: Verlag der Deutschen Technik GMBH, 1941).

4. Luys Collado, *Pratica Manuale di Arteglieria* (Venice, 1586).

5. I do not mean that it is improbable that cannonballs and charges of powder were not measured out with some care; this prudence would demand. I refer only to various hypothetical procedures ancillary to, or involved in, laying the gun.

6. Guilmartin has shown that the range tables often given by the early writers on artillery (some examples of which were printed in the appendix of my book *Ballistics*) are quite impossible and would sometimes require quite unattainable muzzle velocities; all of which strengthens my point.

7. *The Alchemist*, Act II, Scene III, 11.184–202.

8. Robert K. Merton, *Science, Technology and Society in Seventeenth Century England* (*Osiris* 1938; reprint, New York: Harper Torchbooks, 1970), p. 191.

9. B. Hessen, "The Social and Economic Roots of Newton's *Principia*," in *Science at the Cross Roads* (London, 1931; reprint, London: Cass, 1971) p. 166.

10. W. Whewell, *History of the Inductive Sciences* (1837), quoted in Merton, *Science, Technology and Society*, p. 187.

11. Niccolo Tartaglia, *Nuova Scienza* (Venice, 1537); *Quesiti et inventioni diverse* (Venice, 1546).

12. Alexandre Koyré, "La Dynamique de Niccolo Tartaglia," in *La Science au Seizième Siècle*, Colloque de Royaumont 1957 (Paris: Hermann, 1960), 93–113.

13. The *locus classicus* is Pierre Duhem, *To Save the Phenomena: An Essay on the Idea of Physical Theory from Plato to Galileo* [1908] (English translation, Chicago: University of Chicago Press, 1969). See also Edward Grant, "Late Medieval Thought, Copernicus, and the Scientific Revolution," *Journal of the History of Ideas* 23 (1962): 197–220.

14. Galileo Galilei, *Dialogue Concerning the Two Chief World Systems*, trans. Stillman Drake (Berkeley and Los Angeles: University of California Press, 1953), p. 203.

15. Galileo Galilei, *Two New Sciences*, trans. Stillman Drake (Madison: University of Wisconsin Press, 1974), p. 223–229: "I admit that the conclusions demonstrated in the abstract are altered in the concrete, and are so falsified that horizontal motion is not equable, nor does natural acceleration occur in the ratio assumed; nor is the line of the trajectory parabolic and so on." Gregory's work is in "Tentamina quaedam geometrica de motu penduli et projectorum," annexed to William Saunders, *Great and New Art of Weighing Vanity* (Glasgow, 1672); Harriot's is in British Museum Add. 6789, f. 69r. (See D. T. Whiteside, *Mathematical Papers of Isaac Newton* [Cambridge: At the University Press, 1974], 6:7, n. 16 and 17; I am also indebted to Dr. Whiteside for a private communication.)

16. Alexandre Koyré, "A documentary History of the Problem of Fall from Kepler to Newton," *Transactions of the American Philosophical Society* 45 (1955): 329–395; and idem, "Galileo's *De motu gravium*" in *Metaphysics and Measurement* (Cambridge, Mass.: Harvard University Press, 1968), pp. 82–88.

17. Newton, *Correspondence*, 1: 309 (20 June 1674); see also H. W. Turnbull, *Gregory*, p. 282 ff. (note 19 below).

18. See A. Rupert Hall and Marie Boas Hall, *Unpublished Scientific Papers of Isaac Newton* (Cambridge: At the University Press, 1962), pp. 287–292; D. T. Whiteside, *Mathematical Papers of Isaac Newton* (Cambridge: At the University Press, 1974), 6, 65–75.

19. H. W. Turnbull, *James Gregory Tercentenary Memorial Volume* (London: G. Bell & Sons, 1939), pp. 282, 286; H. W. Robinson and W. Adams, eds. *The Diary of Robert Hooke, 1672–1680* (London: Taylor & Francis, 1935), pp. 103, 121.

20. Thomas Birch, *The History of the Royal Society* (London, 1756), 1:20, 33; the experiments were made on 7 April 1661 and presented to the society on 10 July; see also Thomas Sprat, *The History of the Royal Society* (London, 1667), p. 237 and Christiaan Huygens, *Oeuvres Complètes* (The Hague, 1890), 3:323.

21. Birch, *History of the Royal Society*, 2:24 (15 March 1665); *Phil. Trans.* II, no. 26 (3 June 1667): 473–477. It is not impossible that Hooke had some hand in this list.

22. Ibid., 1:20, 33.

23. Birch, *History of the Royal Society*, 1: 281–5, 295, 299, 302–303, 335 (22 July–25 November 1663).

24. Ibid., 1: pp. 425–465 passim (May–September 1664).

25. Ibid. 1:335, 2: 446 ff. (January–February 1667).

26. Ibid. 2:15, 19.

27. Shakespeare, *Henry IV*, Pt. I, I, 3, 11.60–63; D. McKie, "Fire and the *Flamma Vitalis*," in E. A. Underwood, ed., *Science, Medicine and History* (Oxford: Clarendon Press, 1953), I: 469–488; Henry Guerlac, "John Mayow and the Aerial Nitre," *Actes du 7 Congrès d'Histoire des Sciences* (Jerusalem, 1953), pp. 332–349; idem, "The Poet's Nitre," *Isis* 45 (1954): 243–255 (both reprinted in his *Essays and Papers in the History of Science* (Baltimore: Johns Hopkins University press, 1977).

28. E.g., Birch, *History*, 2:468 ff. (13 March 1679 and subsequent meetings) leading into a discussion of Roger Bacon's role in the invention of gunpowder and its use in artillery.

29. Galileo, *Dialogue* (note 14), p. 155; Richard Waller, *Essayes of Natural Experiments* (London, 1684), p. 143; W. E. K Middleton, *The Experimenters* (Baltimore: Johns Hopkins University Press, 1971), pp. 240–243; G. A. Borelli, *De vi percussionis* (Bologna, 1667), p. 169.

30. Birch, *History*, 2:454; Wallis to Oldenburg, 22 November 1670, in A. Rupert Hall and Marie Boas Hall, eds., *The Correspondence of Henry Oldenburg* (Madison: University of Wisconsin Press, 1970), 7:283–284.

31. Birch, *History*, 2:461, 465, 467.

32. Ibid., 1:205 (4 March 1663), 465 (7 September 1664), 474 (5 October 1664).

33. Ibid., 2:339 (14 January 1669).

34. Hall, *Ballistics*, 111–113.

35. Birch, *History*, 2:350, 352–353 (25 February and 4 March 1669). A similar experiment was proposed by Papin, ibid., 4:524–525 (9 February 1687).

36. Ibid., 3:227 (8 July 1675).

37. S. J. Rigaud, *Correspondence of Scientific Men of the Seventeenth Century* (Oxford, 1841), II: 588 (24 August 1674); Wallis here refers to his *Mechanica, sive de Motu* (London, 1669–1671) as a possible first discussion of the projectile's trajectory as "compounded of one accelerated, the other retarded. . . . And practical cannoneers, I am told, find the random of a bullet very different from the parabola, which [Galileo's] hypothesis doth estab-

lish." Cf. Hall and Hall, *Correspondence of Henry Oldenburg*, XI: 108–109 (15 October 1674): "The line of projection in shootings, I have always suspected not to be a Parabola."

38. Birch, *History*, 3:398 (4 April 1678), 400 (18 April). All was not clear, however, for the parabola was to be vindicated again a few days later.

39. Ibid., 2:131–132 (4 December 1666); 344 (1 February 1669).

40. Ibid., 2:469 (9 February 1671); Sir William Petty, *The Discourse made before the Royal Society the 26th of November 1674 concerning the use of Duplicate Proportion in Sundry Important Particulars* (London, 1674), p. 1.

41. Birch, *History*, 3:115 (11 December 1673).

42. Ibid., p. 119 (18 December 1673), p. 136 (29 September 1674). For subsequent returns to the problem of organized experiment, see ibid., p. 309 (6 March 1676), and ibid., 4:6 (29 January 1680), 516 (5 January 1687). Hooke had been censured for failure to perform his duties properly as early as 14 November 1670 (ibid., 2:452). See also Michael Hunter, *Science and Society in Restoration England* (Cambridge: At the University Press, 1981), pp. 56–57.

43. Newton, *Correspondence* (Cambridge: At the University Press, 1975), 5:65–68 (21 September 1710), 70 (30 September 1710). J. Edleston, *Correspondence of Newton and Cotes* (London, 1850), p. 35.

44. Newton, *Correspondence*, 66:103–104 (24 March 1711). I have shortened the story: it was only as a result of Cotes's prodding that Newton finally explained (by the *vena contracta* effect) how the velocity of efflux calculated at $\sqrt{2gh}$—consistent with the height of the jet—was also consistent with a quantity of water flowing as though the velocity at the hole were $\sqrt{gh}$.

45. See A. Rupert Hall, "Newton and his Editors," *Notes and Records of the Royal Society* 29 (1974): 45; Whiteside, *Mathematical Papers of Isaac Newton* (Cambridge: At the University Press), 8:48–58.

46. Because in shooting over sloping ground the range is proportional to $\sin(2a-b)$, where a is the angle of elevation (above the horizontal) and b the angle of slope. Halley's papers are in *Phil. Trans.* XVI, no. 179 (January–February 1686): 3–21, and XIX, no. 216 (March–May 1695): 68–72.

47. See Hall, *Ballistics*, p. 94.

48. Charles Hutton, George Shaw and Richard Pearson, *The Philosophical Transactions*, abridged (London, 1809), 3:265.

49. E. F. MacPike, *Correspondence and Papers of Edmond Halley* (London: Taylor & Francis, 1932), p. 168.

50. Ibid.

51. H. Guerlac, "Vauban: The Impact of Science on War" in his *Essays and Papers in the History of Modern Science* (Baltimore and London: Johns Hopkins University Press, 1977), pp. 413–439.

52. Robert K. Merton, *Science, Technology and Society* (note 8), p. 204 (table). Morris Berman, *Social Change and Scientific Organization* (Ithaca: Cornell University Press, 1978), p. xx. I counted some fifty significant allusions to gunnery and allied matters in rereading Birch's *History*. The four volumes contain 2090 pages; reckoning five subject-allusions per page (surely an underestimation), we may discover altogether 10,000 subject

references, consistently counting each repetition of the same issue at subsequent meetings as a separate subject (thus one topic occurring ten times rates the same as ten topics occurring once). Merton's assessment produces a total of about 5,500 "problems" discussed at Royal Society meetings, presumably counting the total of repetitive allusions to the same "problem" as one: this would indicate a total number of all problems, ignoring repetitiveness, in excess of 10,000. According to my count, then, gunnery topics could not have been more than one percent, at the extreme, of all topics mentioned at meetings. But I am not convinced that the selection and classification of topics alluded to at meetings of the Royal Society can be made rigorous enough to provide more than a very rough guide to the significance of particular interests.

XI

ARCHITECTURA NAVALIS

The subject I have chosen for my discourse is a novel one, I believe, for this Society and certainly for myself. I will tell you how my attention was drawn to it in a moment; my reason for choosing it this evening lies in my opinion that it is important to conceive the history of engineering and technology in broad terms as well as specialist ones. History is not only concerned with electrical engineering or machines or structures; it is concerned with engineering simply. I do not think that the study of naval architecture has ever been discussed here, and though I am far from claiming any competence in that subject, it seems to me to raise as a branch of engineering design historical issues which should be of interest to the members of the Newcomen Society.

The particular occasion I found in a recent book by George S. Emmerson on John Scott Russell, whose association with the younger Brunel is too well known to deserve emphasis. Emmerson tells us how Russell turned from steam road-carriages to steam ships, and in 1835 took up the study of the structure and morphology of the hulls of ships. 'As a result', Emmerson writes, 'he must undoubtedly have become the most scientifically educated naval architect in Britain at this time.' Russell apparently had to start from first principles: no one then knew, he claimed, how fast an engine of a given power would drive a given hull, and similar hulls from the same shipyard behaved in very unlike ways. In the Royal Dockyards, apparently, the shape of neither bow nor stern was supposed to have the least effect on the vessel's performance; questions of buoyancy and stability were total mysteries. Only one scientific study of ship design was known to Russell, it seems, that by the Swede, Chapman, who had reported on model test.[1]

Now this complaint of Russell's of lack of scientific precedent surprised me, because it is well known that there is a large earlier literature on the mechanics of ships, especially from the eighteenth century. Of course, the problems of actually constructing iron vessels were new, and so where the problems of steam propulsion. However, by the time Russell took up these questions there had been some twenty-odd years of experience of putting engines into hulls, and what is perhaps equally important, of proportioning the sizes of engines and boilers to any proposed power. Not long ago I saw advertised in a bookseller's catalogue a table relating power and vessel size attributed to the firm of Maudslay, for example. More to the point—for I am not going into the problem of steam propulsion at all this evening—the study of hull form, buoyancy and stability is in a general sense applicable to all vessels, made of any material and propelled in any manner. It is true that the height and weight of the masts and yards in a sailing ship, together with the pressure of the wind on its sails, cause particular difficulties that do not necessarily present themselves so forcibly in the case of steamship: yet they are by no means irrelevant even to the latter since early steam-vessels were heavily laden with auxiliary sails.

I thought I would like to look into this question, on which secondary literature throws little light, except to establish the fact that a number of modern concepts like that of the metacentre, had been successfully established by the end of the eighteenth century. I make no claim, however, to have made a definitive review of all the early writings nor would time permit their discussion now if I had done so.

A word more on Russell, and then I turn to the first of my themes, the seventeenth century origins of *architectura navalis* as a rational study. Russell's complaint against the ignorance of his own time arose, I think, from two factors: the anti-theoretical conservatism of British shipyards, and the anti-theoretical empiricism of British science. It was well known at the end of the eighteenth century that French ships were generally better sailers than British ones, and their advantage was commonly attributed to the scientific attainments of the French shipwrights. Yet the mathematical and experimental investigations of the forms and mechanical properties of ships made and presumably exploited in France were continually ignored by British shipwrights; and we may suppose that their purely mathematical part would have been distrusted even if known. But let me now try to satisfy you that such investigations were neither recent, nor meagre, nor unintelligible.

The Seventeenth Century

The first point to note is the wide range of engineering problems explicitly or implicitly treated by the shipbuilder: for example, the strength of materials whose study was begun by

Galileo with reference to the Arsenal (or shipyard) of Venice. Other aspects of mechanics and fluid mechanics are obviously prominent also. In general they relate on the one hand to the *hull* (questions of form, strength, buoyancy, stability etc.) and on the other to the *means of propulsion* (oars, paddles or sails). The two set of problems cannot, as is obvious, be considered in isolation for long: for an answer to many problems concerning the mechanical behaviour of a ship it must be considered as a whole. To take an obvious example—the treatment of leeward drift—it is necessary to consider three elements in the sailing-vessel: its sails, its hull characteristics, and its rudder. Moreover, for even mathematical investigations of the ship to have any value at all, they must have some relevance to the actual features of real vessels, even though the mathematician may (for example) consider the ship as powered by one huge sail, as a first approximation to the diverse and divided sailplan of real vessels. Hence the study of naval architecture may properly begin with tools and carpentry and end with calculus applied to fluid mechanics.

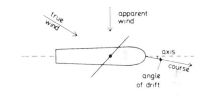

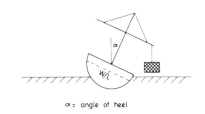

FIG 1. The action of the wind on a sailing vessel.

36

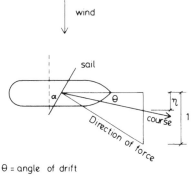

θ = angle of drift

$\eta = \dfrac{\text{bow resistance}}{\text{side resistance}}$

FIG 2. Leeward Drift according to Pardies (1673).

Let us begin in the year 1671, though much had been written before this time,[2] because in that year was published the first of two relevant books, one large and handsome and famous, the second small and largely forgotten.

The title of Nicolaas Witsen's book *Scheeps-Bouw en Bestier* means 'The Construction and Handling of Ships'. Witsen (1641–1717), a notable traveller, was later to be Burgomaster of Amsterdam and to play a small role in the English Revolution of 1688–9. His book is an encyclopaedia of the marine art, covering everything from the history of shipping among all nations to the shipwrights' tools and techniques; it considers all sizes and types of ships and by no means neglects matters we shall hear more of soon. I quote now from the English account of *Scheeps-bouw* in the contemporary *Philosophical Transactions*.[3]

[The book] explains and gives reasons for the several sizes and shapes of the parts of a ship; as why the Masts ought just to be of such a bulk and height? Why some of them must incline backwards, some stand upright? Why a small Rudder can turn a great Ship; and a little Anker stay it? What maketh Ships not feel the Rudder? Why Vessels too broad are weak and prove inconvenient in high winds? Why long and moderately narrow Ships endure the Sea better, than short and broad ones? How the keel ought to be placed? .... Why a Ship is to be broader before than abaft? .... What hinders well-sailing? Why Turkish Vessels are excellent Sailers? And many questions more, considered by this Author.

Witsen's is a book based on commonsense arguments and experience, not mathematics and the science of mechanics. I would say more about it if I could read Dutch; instead, I go on to my second, modest little book, which does initiate the dynamical treatment of ships.

The Jesuit mathematician Gaston-Ignace Pardies (1636–73) is best known for a minor controversy with Isaac Newton. His work published in 1673, two years after Witsen's *Scheeps-bouw*, is entitled *La Statique ou la science des forces mouvantes*.[4] Here Pardies addresses himself to the question I have already raised: the determination of a ship's course in relation to the wind direction. For, he writes; 'the movement of a ship is without doubt one of the finest works of art, and one in which the industry of men seems to pay most respect to the laws of nature'. His argument is that, neglecting any action of the ship's rudder in directing the vessel, its path is determined only by the angle of the sail (or sails, but a single one is considered for simplicity) and the ratio between the ship's resistance to motion in the forward direction, and its resistance to sideways motion, or drift. Call this ratio η. Then, says Pardies, the more acute we make the angle α between sail and wind, the less also the forward speed; while the less we make η, that is to say the longer and narrower the ship, the less again is the angle θ representing the actual drift. You will observe how it is rationally demonstrated by Pardies that the ship's length/beam ratio greatly affects its performance; a point which however, would have become more complicated had he also taken account of the ratio between the area of the sails driving the ship, and the 'dead' area of its hull surface above the water, which largely adds to the drift.

Pardies realises that there is here a problem of optimisation. Blown by a cross-wind, the sails may be so set either that the ship proceeds by the shortest possible path to its destination, or so that the path is longer but the speed greater. As any dinghy-sailor knows, it is a nice matter to balance the two extremes intuitively. Pardies does not solve this problem or others that he raises about, for example, the explanation of a ship's sailing into the wind or the advantage of lateen sails but all are in principle, he maintains, resolvable by the principles of mechanics.

Hence Pardies was rightly regarded by the mathematicians of the next generation as the originator of this branch of applied mechanics. His analysis was simple and did not satisfy for long. Again, I do not mean to pursue the perfection of this matter in detail during the eighteenth century: it is obviously one that presents a nice exercise in the calculus of variations. Let it suffice to mention—because these things seem to have been forgotten—that it was the subject of dispute some years later between Bernard Renau d'Elicagaray (1652–1719), inspector-general of the French Navy and an innovator in methods of shipbuilding that were adopted in the Royal Yards at Brest and elsewhere, and Christiaan Huygens (1627–95), mathematician, physicist, and former leading figure of the Paris Académie Royale des Sciences.[5] It was taken up again by the great Swiss mathematician, Johann Bernoulli (1667–1748)[6] in 1714, and by many more after him.[7]

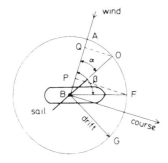

$$a = AB = BO = BF = unit$$

$x = OQ = \sin \alpha$ (α = sail angle)

$p = FP = \sin \beta$ (β = wind angle)

then, $x^4 = a^2 x^2 + \frac{1}{3} p^2 x^2 - \frac{4}{9} a^2 p^2$

gives best sail angle

FIG 3. Huygen's sail angle diagram (1693).

This was, of course, a time in which the French mercantile and military marine were both being built up, almost without regard to cost, and certainly with less respect for tradition than was usual in other countries, in order to outstrip the combined economic and military power of the Netherlands and Britain.[8] Presumably this explains the number of French

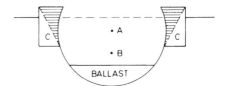

A-centre of circular hull
B-centre of gravity (loaded)

Hoste argued that only the shaded portions of the added 'buoyancy tanks' C,C would increase the stability of the hull

FIG 4. Ship stability according to Hoste (1697).

treatises on shipbuilding and naval arts that appeared during the reign on Louis XIV. I have time to mention only two more. F. Dassié, 'constructeur de vaisseaux pour la marine royale au Havre' was, one supposes, a practical expert; his treatise on *Architecture navale*, Paris 1677, was designed to teach 'infallible ways of making ships sail better, and to find out the just weight of a ships burden, and its true symmetry, and so to bring this Art to perfection'. To rehearse once again the summary in the *Philosophical Transactions*, the book explains[9]

'the use of the Compasses necessary to represent the plan and the proportion of a ship', [that is, in order to draw its lines], 'as also the usual terms of Marine; the Definitions of the several sorts of Vessels; the Proportions and Measures of all the parts of a ship, exhibited in their several figures, a general Description of all the Instruments'.

As with the *Scheeps-bouw*, there is no analysis in terms of theoretical mechanics. Dassie explains the traditional shipyard methods of deriving all the dimensions of a vessel by ratios from the length of the keel, with various *ad hoc* modifications. The complex curves to be given the ribs defining the sections through the ship from bow to stern were similarly derived by established procedures of swinging compass-arcs.[10]

Of the next writer, Paul Hoste (1652–1700) we know a little more as he was a Jesuit mathematician, teaching in various colleges of which the royal seminary at Toulon was the last. His naval experience came from serving as chaplain with the great Anne Hilarion de Cotentin, Comte de Tourville (1642–1701), architect of the French naval organisation that proved victorious at Beachy Head but was in turn destroyed by Admiral Russell at Cape La Hogue (19–23 May 1692). Tourville seems to have used Père Hoste as a mouthpiece for his own systems and ideas with respect to shipbuilding and naval warfare, on which Hoste published several books. His *Théorie de la construction des vaisseaux* (Lyon, 1697) is the first work to approach the subject on the basis of mechanical science. 'What we need,' he says (as many repeat thereafter) 'are sound principles':

At present chance has so great a share in building, that the vessels constructed with utmost care are usually the worst, and those that have been neglected sometimes the best. The ships built today, by intelligent shipwrights following improved methods, are no better than those constructed by carpenters who could neither read nor write.

Hoste himself, however, turns out to have feet of clay. His knowledge of mechanics is elementary, showing no trace of the influence of Huygens, still less of Newton. He argues deliberately *against* the square-law of resistance to motion in fluids, and asserts that resistance increases directly as the velocity. I do not see anything useful that emerges from his discussion of the effect of a solid's shape on its freedom of movement through water. Similarly, his discussions of stability is vitiated by his complete failure to utilize the Archimedean concept of buoyancy as an upward force opposing gravity. His analysis of the effect of girdling a ship—spiking on extra timber round the water-line—accordingly goes all astray. Another incomprehension is revealed by his firm belief that the action of a rudder is different when the ship moves through water in the normal way,

from what it would be if the ship were supposed at rest and the water moving past at the same speed. In fact Hoste's attempt to produce a rationalisation of the behaviour of ships in water is on the whole worthless, and he almost completely fails to generate any useful practical advice, except on such already well known matters as the tendency of vessels with fine bows to pitch more severely than those with bluff bows. Much of Hoste's geometry seems to me quite valueless. On the other hand he did know something about ships: when he tells us that correction of excessive heel in a ship by girdling it will make it a slow sailer, or that correction of the same defect by additional ballast will so increase the roll that the ship may lose her masts, we can believe what he says. He is also correct in the principle that the position of the ship's centre of gravity is vital to an analysis of its behaviour, though he misuses it, and in realising that Galileo's treatment of the strength of materials has real practical relevance. And, above all, he correctly defined the two main problems for the future of theoretical naval architecture: the problems of the relation between shape and resistance, and the problem of stability.[12]

The Resistance of Fluids

Recognition that water resists the passage of boats must be as old as their use; it was formally admitted by Aristotle. In the seventeenth century fluid mechanics rapidly became sophisticated: one could again begin with Galileo, but I will only remind you now of the mathematical investigations of Huygens in Paris and Newton in Cambridge.[13] Both accepted that

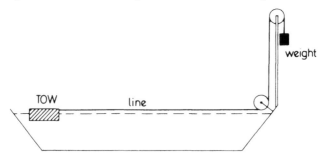

FIG 5. The classical towing tank, Huygens (1669).

as an approximation (at any rate) resistance increased as the square of the velocity of the body moving through the fluid. Newton also made some experiments on freely sinking bodies; Huygens chose for his experiments the 'towing-tank' arrangement that remained classical for centuries (1669). Huygens found that when the pulling weight was doubled the speed of the tow through the water increased by $\sqrt{2}$, confirming the relation he expected.[14] Huygens did not, however, investigate experimentally (or for that matter mathematically) the effect of the shape of the tow on its resistance to motion in the fluid, the topic for which his apparatus was chiefly used by his successors. As is well known, Isaac Newton first tackled the question of the solid of least resistance theoretically in the *Principia* (1687):[15] he there remarked that 'I judge that this proposition will not be without use to the building of ships'—an excessively sanguine expectation.

I shall return to Newton in a moment, but first (because it helps to create the context) let us go back a few years and sink to a less abstract level with Sir William Petty's *Discourse made before the Royal Society the 26 of November 1674 concerning the Use of Duplicate Proportion in sundry important particulars* (London, 1674).[16] Sir William Petty (1623–87) had some small-scale experience of experimental shipbuilding in pursuit of his plan for 'double-bottom' or catamaran construction.[17] His concern in this *Discourse*, however, is to show the generality in nature of what we would call the square-law relation. So, to take an instance, he cites Galileo's theory of beams and points out that

a Ship of 400 tuns, equally strong with one of 50, must have not only 8 times as much Timber in it, but 16 times, which is seldom or never done.[18]

Indeed, he tells us dramatically, no ship in the world is as strong (relatively) as a nutshell. Similarly, Petty argues, because the velocity increases as the applied impelling force, and the resistance increases as the square of the velocity, if the sail area of a ship in increased by a factor of 2 (the wind being constant, obviously) then its speed will be multiplied only by a factor of $\sqrt{2}$. Petty also tried to work out, in what I think is the first attack on this problem, how the resistance of water to the passage of ships with bluff bows differs from that experienced when the bows are sharp. He also makes the econometric point that a vessel having a broad beam and small masts is cheaper to operate than a slender ship of similar capacity having tall masts, because the cost of building the second vessel will exceed, in proportion, the advantage gained from its greater speed of transit. Again, one sees that a problem of optimistation in ship design was perceived here.

Although in other technical contexts the axial symmetry implied in mathematical analyses of the solid of least resistance is maintained, the bows of ships are necessarily non-symmetrical in that sense, and so no attempts, so far as I know, were made to apply that analysis. Johann Bernoulli of Basel did indeed assert hesitantly that 'Perhaps more success in the construction of vessels would be gained if the rules that might be drawn from the solution' of the problem of least resistance were followed, but did not suggest how this might be done.[19] We do owe to him, however, or so I believe, the important realisation that the drift is affected by the form of the bow—the more perfect its form, the less the drift, of course. There is no doubt, he wrote,

that if one followed this method and combined with the blind practice of the artisans the reflections of educated men, one would at length arrive at the highest degree of perfection to which the arts can be borne.

Although Johann Bernoulli anticipated much in the writings of later 18th century theorists especially in the *manoeuvre* or handling of ships, for clearer guidance as to the best form of the bows of ships one has to turn to the celebrated *Traité du Navire* of Pierre Bouguer (1746), who also published eleven years later on the *Manoeuvre des Vaisseaux*. The lines given to a wooden ship had to be at best a compromise between theoretical mechanics, experience, and intuition: the practical shipwright had to consider factors such as strength and seaworthiness not integrable into a theory of resistance.

Pierre Bouguer (1698–1758) is now most often remembered as the 'inventor'—if that is the right word—of photometry; he was an infant prodigy, solemnly recorded as obtaining a royal professorship of hydrography at the age of fifteen: certainly he was awarded three prizes (in 1727, 1729 and 1731) for essays on nautical topics. The *Traité du Navire* was written in the Andes during the long French geodetic expedition to Peru which lasted from 1735 to 1744. His debt to predecessors like Newton and Bernoulli is probably larger than it seems, though he took his overt start from the now discredited studies of Renau and Hoste. Bouguer writes of shipbuilding:

One cannot doubt that posterity will one day be amazed at our behaviour in the past and still at present . . . in a century as enlightened as ours.

Like others, he harped on the inconsistency of principle and practice in naval construction, even after the reforms of 1689 in France, so that there was[20]

no rule, no index even for discovering the truth or for making it known, [constructors] are reduced to reiterating continually the same assertions in place of proofs.

Bouguer first considers mathematically the resistance to motion through water experienced by simple shapes: triangle, semicircle, cone, assuming always that resistance increases as the square of velocity which is, he says,[21]

better confirmed by experience than all others, and serves to account for the enormous effects of which fluids are sometimes capable.

Obviously a cone is least resisted; if the height of the cone is only 50% greater than its diameter, by Bouger's rule the resistance experienced is only one-tenth of that on the flat base. Real ships, he recognised, could not be made so favourable; a bluff-bowed vessel would probably reduce by a factor of six or seven at the most; moreover, the fact that a sailing-ship always has some leeward drift, and hence does not move through the water axially in the direction of the keel, would, as he knew, tend to increase the resistance further. The best one could reasonably hope for would be that a properly designed vessel would attain, at a maximum, some 2/7ths of the speed of the wind—and by the way it was of course known, and mathematically explained, that a wind full aft does not propel a vessel at its highest speed. You will perceive that only a modest speed of say 6 knots would be expected in practice from a nice breeze of say 25 knots, without making any deduction for skin-friction.[22]

Bouguer stated his objective in all this without disguise: it was, he wrote,[23] to treat

'of the construction of ships and of the mechanics of their movements' in such a way as 'to substitute, if possible, exact and precise rules for the obscure and groping practices used in the Navy.'

Mathematical analysis was, in his belief, essential to such an endeavour

'for it is impossible to discover by practice alone, and by trials, however repeated, the properties of a whole surface curve, which is an assembly of an infinity of curved lines and points.

Whether Bouguer succeeded as he claimed, in providing 'a criterion for distinguishing the best bet or for making a careful choice between several designs proposed for the same ship' might be arguable, though later judgment has been favourable to his work. I do not know if any study has been made of the influence of Bouguer's book on French dockyard practices. Certainly one of his recommendations—that ships could well be made of plane surfaces meeting in angles—was never adopted, though one sees something like it in modern 'hard-

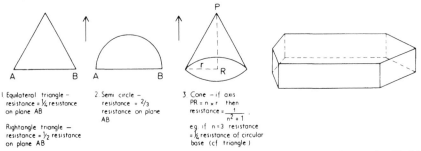

1. Equilateral triangle – resistance = $\frac{1}{4}$ resistance on plane AB

Rightangle triangle – resistance = $\frac{1}{2}$ resistance on plane AB

2. Semi circle – resistance = $\frac{2}{3}$ resistance on plane AB

3. Cone – if axis PR = $n \times r$ then resistance = $\frac{1}{n^2 + 1}$.

eg if n = 3 resistance = $\frac{1}{10}$ resistance of circular base (cf triangle)

FIG 6. Bow shapes in Pierre Bouguer's studies (1746).

FIG 7. Pierre Bouguer's 'plane surface' hull form (1746).

chine' dinghies.[24] It is perhaps unkind to mention the evident fallacies (to modern eyes) in his treatment of the oar, and in his belief that an infinitely long and narrow ship would experience no resistance in moving through the water,[25] and more just to recall his many and valiant attmpts to resolve the complex behaviour of a ship into analytically amenable problems.

To give another example of the mathematical approach let us consider Leonhard Euler's *Theorie complete de la construction et de la manoeuvre des vaisseaux* (1773).[26] Though a Swiss, and no sailor, Euler's attitude to resistance was rather more practical than Bouguer's: that is, he abandoned as useless the idea that theoretical comparisons could be made by considering only the *fore* part of the ship; the aft part too affected its hydrodynamic properties. Accordingly Euler prefers to express the theoretical resistance of a ship (as a fraction of the resistance of its greatest cross-sectional area moving as a plane) in a simple formula which is a function of the bluffness or fineness of the ship's bows. This, he believed, had been sufficiently confirmed by some (unspecified) experiments on ships of the line. All this of course is on the assumption that the vessel is sailing directly before the wind, that is without drift; but Euler has no difficulty in similarly

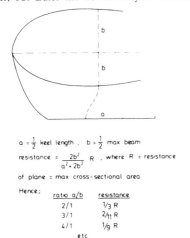

$a = \frac{1}{2}$ keel length ; $b = \frac{1}{2}$ max beam

resistance $= \dfrac{2b^2}{a^2 + 2b^2}$ R , where R = resistance

of plane = max cross-sectional area

Hence;

ratio a/b	resistance
2/1	1/3 R
3/1	2/11 R
4/1	1/9 R
etc	

FIG 8. Leonhard Euler's hull resistance formula (1773).

tabulating the total resistance when the ship, under a side wind, is drifting somewhat to leeward and the resistance is made up of two components. In this case, as he points out, the paradox arises that the least resistance is experienced not when the drift is zero, but when it has a certain small value.[28]

But let us delay no longer over these abstractions or such interesting matters of Euler's treatment of the position of the masts, or his study of the action of oars. I want to press on to the experimental investigation of resistance. Euler himself was well aware that:

from good models in miniature which represent vessels exactly as they are, very important experiments upon the resistance of vessels may be very usefully made and this would be the more significant because the theory of the matter is still very defective.

In this Euler did no more than echo the opinion of Newton in the *Principia* (1687):[29]

if various shapes of ships were constructed as little models and compared one with another, one could test at small cost which [shape] was most suitable for navigation.

Newton went into no details of tank-testing, whereas Euler made a series of practical suggestions about how it should be carried out.

In seems, in fact, that a good deal had been done already. Stoot notes one Samuel Fortree as experimenting on models in a tank during the seventeenth century, later than Huygens but, as Fortree died in 1681, before Newton.[30] To cite something more definite, the Royal Society of Arts as we know it now, founded in 1754, offered in the fourth year of its existence a group of awards for 'ship's blocks' or hull models in order to ascertain by experiments the principles on which a good vessel is founded '. . . . passing through the water with least resistance seeming to be the first quality necessary', although stiffness also was to be investigated. Three years later (1761) a series of quite elaborate trials took place at Peerless Pond near Old Street and later at Snaresbrook Great Pond in Epping Forest, which ended in the award of £100 for a 74-gun ship model, and of £60 for a frigate model. The usual towing tests were made over a distance as great as 105 yards to establish relative resistances; the models were evidently fully made and rigged, since a test was also made of the force required to cause the model to heel, with sails set, through an angle of 20°. Such a test was first proposed by Paul Hoste, was mentioned again by Euler, and was first attempted on a full-size naval vessel in Britain in 1835. The Society of Arts trials must have been fun to watch, and were repeated in 1762.[31]

Yet they probably produced no useful effect, since the first body in Britain systematically to concern itself with rational shipbuilding was the Society for the Improvement of Naval Architecture founded in 1791, which lasted about ten years. Models resistance experiments were, however, so well known that they were naturally taken up by Benjamin Franklin when investigating the passage of boats through shallow canals, in 1768.[32]

Far better known than these early English experiments were the elaborate trials carried out by Frederick af Chapman in Sweden and three French scientists in Paris. Chapman (1721–1808), whose father was a Yorkshire-born carpenter, became one of the most famous shipbuilders of all time, and an Admiral in the Swedish Navy. I shall not discuss his career as a constructor; his publications, of which I must say something, are rather complex. First came a Latin title, *Architectura navalis mercatoria*, published at Stockholm in 1768. This is an ordered series of 62 plates of merchant (and privateer) ship-designs, with outline specifications, but no other text. Then, seven years later, Chapman published his chief work, *Tractat om Skepps- Byggeriet* or treatise on shipbuilding. This contains a first brief account of model tests. This book was twice translated into French in 1779 and 1781. The second of these versions, by Vial du Clairbots, was translated into English (1820) by the Rev. James Inman, principal of the short-lived School of Naval Architecture at Portsmouth, for whose students it was chiefly intended. This book is not in the British Library, nor was it well known in the early nineteenth century since a writer of 1822 expressly declared that Chapman's works,[33]

if translated, would be of little value to this country, since they are not to be understood without a previous acquaintance with the higher branches of mathematics, of which little is known among our artists.

Finally, it is said (in the recent reprint of his *Architectura navalis*) that more than 2,000

Weight of the bodies		N°. 1 27 pounds	N°. 2 27 pounds		N°. 3 27 pounds		N°. 4 22 pounds		N°. 5 19 3/4 pounds		N°. 6 16 1/4 pounds		N°. 7 12 pounds	
Form of the bodies		 *A A*	*B*	*C*	*D*	*E*	*F*	*G*	*H*	*I*	*O*	*P*	*R*	*P*
Moving weights	Retarding weights	Time the bodies have been describing the space of 74 feet, in seconds												
		Seconds *A*	Seconds *B*	*C*	Seconds *D*	*E*	Seconds *F*	*G*	Seconds *H*	*I*	Seconds *O*	*P*	Seconds *R*	*P*
3/4 the weight of the body	1/2 the weight of the body	25 1/2	26 1/4	24 3/4	27 3/4	26 1/2	25 3/4	25 1/2	27 1/4	24 1/4	30	29 3/4	45	29 1/2
The weight of the body	1/2 the weight of the body	14	14	14 1/2	14 1/2	16 1/2	13 3/4	13 3/4	15	16	24 1/2	24 1/4	38	24
1 1/2 weight of the body	1/2 the weight of the body	11	10 1/2	11 1/2	10 1/2	13 1/2	11	11	10 1/4	11 1/2	12 1/2	17 1/2	30 3/4	19 1/4
37 pounds in all	12 lb. and 1/3 in all	12 1/2	lost		11	14	10 3/4	11	10	11 1/4	12	16	—	—

FIG 9. Fredrik af Chapman's model tests in Sweden (1775).

model tests form the basis of his *Fysiska rön om det motstand Kropper lida, some föras rätt fram genom vattned* ('Physical data concerning the resistance to which bodies are subject when moved straight forward through water') published in 1795.

It is unjust to a pioneer to summarize his achievements in a few words, perhaps not fully perceptive. Chapman seems to have begun, like the French mathematicians, from dissatisfaction with the state of his art in which 'the construction of a ship with more or less good qualities is a matter of chance and not of previous design' which led him to the belief that 'a good theory' was needed to attain 'any greater perfection than [shipbuilding] possesses at present'. In fact he takes as his mathematical basis Newton's solid of least resistance but like Beaufoy twenty years later he recognised that water is not an ideal fluid—its viscosity and inertia are manifested in the bow wave and stern turbulence. These are caused by the continual displacement of water equal to the ship's submerged volume as it moves forward. He deduced mathematical expressions relating resistance to size and also fixing the positions of the masts in the vessel, and of the maximum beam. He seems not to have been interested in developing a 'perfect' shape for the bow. His experiments (in the *Tractat*) were directed to comparing the performance of obtuse and acute angled shapes: his results indicated that the obtuse end should precede when the velocity is high, but the acute angled end should lead when the velocity is low. But the experimental differences were not very decisive.

The trials of the three French scientists were made in 1775 at the Ecole Militaire in Paris and reported in *Nouvelles Experiences sur la Resist-* *ance des Fluides* in 1777. Charles Bossut (1730–1814) who receives chief credit for them—though his colleagues D'Alembert and Condorcet were men of greater weight—occupied the chair of hydrodynamics in Paris created by the great Minister Turgot, with practical applications particularly in mind; which did not however, make Bossut more popular with French naval men. At Turgot's request Bossut made 300 trials with 12 models. His main concern was to test the then ordinary hydrodynamical principles—that resistance is equivalent to a corresponding static pressure, that it accordingly increases with the square of the velocity, and that the resistance to a plane inclined to the direction of motion is as the sine of the angle. His results confirmed this theory, *grosso modo*, but (like Franklin) he found that the resistance was markedly increased in narrow, shallow channels. Bossut was mistaken in concluding that the frictional resistance of a hull is 'nil in comparison with that arising from the inertia' of the water particles, on the other hand he remarked upon the elevation of the water before the stem of a vessel and its depression at the stern, with consequential modification of the solid of resistance.[34]

Finally, as is well known, very elaborate experiments on fluid resistance—10,000 trials in all, it is said, costing £50,000—were made between 1793 and 1798 by Mark Beaufoy (1764–1827) of which a painfully full account was given posthumously in his *Nautical and Hydraulic Experiments* (London, 1834). This is a huge and maddening book. With its endless undigested tables of experiments which (because of changes in method) cannot be directly compared one with another, its five and six figure decimals, its mis-copying and misprints,

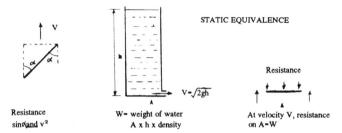

STATIC EQUIVALENCE

Resistance
sinα and v^2 W= weight of water
A x h x density   At velocity V, resistance
on A=W

$V=\sqrt{2gh}$

FIG 10. Charles Bossut's use of 'ordinary hydrodynamical principles' (1777).

its failure to analyse and digest, it is almost totally incomprehensible in detail. Beaufoy studied strictly the mechanics of resisted motion: he did not try to discover what form of hull might show the least resistance nor did he investigate the behaviour of the fluid flowing round the moving body. He had been preoccupied with this topic since boyhood, drew up his first reflections at the age of twenty, and was a principal activist in the Society for the Improvement of Naval Architecture, no doubt with an eye to carrying out those large-scale experiments at Greenland Dock which are its chief memorial. Something of his work was published in his lifetime, for example in the Society's 1794 *Report* on its experiments and also in a *Marine Dictionary* of 1815, so it is not wholly true that Beaufoy's work remained buried and unknown for over thirty years, until his son Henry rescued it.[35]

What general ideas emerge from Beaufoy's investigation? Firstly, with reference to the mathematicians, he found that resistance did not vary *exactly* as the second power of the velocity. Sometimes the index was slightly greater than two, sometimes less, depending on the particular type of experiment. Beaufoy did not think the discrepancies due to experimental error. More important, he showed clearly that skin-friction or drag is an important component in total resistance whatever the speed (which was from one to eight knots). This friction of course increased with length or more strictly surface area and therefore the mathematical assumption that an infinitely long, infinitely narrow ship would have no resistance was false. Secondly—if I understand him rightly—he discovered and measured negative pressure, previously only hinted at, which he called 'minus pressure', understanding that the excess pressure of water at the bows (or bow wave) must be balanced by an equally adverse defect of pressure at the stern: or as he put it, as early as 1784,[36]

if the velocity of the [moving] solid be greater than that with which the water rushes into the opening [created by the motion forward] the resistance made to the solid will be increased by the weight of water resting upon it in consequence of the vacuum . . . .

Accordingly Beaufoy advocated as much attention to the shaping of the lines at the stern of the vessel as to the lines of the bow, and condemned the practice of cutting off the stern

in a square plane. In his view, there were three components in the total fluid resistance: excess pressure at the bow, friction along the sides, and minus pressure at the stern. Only the first of these had been considered by mathematical theorists. It is a great pity that Beaufoy, who devoted such fantastic effort to his research, and aimed at practical benefits, should have failed so signally to express his 'lessons' for shipwrights in any clear and positive message. He seems to have been content to compile raw data.

Stability

From a practical point of view only marginal gains could be looked for, probably, from improvements in the bluff, stocky shape of traditional wooden ships, made with an eye to facilitating passage through the water. Sailing qualities, as was well known, demanded a length/beam ratio of not much more than about three to one, and forbade excessive 'fineness' in the bows. Rationalization—in the reduction of excessive overhang and top weight—could be and was effected. But gains in speed of more than a few % were unattainable, and so luck might still be more important than design.

When we turn to questions of stability the situation is very different; it is essential to all ships at all times in a way in which speed is not. A slow ship may be tedious, an unstable ship may cost hundreds of lives, as with the *Mary Rose* at Portsmouth in 1545 or the now raised *Wasa* at Stockholm in 1628. Many ships proved too tender when first commissioned, to be remedied by girdling, shortening the masts, and similar detrimental expedients. Others did not realise design expectation as to their water-line level or burden. The eighteenth century mathematicians devised useful new methods for working out the volume of a ship, and assessing its static properties.

Time will not permit me to consider all the interesting analysis devoted to problems of pitch and roll, much aided of course by the understanding of the dynamics of oscillation gained in the seventeenth century. However, I cannot altogether pass over Bouguer's analysis of the position of the ship's centre of gravity, and its effect on stability. So far as I am aware he was the first to point out that there is a centre of gravity for the immersed part of the hull, which is also the centre of thrust of the force of buoyancy exerted by the water, and a second centre of gravity of the whole ship, which is necessarily higher than the former. Further, he defined on the common axis of these two, which is the vertical axis of the ship, a third point later called the metacentre. If the centre of gravity of the whole ship lies above this point, then it is inherently unstable; the greater the metacentric distance, the greater the stability, and the longer the period of roll. Bouguer showed how the position of the metacentre could be found from a test revealing the force required to heel the ship through some measured angle. Metacentric height is increased by greater breadth and heavier ballast, reduced by height and 'top-hamper' of all kinds, particularly the weight of cannon on the upper gun-decks. Bouguer's general expression defining the metacentric distance HQ is

$$HQ = \frac{2}{3}\int y^3 \frac{dx}{p}$$

where p is the immersed volume, and x and y are the elements of length and breadth respectively. To repeat Bougher's own example of Noah's Ark—of uniform section 50 by 30

cubits, drawing say 10 cubits of water—we can simplify to

$$AH = \frac{\frac{1}{4}\,(\text{widths})^2}{3 \times (\text{depth})} = \frac{\frac{1}{4}\,(50)^2}{3 \times 10}$$
$$= 20\,\frac{5}{6}\text{ cubits}$$

this gives a metacentric height of $5\frac{5}{6}$ cubits, so that the Ark is highly stable.[37] Somewhat similarly, the rolling period of the Ark can be calculated as being about 2½ seconds.

Conclusions

It is now time to offer you my conclusions, based on rather more than this brief review of only a small fraction of the evidence. I hope I have succeeded in convincing you that the rational and experimental study of naval architecture and hydrodynamics has a long and respectable history before 1800, and have at least provided hints that a comprehensive study of the relevant literature made, say, in 1830 would have been a matter not of days or weeks but of months or years. Despite a good deal that was ephemeral, this literature contained much of permanent value relating to the strength of the ship as a structure, to its stability and seaworthiness, to the calculation of its

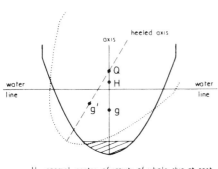

H = normal centre of gravity of whole ship at rest
g = normal c g of immersed hull at rest
g' = c g of immersed hull, heeled
Q = metacentre - intersection of axes

$$HQ = \frac{2}{3}\int y^3 \frac{dx}{p}$$

p = immersed volume ; x,y = ordinates of length and breadth

$$AQ = \frac{\frac{1}{4}(\text{width})^2}{3\,(\text{depth})} = \frac{\frac{1}{4}(50)^2}{3 \times 10} = 20\,\frac{5}{6}$$

Ag = 5 cubits
AH = 15 "
AQ = 20 ⁵⁄₆ "

FIG 11. Pierre Bouguer's analysis of ship stability and his example from Noah's Ark (1746).

weight and displacement, and so forth. In the hands of those who could master this literature shipbuilding could become a more definite and rigorous combination of design and craft than it had formerly been.

Thus, to revert to John Scott Russell, if he really made a claim that nothing worthwhile of a theoretical or scientific kind had been done in the past towards the improvement of shipbuilding, he was mistaken. If he had been making the rather different claim that what had been done was unknown to or ignored by *British* shipbuilders, then this modified claim might have been just, or at least in conformity with an ordinary view of the historians of shipping; I quote a recent book:[38]

In France, following research into the resistance of solid bodies to water and the action of waves, particular attention was given to the design of the underwater hull form of warships. In consequence, the French warships in general had better sailing qualities and were faster than English ships of a similar class.

(I do not myself think this is quite correctly put and when the authors hint that the chief advantage of having faster ships was that the French navy had the option of running away they are perhaps being less than generous.) But I think Russell meant more than this. In one of his British Association Reports (1842) summing up the long series of experiments (more than 20,000, he claimed) on ship models and vessels 'up to 1000 tons burden' Russell presented his sense of the deficiency of the experiments previously made: 'they had not been conducted with an adequate knowledge of the wants of the constructor'; they had used shapes such as were never given to ships; their scale had been too small; the law by which 'the results of experiments on one scale of magnitude are to be transferred to a different scale' had not been discovered; and finally the apparatus used by earlier experimenters had been defective. His own trials, Russell asserted, had been aimed to remove these objections.[39] Clearly—if we may believe the B.A. abstract—Russell did not claim to be a pioneer in the experimental study of naval hydrodynamics, and he should not have claimed that all his results (such as the influence of the channel on the resistance which a ship moving in it experiences) were new in principle. I think, however, that he could justly affirm that his experimental

study 'of the value of giving proper FORM to ships, altogether independently of proportion and dimension' was something new, though clearly foreseen by such mathematicians as Newton and Euler.

In fact, Russell started not from ignorance of theory, but from dissatisfaction with it. The discrepancy between the theory and the experiments already reported was, he wrote, such as to 'render the principles of the theory exceedingly false guides, when followed as maxims of art.' Avoiding models as dubious, his early experiments were made on sizeable canal-boats. This circumstance led him to a number of considerable discoveries concerning the motion of vessels in confined waterways. Particularly he examined wave formation and turbulence, noting that the speed of the bow-wave associated with the vessel depended on the depth of the channel, not on the speed of the vessel; and that wave, turbulence and resistance all increased disproportionately as the speed of motion approached the speed of the wave. When the boat rode on its crest, or exceeded the natural wave-speed for the channel, it moved smoothly and with less resistance; thus the boat 'on the summit of the wave' was pulled 'with greater ease at 10 or 12 miles an hour, than at 6 or 7'.[40] This suggested an important lesson for canal navigation. But observation of the 'large, solitary, progressive wave' also led Russell to redesign the hull form, adopting the concave in place of the convex bow. In truth, concave lines were already, it seems, to be seen on wherries and perhaps already on the early U.S. 'proto-clippers'; that need not concern us, it is enough that Russell showed, in an unprecedentedly exact way, that very similar hulls, of the same principal dimensions and displacements, and all good sea-boats, could differ 'so much in resistance, that the one had nearly double the resistance to another'. The concave form which he gave to the *Wave* proved to suffer the least resistance and it was (he adds proudly) 'also the best sea boat, the easiest and the driest.'[41]

There is, of course, something wrong with Russell's theory as regards oceans and ships in general. In a canal 5 feet deep Russell found his wave to travel at about 7½ m.p.h. in accord with his theory; if the same theory were applied to an ocean 1000 ft. deep, the corres-

FIG 12. Waves from a boat moving in a confined channel, Scott Russell, 1842.

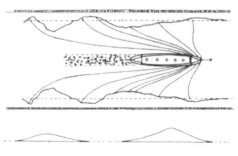

FIG 13. Vessel moving among waves, Scott Russell, 1842.

maximum disturbance and resistance
when boat slightly slower than wave

disturbance and resistance diminish
with coincidence of velocities and
boat rides smoothly on wave

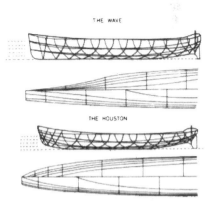

THE WAVE

THE HOUSTON

Dynamometer measurements of the resistance of dimensionally similar but diversely shaped hull forms (No. 1. is *Wave*). Speed = 7¼ mph.

No. 1 Form	56.6 lbs. resistance
No. 2 Form	138.5 ..
No. 3 Form	102.7 ..
No. 4 Form	90.2 ..

FIG 14. Two of the hull forms tested by Scott Russell, 1842.

XI

ponding wave speed would be 122 m.p.h. Even in deep channels a canal boat could not possibly be hauled at the speed of the wave. And I cannot myself credit that with his new lines all vessels at all speeds would behave like *Wave* at 17 m.p.h.:

there is no spray, no foam, no surge, no head of water at the prow, but the water is parted smoothly and evenly asunder and quietly unites after the passage of the vessel.[42]

But I must regretfully leave Russell to offer you my main historical conclusion, which turns on the nature of high technologies and industrial revolutions. It has often been supposed that scientific knowledge made a direct contribution to the changes in technology that were a feature of and a factor contributing to the industrial revolution of eighteenth century Britain. The 'scientist', or natural philosopher and mathematician, it is urged, assisted the entrereneur and the inventor to unbind Prometheus. Historians have examined the Lunar Society of Birmingham, Watt's indebtedness to the heat theory of James Black, the scientific attainments of Josiah Wedgwood, and so forth. Science and innovation, some have argued, went hand in hand.[43]

In the history of naval architecture and shipbuilding, it seems to me, we have a very different state of affairs. The wooden ship, at the hands of a number of mathematicians, philosophers, and experimenters received a vast deal of study and intellectual effort. Even the toil of the galley-slave was measured by the differential calculus. All this study of which I could set so pitifully little before you today was by no means without effect; it made French ships, by 1780 or so, a little faster than English ones. It did not, by itself, enable Villeneuve to win the Battle of Trafalgar. And clearly all the ink and experiments from Galileo to Beaufoy did nothing to change the obvious and fundamental characteristics of the wooden sailing ship, or of the shipyards from which it came. In a very real sense, through all the changes from clinker to carvel build, from single decks and single masts to three-deckers with multiple masts, for square sails to fore-and-aft rig, *Victory* and her contemporaries were of the same

stock as ships of 2,000 years before and cousins to the sailing barge and fishing lugger. We know where the great change in the art of the ship came from: it came from John Wilkinson and James Watt via such pioneers as Symington and Fulton. The new ship came not like Venus from the sea foam but from the mines of iron and coal. The industrial revolution in shipping, if I may so term it, owed nothing to Newton and Bouguer, Beaufoy or Chapman. The iron steamship was a new created thing; it was an offspring, doubly, of the riveted steam boiler.

Obvious analogies could be drawn. The theory of structures, as it developed in the eighteenth century, had very little relevance to the structural use of cast and wrought iron which was the chief revolution in practice towards the close of the century. It might almost seem that the most able theoreticians were suspicious of the new materials. Indeed, analysis and innovation are very unlike activities.[44] Of course I am not saying that they are opposed activities. In the end they have to combine and reinforce each other. Think of the great steel barques of eighty or so years ago— surely the most efficient sailing carriers ever devised. But my point is that analysis, which is science, though it may sometimes be a necessary factor for engineering innovation, cannot by itself be a sufficient factor; and that I think was Scott Russell's real criticism of his predecesors in the rational and experimental study of naval architecture. They had sought and in a measure found analysis; they had not sought or discovered innovations in design. To my mind, this is a generally valid and and useful way of considering the British industrial revolution; the men who brought it about were not analysts, like Bouguer or Beaufoy, they were men looking for new engineering and productive principles. Calculus has very little to do with inventing the steamship, though it has a great deal to do with perfecting it. It was not the case that rational shipbuilding made the iron steamship possible: quite the contrary, it was the iron steamship that made rational shipbuilding possible.

References

1. George S. Emmerson, *John Scott Russell*, London, John Murray, 1977, p. 13. Compare J. B. Caldwell, 'Three Great Ships', in Alfred Pugsley (ed.), *The Works of I.K. Brunel*, London and Bristol, 1976, pp. 137–62 where only 'Atwood, on stability, and Beaufoy, with his systematic experiments on ship resistance' are noted as standing out for their 'scientific contributions'.

2. The classic early printed book on ship-construction is Bartolomeo Crescenzio, *Nautica mediterranea, nelle quale si mostra la fabrica delle galèe* (the colophon dated 1602, perhaps first issued at Rome, 1607). The classic modern study is F. C. Lane, *Navires et Constructeurs à Venise pendant la Renaissance*, Paris 1965.

3. *Philosophical Transactions*, **VI**, no. 77, 20 November 1671, p. 3011. Cf. A. R. and M. B. Hall, *The Correspondence of Henry Oldenburg*, University of Wisconsin Press, Madison, Milwaukee and London, **VIII**, 1971, Index S. V. 'Witsen'. I have examined a facsimile reprint of the second edition.

4. I have used the second edition of *La Statique*, Paris, 1674, see pp. 219–32.

5. Renau's *De la théorie de la manoeuvre des vaisseaux*, published at Paris 'De l'exprès commandement de sa Majesté in 1689 was criticised by Huygens in print in 1693: he thought Renau's treatment could lead pilots 'into great and serious errors' (*Oeuvres Complètes de C. Huygens*, X, 1905, 523–31). The dispute continued into 1694 (*ibid.* 588–96, 654–8, 690–93). Huygens's criticism was directed against Renau's taking the ship's velocity as proportional directly to the force applied, rather than the square root of the force (being resisted).

6. Johann Bernoulli, *Essay d'une nouvelle théorie de la manoeuvre des vaisseaux*, Basel 1714 (in *Opera omnia*, Lausanne & Geneva 1742, **II**, 1–167).

7. Such as H. Pitot, *Théorie de la manoeuvre des vaisseaux reduite en pratique*, Paris, 1731. (English translation, 1743).

8. See C. W. Cole, *Colbert and Century of French Merchantilism*, New York, 1939.

9. *Philosophical Transactions*, *XII*, no. *135*, *26 May 1677*, *pp. 879–883*; the quoted words are directly translated from Dassié's introduction.

10. Cf. Sir Westcott Abell, *The Shipwright's Trade*, Cambridge, 1948.

11. Hoste, *Théorie* (1697), Preface.

12. *Ibid.*, 46–8, 68–82 (on pitching and rolling), 84–5.

13. Further details may be found in my book, *Ballistics in the Seventeenth Century*, Cambridge, 1952) especially in Chapter 5.

14. Christiaan Huygens, *Oeuvres complètes*, **XIX**, 122, 124–5.

15. *Principia*, 1687, Book III, Scholium to Prop. XXXV (pp. 326–7). I. Bernard Cohen argues (in R. S. Cohen *et al. For Dirk Stuik:* Boston Studies on the Philosophy of Science, **SV**, 1974, 169–87) that Newton's words quoted in the text apply only to the paragraph in which they appear, and not to the preceding paragraph of the Scholium where Newton has considered resistance to a conical shape. But as the second paragraph relates to the first (for Newton states that an oval figure terminating in a cone is less resisted than the oval alone) they can hardly be considered as independent.

16. The imprimatur is dated 10 December 1674. See also the Royal Society's Register-Book, **IV**, 246–67.

17. The Marquess of Landsdowne, *The double-bottom or twin hulled ship of Sir William Petty*, Oxford 1931.

18. Petty, *Discourse* (1674), 17, 39, 43–45.

19. Johann Bernoulli, *Nouvelle Théorie de la Manoeuvre des Vaisseaux (Basel, 1714)*, 54.

20. P. Bouguer, *Traité de Navire* (Paris, 1746), xvii, xviii. He referred to the ordinance of the Arsenal 1689, laying down the principal dimensions of French naval vessels.

21. P. Bouguer, *De la manoeuvre des Vaisseaux, ou Traité de Méchanique et de Dynamique*, Paris, 1757, 183; cf. *Traité de Navire*, 355.

22. *Traité de Navire*, 364–379, 418, 421. Bouguer, of course, understood that, as the speed of the vessel increases, the pressure of the wind on the sails diminishes (445).

23. *Ibid.*, 1, 63, 679.

24. *Ibid.*, 663–4.

25. *Ibid.*, 105, 565.

26. I have used the *Nouvelle Edition corrigée et augmentée* of 1776. The original (St. Petersburg, 1773) is reproduced in Euler's *Opera Omnia*, Ser II, 21, 1978, 82–213. This work was published in English (London, 1776) the translator being Henry Watson (c. 1737–86), formerly chief Engineer of Bengal. Dedicating his work to the First Lord of the Admiralty Watson hoped that it might 'excite the Navigators and Artists of this Kingdom to render their Theory more perfect, and to become as eminently skilled in the scientific as they now confessedly are in all the practical Braches of their Profession'. Euler's work was also translated into other languages. It is a condensation of his *Scientia navalis* (1749; *Opera Omnia*, Ser. II, 18, 1967, 19, 1972).

27. *Theorie complete* (1776), 91–4.

28. *Ibid.*, 116.

29. *Ibid.*, 231–7. Newton, *Principia*, 1687, 354. This passage was omitted from later editions.

30. W. F. Stoot, 'Some Aspects of Naval Architecture in the Eighteenth Century', *Trans. Inst. Naval Architects*, Vol. 101, 1959, 35 quoting G. S. Baker in *Trans. North East Coast Inst. of Shipbuilders and Engineers*, *54, 1938, 33*–6. The Fortree MS. is in the Pepysian Library, Magdalene College, Cambridge.

31. D. Hudson and K. W. Luckhurst, *The Royal Society of Arts, 1754–1954*, 143; Minutes of Mechanics Committee, 27 July 1761.

32. B. Franklin, *Experiments and Observations on Electricity*, London 1769, 492–6. *Papers of Benjamin Franklin*, 15, 1972, 115–8. He simulated the canal by a trough 14 feet long by six inches square. The experiments were made while Franklin was visiting London.

33. D. Steel and J. Knowles, *Naval Architecture*, London, 1822, Preface, quoted by Geoffrey Dyson, 'The De-

50

velopment and Instruction in Naval Architecture in 19th century England', unpublished M.A. dissertation, University of Kent, Canterbury, 1978. Chapman's work may now be conveniently examined in the reprint of *Architectura navalis mercatoria* (Rostock, 1968; English version, London 1971) containing part of the 1775 *Tractat* in Inman's translation (*A Treatise on Shipbuilding . . . .* Cambridge, University Press, 1820; it contains Vial du Clairbot's notes).

34. D'Alembert, Condorcet and Bossut, *Nouvelles Experiences sur la resistance des fluides*, Paris 1777, especially pp. 131, 147–8, 173–4.

35. *Report of the Committee appointed to manage the experiments of the Society [for the Improvement of Naval Architecture]*, London 1794; William Falconer, *Universal Dictionary of the Marine* [first published London 1769] modernized and enlarged by W. Burney, London, 1815.

36. *Nautical and Hydraulic Experiments*, xxiv-xxv. I find Beaufoy's formulation of this issue clearer than Chapman's. But see also P.L.G. due Buat, *Principes d'Hydraulique* (Paris, 1779) where the formation of high water pressure at the bow end and low pressure at the stern was studied. Mention should be made also of the Rochellois Charles Romme, *L'Art de la Marine, ou principes et précéptes généraux de l'art de construire, d'armer, de manoeuvrer et de conduire des Vaisseaux*, La Rochelle 1783, which is said to report model-tank experiments.

37. P. Bouguer, *Traité du Navire*, Paris 1746, 257–91.

38. D. MacIntyre and B. W. Bathe, *The Man-of-War*. London 1968, 61.

39. *Report of the 12th Meeting of the British Association . . . . Manchester, 1842*, London 1843, 104–5.

40. J. S. Russell, 'Experimental Researches into the Laws of certain Hydrodynamical Phenomena' *Trans. Roy. Soc. Edinburgh*, 14, 1840, 47–99. This paper must be used with care—it is full of mistakes.

41. *Report* (note 39 above).

42. 'Experimental Researches', 99.

43. See A. E. Musson and Eric Robinson, *Science and Technology in the Industrial Revolution* Manchester 1969; R. E. Schofield, *The Lunar Society of Birmingham*, Oxford, 1963; P. Mathias, *The Transformation of England*, London, 1979, Chs. 3 and 4. N. McKendrick, 'The role of Science in the Industrial Revolution: a study of Josiah Wedgwood' in M. Teich and R. M. Young (eds.) *Changing Perspectives in the History of Science*, London, 1973. My own views were last stated in 'What did the Industrial Revolution owe to Science?', in N. McKendrick (ed.) *Historical Perspective, Studies in English Thought and Society in honour of J. H. Plumb*, London, 1974.

44. A. W. Skempton, 'The Origin of Iron Beams', *Actes du VIIIᵉ Congrès Int. d'Hist. des Sciences*, **III**, 1958, 1029–39; J. G. James, 'Iron Arched Bridge Designs in Pre-Revolutionary France', *History of Technology 1979*, 63–99.

This article is reprinted by permission of the Council of the Newcomen Society. Their co-operation is much appreciated

Ed.

XII

Hooke's Micrographia, 1665-1965

THE twentieth of January, 1665, was a typically busy day for Samuel Pepys, with public and private business, dinner at the Swan tavern in Westminster (followed by kisses from the maid there), and purchase of a hare's foot for its medical virtue; it was on this day that he took home, bound, 'Hooke's book of microscopy, a most excellent piece, and of which I am very proud'.[1] He had seen it at the bookseller's on the second day of the year, and found it too pretty to resist. This was not Pepys' introduction to the world of the very small, however, for he had already read Henry Power's book on the same subject in the preceding summer of 1664: 'very fine and to my content'. Nor was he yet a Fellow of the Royal Society himself; this he was to become the following month (on 15 February), when he for the first time attended a session of the Society at Gresham College; that afternoon, I imagine, he also met Robert Hooke face to face for the first time, and was so impressed that he wrote of him that he 'is the most, and promises the least, of any man in the world that ever I saw'. It seems that Pepys was never disenchanted, though in later life he was also to become a friend of Hooke's arch-enemy, Isaac Newton.

Perhaps it would have been still more appropriate to have celebrated the tercentary of the *Micrographia* at this time a year ago, for the book was completely printed by October, 1664, and was given its *imprimatur* by Lord Brouncker, the President of the Royal Society, on 23 November. Hooke himself explained that there was a delay in publication because some of the Fellows insisted on reading the book before it was approved. So it is likely that *Micrographia* was on sale in London before the end of 1664, the booksellers of those days often putting next year's date on a book that actually came out in the late autumn.

In any event, the way for *Micrographia* was well prepared. For

---

[1] This lecture was arranged by the Faculty of Zoology, University of London; Professor G. P. Wells took the Chair. An exhibition of Hooke's published works from the Library of King's College, London, was made possible by the kindness of the University Librarian and the Librarian of King's College.

example, Henry Oldenburg the Secretary of the Royal Society spoke of it as forthcoming in a letter to the Danzig astronomer Hevelius of mid-November; and no doubt also wrote of it to others. More important, Hooke himself had been presenting his microscopical descriptions to the Society for some years; for on 25 March, 1663 Hooke was desired to 'prosecute his microscopical observations, in order to publish them', while at the meeting of the following week he was instructed to produce at least one such observation before the Society at each of its sessions. Accordingly in April he began to reveal some of the observations, and no doubt to produce for inspection some of the drawings, that were later to appear in *Micrographia*. In fact it would seem that Hooke had been engaged upon microscopy in his Oxford days with Robert Boyle, long before he came to London to work for the Royal Society as its Curator of Experiments.

Something has been said of the genesis of the famous *Micrographia*; what of its author? When *Micrographia* appeared Robert Hooke was just thirty years and six months old; a mature man for that period. *Micrographia* was his first considerable achievement in science, though not actually his first publication; if he was not—as some have supposed—actually the author of a continuation of Bacon's *New Atlantis* that appeared in 1660, he was certainly the 'R.H.' who published *An Attempt for the Explication of the Phaenomena, observable in an Experiment Published by the Honourable Robert Boyle* in 1661. The phenomenon was that of capillary attraction. As this publication indicates, it was Boyle who brought Hooke firmly into the current of seventeenth-century science, as it was Boyle also who secured for Hooke his position of Curator of Experiments to the Royal Society. Hooke's early life, before his fruitful association with Robert Boyle began at Oxford in 1658, is rather a series of legends than of well-ascertained facts. His origins were humble, and his prospects were not improved by the early death of his father, curate of Freshwater in the Isle of Wight. The boy, it seems, was sent to London with a sum of money, and is said to have spent some time with the portrait-painter Peter Lely and the miniaturist Samuel Cooper. He is also associated—though in no very regular way—with Westminster School. However all this may be, it is clear that in 1653—at the age of eighteen which was advanced for those days—Robert Hooke was at Christ Church, though how he lived and what he did is obscure. He may have been a sizar, as Newton was at Cambridge; he may have acted as lab-boy to Thomas Willis, the great anatomist and physician; but he did not go through the customary exercises for the degree, and Hooke himself speaks, at this time, of his acquaintance with the celebrated Dr. John Wilkins, Warden of Wadham College, and of his great preoccupation with the problem of mechanical flight. It is obvious enough, however, that Hooke was much interested in other mechanical matters, in astronomy, and in geometry, though he was never to attain the highest level of theoretical proficiency in either of the latter studies.

As everyone knows, Oxford was particularly rich in eminent or potential scientists during the late 1650s. Besides Wilkins, the great organiser (for he was not actually doing anything himself), Willis, and Boyle, there was John Wallis the great mathematician, Seth Ward the astronomer, Jonathan Goddard the chemical physician, William Petty the statistician, and others; while among a younger generation were to be found, besides Hooke, Christopher Wren— potentially perhaps the most brilliant man in Oxford—Richard Lower, John Mayow and quite a few more. Such men provided an extraordinarily stimulating environment; unfortunately, we have no real knowledge of Hooke's place in it. The most important of his acquaintanceships in the present context was that with Christopher Wren, for it seems clear that these two began their studies in microscopy together. Henry Power—of whom I shall speak again in a moment—predicted to his readers in 1664 that 'you may expect shortly from Dr. Wren and Master Hooke, two Ingenious Members of the Royal Society at Gresham, the Cuts and Pictures drawn at large, and to the very life of these and other Microscopical Representations'. And it was Wren, says Sprat in his *History of the Royal Society*, who was the 'first Inventor of drawing Pictures with Microscopical Glasses'. If we may rely on a letter from Sir Robert Moray to Wren of August, 1661, as printed in *Parentalia*, it was when Wren himself begged off the task of making microscope drawings of insects for Charles II that Hooke was invited to take up the business. This story both agrees and disagrees with Hooke's own words in the Preface to *Micrographia* that it was with reluctance that he followed in the footsteps of Dr. Wren 'who was the first that attempted any thing of this nature; whose original draughts do now make one of the Ornaments of that great Collection of Rarities in the King's Closet'. It has even been suggested that some of the figures in the *Micrographia* were drawn by Wren; for these famous plates are unsigned, and Hooke did not explicitly claim them as his own. As with the woodcuts in Vesalius' *De humani corpus fabrica* the credit for the illustrations that contribute largely to the success of the work must be left unassigned; at this time it was, of course, quite usual to employ professional artists and engravers, like Faithorne, to make cuts for scientific books, without assigning them any credit for their work.

Thus, in his last years at Oxford with Robert Boyle, who first introduced Hooke into disciplined scientific research, Hooke was probably already busy with microscopy, as well as with pneumatics and, one presumes, astronomy and mechanics as well as the chemistry which was always Boyle's chief interest. Combustion is discussed in *Micrographia*. Moreover, in 1664, Boyle was to publish his book on *Colours*, which although largely concerned with colour-chemistry does touch a good deal on light and optics; it may therefore be that it was Boyle who stimulated Hooke to the interest in optical matters that is also displayed in *Micrographia*. But it was only when he left Boyle in 1662, to become Curator of Experiments to the infant Royal Society, that Robert Hooke's interests, activities, and powers as a scientist come clearly into the light—partly in his own publications beginning with the *Micrographia*, and still more in the records of the Royal Society largely extracted by Thomas Birch in his *History* (1756–7).

Especially in the two and a half years of his service before the Plague struck London in the summer of 1665—that is, through the period of the preparation and publication of *Micrographia*—Hooke was exceptionally busy. He was never, I think, to give so much time to science again, partly because of his occupation as a surveyor and architect after the destruction of London by the Fire in 1666, and partly for more personal reasons. This was the peak period of his activity, the one which determined much of the later course of his life, and of course its major product, *Micrographia*, was the greatest single achievement of Hooke's career.

But Hooke was not the first microscopist, and *Micrographia* was not the first book devoted to microscopy; it was no more unprecedented than Vesalius' *Fabrica*. Rather, like the *Fabrica*, it outclassed its predecessors, raising the whole business to a new level of performance and criticism. The microscope was a slow starter. Its early history is even more obscure than that of the telescope; and whereas the telescope became (one may well say) notorious in the hands of Galileo in 1609—so that no one could miss its implications—the microscope long remained simply a toy. Its early use or rather non-use is an excellent illustration for the argument that science can only advance by answering questions. At first no one could see what kind of questions could be answered by close-range optical magnification; or rather, that is not quite true: for at a time when everyone believed that the minute structure of things was of great significance for gross phenomena, it was clear enough that a microscope revealing the ultimate particles of bodies would be of enormous importance in physics and chemistry; but such an instrument was then unattainable and no one, before the 1660s, seems to have clearly got hold of the idea that there were *biological* structures everywhere awaiting investigation, and that existing instruments and quite low magnifications were suitable for such uses.

In fact the earliest microscopes were simply 'flea-glasses'. It was astonishing (as Galileo remarked in a letter of 1619) to see a fly appear as big as a lamb; instructive, too, to observe the hairs on insect legs, their compound eyes, and the articulation of their jaws and limbs. Perhaps if entomology or parasitology had been as flourishing subjects in 1609 as astronomy was, the microscope would have been hailed as an instrument of revolutionary significance, and in fact the first pictures ever printed from a microscopic observation were those of the bee, published by Francesco Stelluti in 1625. But there was little interest at this time in insects. Perhaps even more surprising was the almost complete lack of evidence, during a whole generation, of any attempt to apply the microscope to anatomy, or embryology, or medicine; the only exceptional name one might mention is that of Athanasius Kircher, whose observations, like his theories, were remarkable for imagination rather than accuracy. Nevertheless, a subject began to take shape: Hodierna of Palermo, in 1644, had studied the compound eye in thirty-four species of insects, while Fontana of Naples, two years later, gave four pages in his 'New Observations of the Things of Heaven and Earth' to the microscopy of insects.

To Pierre Borel (1620–71), however, belongs the honour of

issuing the first separate treatise on microscopy; his *Centuria Observationarum microscopicarum* (1651). We are now treading hard on the *Micrographia*, in content as well as in style, for Borel went beyond entomology to examine such subjects as the hairs of insects and of man, and vegetable fibres (all, he asserted, were hollow), the eyes of slugs growing on the ends of their horns, fat globules in milk, the markings on leaves, and that perennially fascinating exhibit, the nematode inhabitants of stale vinegar.

I will only mention two more predecessors of Hooke. One of them deserves a separate celebration on another occasion: I mean Marcello Malpighi, the great Italian microscopist. It was in 1661 that he issued at Bologna in the form of a letter to Giovan Alphonso Borelli his first and seminal work, *De Pulmonibus Observationes Anatomicae*, an extremely rare folio pamphlet. This was the first application of the microscope to anatomy—and hence to physiology—which Malpighi was to take farther forward with his *Tetras Anatomicarum Epistolarum* of 1665. But I must now move on hastily from Malpighi to Henry Power, whose name I mentioned in passing before, because Power was the first Englishman to publish observations with the microscope, in his *Experimental Philosophy* of 1664. It has been suggested that Power followed Hooke's example, but it was clearly not so, since Power's papers disclose his preoccupation with microscopy over several years, long before he could have known of Hooke's work. Power's was not a colossal contribution to the development of this new branch of science, perhaps, but it was far from despicable; many of his descriptive passages can be quoted alongside the parallel ones from *Micrographia* without disparagement.

This is the point, perhaps, where I should put in something about Hooke's relation to the development of the instrument itself. Henry Power's microscopes, we know, were made for him by the London optical-instrument-maker Richard Reeve, with whom Hooke also was very well acquainted. Pepys bought a microscope from the same maker after reading Power's book. These must have been microscopes of a typical English style for this period, of which a number of examples survive in our museums: they had three lenses, the little objectives being usually interchangeable, screwed into turned wood mounts glued to pasteboard tubes. A small tripod base had a female thread, in which a long screwed piece or 'nose' of the

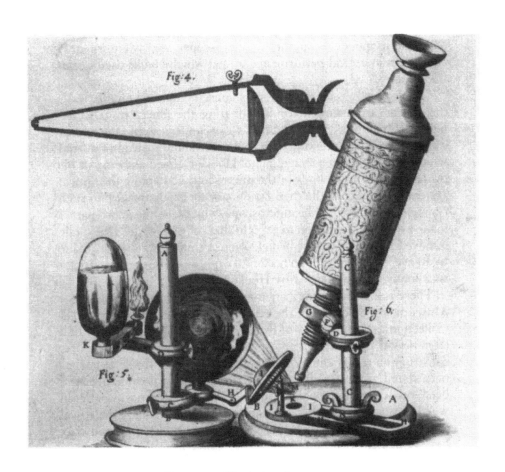

PLATE I: Hooke's microscope.

microscope worked to provide for focusing by an inconvenient rotation of the whole body of the instrument. Malpighi's instrument, again, was constructed on similar principles—probably, it seems, by the famous instrument-maker of Bologna, Giuseppe Campani. It was certainly quite possible, in London, Paris, Rome, Bologna and elsewhere to obtain for relatively large sums—say £100-£200 of modern money—compound microscopes that were roughly equivalent in optical performance to that which Hooke used for *his* microscopy.

This is chiefly remarkable for its construction and auxiliaries—there is an arrangement for illuminating the object by reflected light with a globe of water and bull's-eye condenser, and the microscope body is mounted upon an inclinable stand instead of a fixed tripod (Pl. 1). These arrangements Hooke claimed as being of his own devising. Of the body of the microscope containing the optical system—which is of the then customary form—he makes no such claim, saying only, 'The microscope which for the most part I made use of . . .' going on to explain that for most critical vision he removed the intermediate or field-lens. I think the tube was made by Reeve, and I know of no reason to suppose that its optical design was in any way influenced by Hooke himself.

There is a well-known instrument still surviving in the Science Museum and described in the original Catalogue of the George III Collection as the property of Robert Hooke. This is obviously of a later period than *Micrographia*; as well as being larger it is of a construction rather different from that described there. If, as has been stated, it was the work of Christopher Cock, it could not have been made before 1668 or 1669. I have sometimes wondered whether this was not the 'great microscope' bought by the Royal Society from Cock, which passed into Hooke's keeping and may have remained with him.

None of these instruments has been well examined, so I can offer no comments on their resolving-power. But it is known that the highest magnifications and resolutions attained at this period—and indeed until the mid-nineteenth century—were provided by the simple microscopes of Antoni van Leeuwenhoek, from about 1674 onwards. Hooke himself was well aware of the virtues of a single refraction (as he misleadingly put it), describing how a tiny glass ball could be fused upon a thread of glass, and ground off smooth. If, he says in *Micrographia*, 'one of these be fixt with a little soft wax against a small needle hole, prick'd through a thin plate of brass, lead, pewter or any other metal, and an object, plac'd very near, be look'd at through it, it will both magnify and make some objects more distinct than any of the great microscopes' (Preface). But the device was, he said, very troublesome to use; as without any means of focus it would be.

I do not think that it could be said that, compared with his contemporaries and immediate successors, Hooke had any conspicuous advantage in the instrument he used. But it should be said at once that Hooke's book, the *Micrographia*, had one splendid advantage, almost one might say an original invention: its engravings. Here again a comparison with Vesalius is inevitable. How can one explain the force—and hence the fame, and today the price—of a really well-illustrated book, one that reveals a whole new visual world? This Hooke's did in a way anticipated by none before him, not even Malpighi, though Malpighi was a far greater biologist than Hooke. It was Hooke who showed the way which his contemporaries and successors—Malpighi, Grew, and Swammerdam—were to follow, though not Leeuwenhoek, who seems never to have mastered the art of sketching a microscopic object. It is curious that although the merits of anatomical illustration had been recognised for a century and a half, the first generation of microscopists remained completely blind to them, until *Micrographia* showed what could be done. In place of the miserable little postage-stamp sketches that had been offered before, Hooke printed thirty-two plates of observations, all of them exquisitely executed, many of them very beautiful, and a few of almost gigantic size. It is hardly astonishing that it was Hooke's elephantine flea that impressed itself on William Blake's imagination or that—just as Vesalius's figures adorn the article on anatomy in Diderot's *Encyclopédie*, so Hooke's served in manuals of natural history until the nineteenth century.

There can be no doubt that a very great deal of the fascination of *Micrographia* has always lain in its illustrations. The design of the book—I mean of the series of microscopical observations, for I leave its other interesting content aside for the moment—is not after all very original: We can digest the whole range of visible objects considered as follows:

Nearly half the observations, and perhaps rather more of the space, were given to insects, the traditional subject for microscopy. In fact very few of Hooke's subjects for examination were quite unprecedented. What is new, rather, is the accuracy and completeness of Hooke's description. An entomologist, or a naturalist, would be able to do much more justice to this than I can, but here is the beginning of Hooke's examination of the eye of a 'large grey drone-fly' (Pl. 2).

First, that the greatest part of the face, nay of the head, was nothing else but two large and protuberant bunches, or prominent parts, ABCDEA, the surface of each of which was all cover'd over, or shap'd into a multitude of small Hemispheres, plac'd in a triagonal order, that being the closest and most compacted, and in that order, rang'd over the whole surface of the eye in very lovely rows, between each of which, as is necessary, were left long and regular trenches, the bottoms of every of which, were perfectly intire and not at all perforated or drill'd through, which I most certainly was assured of, by the regularly reflected Image of certain Objects which I mov'd to and fro between the head and the light. And by examining the Cornea or outward skin, after I had stript it off from the several substances that lay within it, and by looking both upon the inside and against the light.

Next, that of those multitudes of Hemispheres, there were observable two degrees of bigness, the half of them that were lowermost, and look'd toward the ground or their own leggs, namely, CDE, CDE being a pretty deal smaller then the other, namely, ABCE, ABCE, that look'd upward, and side-ways, or foreright, and backward, which variety I have not found in any other small Fly.

Thirdly, that every one of these Hemispheres, as they seem'd to be pretty neer the true shape of a Hemisphere, so was the surface exceeding smooth and regular, reflecting as exact, regular, and perfect an Image of any Object from the surface of them, as a small Ball of Quick-silver of that bigness would do, but nothing neer so vivid, the reflection from these being very languid, much like the reflection from the outside of Water, Glass, Crystal, &c. . . .

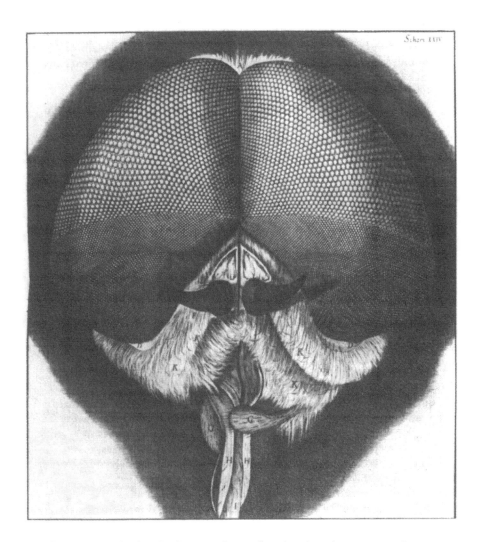

PLATE 2: The head of a grey drone-fly, showing the compound eye.

Fourthly, that these rows were so dispos'd, that there was no quarter visible from his head that there was not some of these Hemispheres directed against; so that a Fly may be truly said to have an eye every way, and to be really circumspect. And it was further observable, that that way where the trunk of his body did hinder his prospect backward, these protuberances were elevated, as it were, above the plain of his shoulders and back, so that he was able to see backwards also over his back (pp. 175–6).

If we continue to read this one section, we shall learn that Hooke was no mere cataloguer of verbal appearance; he is aware of the significance of *comparison* (as when he points out that 'the eyes of Crabs, Lobsters, shrimps and the like . . . are hemispher'd, almost in the same manner as these of flies are'; he is aware that the microscopic anatomist must be concerned with *function*, as when he enters on a consideration of the compound eye as a true seeing eye, raising the question of the combination of the multiple retinal images, and pointing out that there is, as it were, a kind of balance in the fact that while such eyes have so wide a visual angle, the creatures bearing them have little freedom to revolve their heads; which last, in turn, he can as a good Christian naturalist regard as a fair provision of God's balanced design for living things.

Or again, let us consider Hooke's description of an almost heroic investigation of the sting of a nettle, with an experimental demonstration of its action (Pl. 3). Each needle on the stinging-nettle, he says,

consists of two parts very distinct for shape, and differing also in quality from one another. For the part A, is shaped very much like a round Bodkin, from B tapering till it end in a very sharp point; it is of a substance very hard and stiff, exceedingly transparent and cleer, and, as I by many trials certainly found, is hollow from top to bottom. This I found by this Experiment, I had a very convenient Microscope with a single glass which drew about half an inch [that is, a simple microscope magnifying about twenty times], this I had fastened into a little frame, almost like a pair of Spectacles, which I placed before mine eyes, and so holding the leaf of a Nettle at a convenient distance from my eye, I did first, with the thrusting of several of these bristles into my skin, perceive that presently after I had thrust them in I felt the burning pain begin; next I observ'd in divers of them, that upon thrusting my finger against their tops, the Bodkin (if I may so call it) did not in the least bend, but

I could perceive moving up and down within it a certain liquor, which upon thrusting the Bodkin against its basis, or bagg B, I could perceive to rise towards the top, and upon taking away my hand, I could see it again subside, and shrink into the bagg; this I did very often, and saw this Phenomenon as plain as I could ever see a parcel of water ascend and descend in a pipe of Glass. But the basis underneath these Bodkins on which they were fast, were made of a more pliable substance, and looked almost like a little bagg of green Leather, or rather resembled the shape and surface of a wilde Cucumber, or *cucumeris asinini*, and I

PLATE 3: Part of the leaf of a stinging-nettle.

could plainly perceive them to be certain little baggs, bladders, or receptacles full of water, or as I ghess, the liquor of the Plant, which was poisonous, and those small Bodkins were but the Syringe-pipes, or Glyster-pipes, which first made way into the skin, and then served to convey that poisonous juice, upon the pressing of those little baggs, into the interior and sensible parts of the skin, which being so discharg'd, does corrode, or, as it were, burn that part of the skin it touches; and this pain will sometimes last very long, according as the impression is made deeper or stronger (pp. 142–3).

I believe that this clear description is quite new; its mechanistic quality is, of course, typical.

In this 'observation' also we find a characteristic typical of Hooke as a scientist; his inability to stick to one narrow line, his eagerness to follow every analogy, every suggestion of a fertile imagination. For the discussion of nettle-rash leads Hooke to suggest the reason why acids burn, and salt kills fish and frogs, why bathing in mineral waters is a sovereign remedy for so many distempers, especially chronic ones, the virtues of the method of

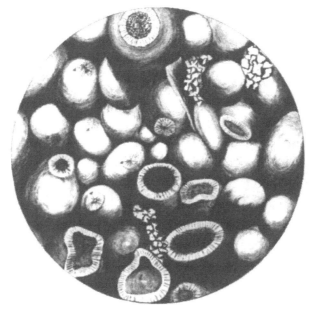

PLATE 4: 'Kettering' stone.

injecting medicines into the veins, and more of the same kind. An observation of snow-crystals conduces, perhaps more plausibly, to a note on crystallography. The microscope is turned to Kettering stone (as Hooke calls it, I think he meant Ketton stone, which is still quarried), which is an oolitic limestone (Pl. 4); here is the description:

It is made up of an innumerable company of small bodies, not all of the same cize or shape, but for the most part, not much differing from a Globular form, nor exceed they one another in Diameter above three or four times; they appear to the eye, like the Cobb or Ovary of a Herring, or some smaller fishes, but for the most part, the particles seem somewhat less, and not so uniform; but their variation from a perfect globular ball, seems to be only by the pressure of the contiguous bals which have a little deprest and protruded those toucht sides inward, and forc'd the other sides as much outwards beyond the limits of a Globe; just as it would happen, if a heap of exactly round Balls of soft Clay were heap'd upon one another; or, as I have often seen a heap of small Globules of Quicksilver, reduc'd to that form by rubbing it much in a glaz'd Vessel, with some slimy or sluggish liquor, such as Spittle, when though the top of the upper Globules be very neer spherical, yet those that are prest upon by others, exactly imitate the forms of these lately mention'd grains (pp. 93–4).

Hooke has no idea of the organic origin of the limestone; he has no notion that his little balls were once shells. Rather, he speculates, 'it seems to me from the structure of it to be generated from some substance once more fluid, and afterwards by degrees growing harder'. But—this is what I wish to point out here—Hooke finds here occasion (in speaking of the discernible pores in bodies, and so of the supposed pores by which light passes through transparent ones) to speak at some length of the medium which fills such pores, or aether. Similarly we find him digressing into the fabrication of a hygrometer from the beard of a wild oat, into the colouring of marble and the making of artificial mother-of-pearl, with much more of the same kind to add to the interest of the reader.

But what of Hooke as a biological microscopist? The first point which is obvious is that a large part of his description is devoted to externals: superficial appearance, skin, scales, feathers, hairs, limbs and feet, eyes, and so on. He has a consistent interest in surface structure; here is an account of 'the curious texture of sea-weeds' (Pl. 5):

the whole surface of the plant is cover'd over with a most curious kind of carv'd work, which consists of a texture much resembling a Honeycomb; for the whole surface on both sides is cover'd over with a multitude of very small holes, being no bigger then so many holes made with the point of a small Pinn, and rang'd in the neatest and most delicate order imaginable, they being plac'd in the manner of a Quincunx, or very much like the rows of the eyes of a Fly, the rows or orders being very regular, which way soever they are observ'd (p. 140).

He finds a certain common texture among sponges, mushrooms, and skin (or rather leather):

As for the skin the Microscope discovers as great a difference between the texture of those several kinds of Animals, as it does between their hairs; but all that I have yet taken notice of, when tann'd or dress'd, are of a Spongie nature, and seem to be constituted of an infinite company of small long fibres or hairs, which look not unlike a heap of Tow or Okum; every of which fibres seem to have been some part of a Muscle, and probably, whil'st the Animal was alive, might have its distinct function, and serve for the contraction and relaxation of the skin, and for the stretching and shrinking of it this or that way (p. 160).

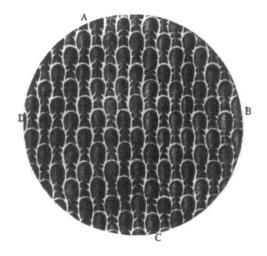

PLATE 5: Structure of the surface of a sea-weed.

And he finds it hard to imagine how an animal's skin could be so elastic unless it possessed this sort of structure.

I need hardly emphasise the fact that seventeenth-century science was in general extremely interested in the structure of things, and the relation of structure to properties and phenomena, so that Hooke's concern is characteristic for his time. In his remarks on feathers he goes farther than he did in what I have just quoted, in actually offering an explanation of a peculiar feature of some living things—their bright coloration—in terms of physical structure and the physics of light; for, he says, such specially glorious, vivid and clear colours like those of silk or feathers always arise when light is refracted through very thin, transparent bodies. We would speak of interference; but more of this later.

There are other traces of the fact that Hooke was a 'mechanical philosopher', one of that Royal Society which found 'some reason to suspect, that those effects of Bodies which have been commonly attributed to qualities, and those confessed to be occult, are performed by the small machines of Nature, which are not to be discerned without these helps, seeming the mere products of motion, figure, and magnitude' (Preface). These are Hooke's own words. A simple illustration of them is his correlation (p. 172) of the note of an insect's buzz or hum with the frequency of vibration of its wings. Then the rising of sap in plants he regards as a purely mechanical phenomenon, as much as the rising up of water to the cisterns from the pumps at London Bridge (p. 162) while 'mushrooms' (moulds) are to Hooke so completely mechanical that he places them immediately above crystals in the great chain of being.

'As far', writes Hooke,

as I have been able to look into the nature of this Primary kind of life and vegetation, I cannot find the least probable argument to perswade me there is any other concurrent cause then such as is purely Mechanical, and that the effects or productions are as necessary upon the concurrence of those causes as that a Ship, when the Sails are hoist up, and the Rudder is set to such a position, should, when the Wind blows, be mov'd in such a way or course to that or t'other place (p. 130).

Naturally enough, Hooke links this view of the humblest kind of plant with the doctrine of spontaneous generation. There can be little doubt that in 1665 Hooke was a believer in this ancient notion, though he grew progressively more dubious. His explanation was that disjointed elements of a life-force might remain in putrefying matter, which could not re-create the *original* organic form but which yet, when (as he put it 'wound up like a spring by the heat of the sun') would reassemble themselves into a new form, a new living thing (p. 134). It is a quasi atomic theory; but as regards insects Hooke seems to speak always of their contributing a 'seed' to the putrefying substance as a 'womb'.

In such a remark, of course, Hooke has moved over altogether from the language of mechanism to the language of Aristotle;

so, perhaps, he does not further increase inconsistency by also speaking of an *anima* in plants, a 'vegetative faculty' or 'forma informans', which 'does contrive all the structures and mechanisms of the constituting body, to make them subservient and useful to the great work or function they are to perform'. Having said that, Hooke moves on smoothly and easily in the next sentence to confess that it is all a matter of divine design, after all: 'And so I guess the pores in wood and other vegetables, in bones and other animal substances, to be so many channels provided by the great and all-wise Creator, for the conveyance of appropriated juices to particular parts' (p. 95).

Whether or not Hooke compiled such passages because he felt his readers expected them, many of his chapters end on just such a pious note; as again when he so far runs counter to his basic philosophy as to write:

> And to conclude, we shall in all things find, that Nature does not only work mechanically, but by such excellent and most compendious, as well as stupendious contrivances, that it were impossible for all the reason in the world to find out any contrivance to do the same thing that should have more convenient properties. And can any be so sottish as to think all those things the productions of chance? Certainly, either their ratiocination must be exceedingly depraved, or they did never attentively consider and contemplate the works of the Almighty (pp. 171–2).

And such care in design, Hooke notes, is as evident 'in these small despicable creatures, Flies and Moths, which we have branded with a name of ignominy, calling them vermin, [as] in those greater and more remarkable animate bodies, Birds' (p. 198).

Now to return to my remark a moment ago, that the *Micrographia* is (so far as it is concerned with microscopy) largely concerned with externals, I must qualify by pointing out that Hooke shows some awareness of insect anatomy. Of the snail, for example, whose 'teeth' he has examined, he says,

> the Animal to which these teeth belong, is a very anomalous creature, and seems of a kind quite distinct from any other terrestrial animal or insect, the anatomy whereof exceedingly differing from what has been hitherto given of it I should have inserted, but that it will be more proper in another place (p. 181).

Schem: XXVI.

Fig: 1.

Fig: 2.

PLATE 6: A blue-bottle fly, and its wing.

This, however, was to remain for Swammerdam. Similarly, Hooke cut open a blue-fly (Pl. 6), seeing 'much beyond my expectation' an abundance of branching of milk-white vessels, no less curious than the branchings of veins and arteries in bigger terrestrial animals, and he asks whether such vessels may be analogous to the *vena porta*, or mesenteric vessels, or rather to blood-vessels, or even to the lacteals; but this, 'not having made sufficient enquiry' he could not determine (p. 184). And a little farther on Hooke remarks of the water-insect or gnat (Pl. 7) that he could see several internal motions, including a peristaltic motion, leading to a comment on the benefit of the microscope

for the discovery of Nature's course in the operations perform'd in Animal bodies, by which we have the opportunity of observing her through these delicate and pellucid teguments of the bodies of insects, acting according to her usual course and way, undisturbed, whereas when we endeavour to pry into her secrets by . . . dissecting and mangling creatures while there is life yet within them, we find her indeed at work,

but put into disorder by the violence offered (p. 186). Not surprisingly, analogical language comes readily to Hooke's pen in passages where he speaks of anatomy or function. He is amazed at the digestive power of the silver-fish, having so violent a fire in its stomach nourished by its lungs (p. 210); insects are assumed to respire, perhaps by passages like the gills of fish (p. 173); and what he calls the 'pores of bodies' (that is, of plants) 'seem to be the channels or pipes through which the *succus nutritius* or natural juices of vegetables are convey'd, and seem to correspond to the veins, arteries and other vessels in sensible creatures' (p. 114). It would seem indubitable that Hooke here, and elsewhere, gave support to the seventeenth-century belief that an analogy would be found between the circulation of animals and the motions of the sap in plants, a notion which prompted many investigations.

This leads me to discuss the passages that are most remarkable of all in *Micrographia* from a biological point of view, which have given grounds for the assertion that Hooke discovered the cell. Before discussing what this might mean, permit me to quote again the salient passages from *Micrographia*, familiar though these are. Observation XVIII is headed: 'Of the Schematisme or Texture of Cork, and of the Cells or Pores of some other such frothy Bodies';

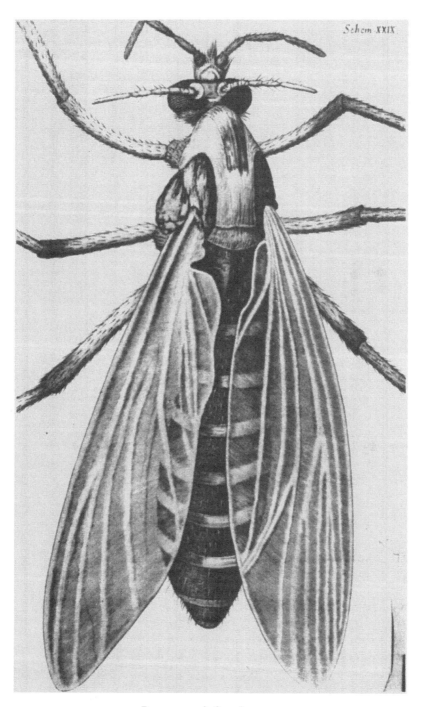

PLATE 7: A female gnat.

and Hooke relates how, examining a thin slice of cork by reflected light (Pl. 8), 'I could exceeding plainly perceive it to be all perforated and porous, much like a honey-comb, but that the pores of it were not regular' (p. 113). The walls of the cells—the word *cell* had of course long been applied to the honey-comb structure— were very thin, while these 'pores or cells were not very deep, but consisted of a great many little Boxes, separated out of one long continued pore by certain diaphragms'. He claims that these were

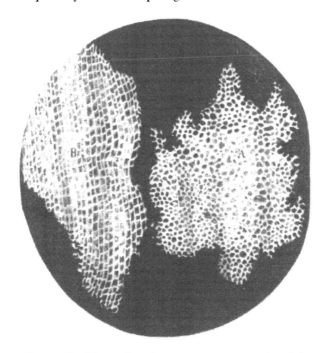

PLATE 8: Slices of cork, transverse cut on the right.

the first microscopical pores ever seen, but he has no idea of their organic significance. Rather, it seems, the cell is subordinate to the 'continued pore', which as we have just seen had in Hooke's view a possible vascular function. What he does perceive is that cellular or porous structure has a physical significance in explaining why such bodies as cork are light and elastic. He is aware that the structure appears variously to the eye as it is cut in the plane or transversely (A in the figure (*right*) is what Hooke calls a transverse cut through the pores); and he discovers also that the microscope yields a similar appearance in the pith of an elder, or almost any other tree, the inner pith or pulp of hollow plant-stalks, and also in the pith of a feather, though he likens this last rather to a congealed froth.

It seems to me that Hooke uses the term 'cell' in a simple descriptive sense, and no more than that; it is not for him a technical term, and therefore it is misleading to say (as Arthur Hughes does)[1] that Hooke used the word 'cell' in our sense of the word. For he nowhere hints that the cell is some sort of a living unit, or that it is the basic structural form of living things. As I understand it, the word 'cell' as employed by biologists today—and naturally the concept undergoes a continuous evolution in meaning—is a conceptual term defined by the content of cell-theory, which was evolved by Schwan and Schleiden. It would be as impossible to understand the meaning of 'cell' without cell-theory as to understand momentum without a science of mechanics. The actual *seeing* of cells is only the first step—though obviously an indispensable first step—to entertaining the *concept* of cell; and it is interesting that in his figure of the nettle-leaf Hooke actually depicts cell-walls which he does not recognise and on which he makes no comment. Similarly the rainbow was observed by countless generations without their divining the nature of light. In so far as Hooke set his visual observation of cellular structure into a theoretical context, it was the context of a vascular physiology; even his failure to discover passages from cell to cell could not convince him that there were 'none such, by which the *succus nutritius* or appropriate juices of vegetables, may pass through them; for in several of those vegetables whilst green I have with my microscope plainly enough discover'd these Cells or Pores fill'd with juices, and by degrees sweating them out' (p. 116).

It is by no means surprising that Hooke should fail to comprehend the nature of the cell, or that he should have been puzzled by the relation between the true cells and what he called the 'pores', the true *vessels* in the plant tissue; neither Grew nor Malpighi was completely successful in clearing up the matter. But we should try to understand what Hooke did see, and what he tried to express. There was, in short, no seventeenth-century cytology.

[1] *A History of Cytology*, Abelard–Schuman, London and New York, 1959, p. 3.

It would be unjust to Hooke's scientific versatility to write of *Micrographia* as though it were wholly (in the words of its sub-title): 'Physiological Descriptions of Minute Bodies made by magnifying Glasses.' In fact the first hundred pages of the book and the last thirty are concerned not at all with biology and little with magnifying glasses. Hooke was much more a physicist and astronomer than he was a biologist (though his lifelong interest in respiration must not be overlooked), and the *Micrographia* is enriched with observations on the strange phenomena of capillary attraction—a reprint of the tract of 1661—on the possible causes of the appearances seen on the surface of the Moon, on various properties of the atmosphere, and on other matters rendering the book hardly less of a landmark in the history of the physical sciences than it is in the history of biology, for it contains the enunciation of at least three theoretical principles of great importance.

The first is Hooke's idea that each primary planet and each satellite of a primary in the solar system has its own gravitation. In the case of the Moon, says Hooke, this seems to be evident from its spherical shape and from

the shape of the superficial parts [which] are, as it were, exactly adapted to suit with such a principle . . . for I could never observe, among all the mountainous or prominent parts of the Moon (whereof there is a huge variety) that any one part of it was plac'd in such a manner, that if there should be a gravitating, or attracting principle in the body of the Moon, it would make that part to fall . . . (p. 245).

This, he appreciated, was a point of some fundamental significance since the generally prevailing Cartesian theory of gravitation attributed it to the Earth alone, and found its cause in the rotation of the terrestrial vortex. If the Moon possessed gravity, and was not at the centre of a vortex (being rather embedded in that of the Earth), then the cause of gravity could not be found in the vortex: 'and therefore some other principle must be thought of, that will agree with the secondary as well as the primary Planets' (p. 246).

Further, *Micrographia* contains Hooke's well-developed thoughts on heat and combustion. They had a long history behind them; the kinetic or mechanical theory of heat went back through Francis Bacon to remote origins, while Hooke's chemical theory of burning derives from the Paracelsan tradition. If what Hooke wrote on these points was not novel to his age, it was clear, forceful and so in turn influential upon the readers of *Micrographia*. 'Heat', he declares, 'is a property of a body arising from the motion or agitation of its parts' (p. 37), and he will have no truck with fire-atoms or figments of that kind, seizing the opportunity to criticise Descartes in this connection for insufficient attention to experiments (p. 46). As for the heat that accompanies fire, this too is a manifestation of a rapid agitation of the particles of matter, whose nature Hooke defined in the following propositions:

First, *that the Air* in which we live, move and breathe, and which encompasses very many, and cherishes most bodies it encompasses, that this Air is the *menstruum*, or universal dissolvent of all *sulphureous* bodies.

Secondly, *that this action* it performs not, till the body be first sufficiently heated . . .

Thirdly, *that this action* of dissolution, produces or generates a very great heat, and that which we call Fire; . . .

Fourthly, *that this action* is perform'd with so great a violence, and does so minutely act, and rapidly agitate the smallest parts of the *combustible* matter, that it produces in the *diaphanous medium* of the Air, the action or pulse of light . . .

Fifthly, *that the dissolution* of sulphureous bodies is made by a substance inherent and mixed with the Air, that is like, if not the very same, with that which is fixt in *Salt-peter*, which by multitudes of Experiments that may be made with *Saltpeter*, will, I think, most evidently be demonstrated (p. 103).

There are twelve principles in all, the burden throughout being the definition of combustion as a normal chemical process in which a soluble body is dissolved in a solvent exothermically. One major branch of seventeenth-century chemical theory taught Hooke that this was in the class of reactions between 'sulphurs' and 'salts' (two of the three great Paracelsan principles) so that he could regard combustibility as a typical 'sulphureous' characteristic and solvent action as a typical characteristic of a 'salt'—here, for good historical and experiential reasons, nitre or saltpetre. Hooke did not, of course, hold that nitre is dispersed in the air, any more than he held that the air itself participated in combustion; rather something unknown in the air was the 'solvent' of combustible bodies which was also 'fixed' in (that is, a component of) saltpetre, a substance whose compound nature had been demonstrated by Robert Boyle. (Boyle's own ideas on combustion resembled Hooke's, but were less dogmatic.) With Boyle, Ent, Mayow, Lower, Willis, Thruston and others—some before him, some later —Hooke was in fact working towards a double objective: the more limited one of comprehending the scientific rôle of the air (and substances contained in the air) in a number of diverse phenomena, and the major, conceptual object of bringing what was known of chemistry and physiology within the scope of the mechanical philosophy. For one cannot understand Hooke's treatment of combustion except by bearing its mechanical articulation constantly in mind.

Finally, this is even more true (if possible!) of Hooke's discussion of light in *Micrographia*. Curiously, the study of optics in relation to the improvement of instruments seems to have concerned him little, but what he did open up here was the investigation of interference-colours, as we should now call them, especially the so-called Newton's rings: 'wheresoever you meet with a transparent body thin enough, that is terminated by reflecting bodies of differing refractions from it, there will be a production of these pleasing and lovely colours' (p. 53). Again, this new cause of colour appeared inexplicable by conventional theories, especially the Cartesian, and Hooke naturally enough tried to embrace it under a general concept of light and particularly to make these new colours analogous to the more familiar ones produced by refraction, as in a prism. As for the former, Hooke declared (pp. 54–6) that light consists in a very short, rapid, vibratory motion of transparent media through which it can travel, by which he meant in effect that light consists of a stream of pulses emanating from the source. Normally light is white; in refraction, however, according to Hooke's study of the geometry, the line of the light-pulse ceases to be normal to the direction of the beam, and it is this disturbance of the simplicity of white light that causes the colours to appear; hence '*Blue is an impression on the Retina of an oblique and confus'd pulse of light, whose weakest part precedes, and whose strongest follows. And, Red is an impression on the Retina of an oblique and confus'd pulse of light, whose strongest part precedes, and whose weakest follows*' (p. 64).

Here was to lie the root of the later difference between Hooke and Isaac Newton, for the latter's experiments led him to assert that white light is compound, not simple, and that for its (normal) composition the whole spectrum of colours is required, not red and blue alone. Newton's famous letter on light of 1672 seems to leave little intact of Hooke's theory concerning refraction and colour. But it must be remembered that this theory in *Micrographia* was above all designed to explain interference-colours, such as those seen in the thin coats and transparent particles or fibres of natural bodies, and that when Newton himself came to deal with such colours (in 1675), he too was compelled to introduce an effect of periodicity which is not dissimilar from Hooke's train of pulses. As Hooke himself was not slow to remark.

In Hooke's time *Micrographia* was as much read for such hypotheses as these, as for its microscopical observations. The devoted band of microscopists was small and did not notably increase; near the end of the century Hooke himself was to lament the fact that only Leeuwenhoek was left to continue such researches. Yet by then microscopy had gone a good deal beyond the level of *Micrographia*; the art had extended to histology, to the anatomy of insects and the structure of plants, to comparative anatomy, to embryology and entomology, to protozoology and even bacteriology; only a part of this great potentiality had been suggested by Hooke himself—who remained above all a naturalist with the microscope; we cannot infer that these developments would have been lacking had *Micrographia* not appeared, though Leeuwenhoek at least may have derived some inspiration from its pictures (he could not read its English text). But we can confidently say two things: first, that the *Micrographia* is one of the major English contributions to seventeenth-century science and that, in particular, it was the first book in the history of science to treat microscopy as a serious aspect of biology. No one could, or can, read *Micrographia* without the sense that here is a powerful, exciting new technique of scientific investigation. Hooke himself continued to use the microscope for many years, though his further, desultory observations have never been collected. Thus it was his historic rôle if not to complete, still to initiate and introduce the great age of classical microscopy. If one may hazard the generalization that the microscope has been the indispensable tool of modern biology, then one might say that it was with *Micrographia* that modern biology began.

XIII

ENGLISH MEDICINE IN THE ROYAL SOCIETY'S CORRESPONDENCE: 1660–1677*

I MUST CONFESS at once that since I have the honour of addressing you this afternoon as your Sydenham Lecturer there is some lack of propriety in speaking of the Royal Society. Thomas Sydenham was never a Fellow, unlike so many of the outstanding English physicians of his time. We may presume I think that this fact implies some disinclination on his part to be elected. All but one of his closest colleagues, Dr. Dewhurst tells us, were Fellows and he was at least casually acquainted with many more. However, it is clear that Sydenham's cast of mind was wholly opposed to the kind of scientific medicine that the Royal Society was endeavouring to foster. 'That anatomie is like to afford any great improvement to the practise of physic, or to assist a man in the findeing out and establishing a true method, I have reason to doubt', he wrote. He had an equal distrust of the attempt to draw lessons from pathology and an even greater aversion to microscopy, which he regarded as a complete waste of time. Thus although I am not aware of Sydenham's actually being hostile to the Royal Society he must have regarded its pursuits as quite vain, if not indeed damaging to the advance of medicine by clinical observation.[1]

As might be expected, therefore, Sydenham's name appears seldom in the early correspondence of the Royal Society; even in the writings and correspondence of his friend Robert Boyle he has left very little trace. Sydenham is mentioned as being convinced, against his initial prejudice, of the reality of the cures wrought by Valentine Greatorix, the Irish stroker.[2] There are records of two copies of *Methodus curandi febres* being despatched by request to the continent.[3] Finally, there is the odd fact that when the Secretary of the Royal Society was thrown into the Tower on suspicion of unpatriotic correspondence with the enemy during the second Dutch war, Sydenham was (he claimed) the only person to speak ill of him; and he declined further acquaintance.[4]

Now this Secretary of the Royal Society was, of course, Henry Oldenburg; he was appointed to that office in the first of the Society's Charters, of 1662, and he continued to be re-elected into it each year until his death, aged about sixty, in 1677. It is upon his correspondence, partly personal but increasingly conducted on behalf of the Royal Society as the years roll by, that this lecture is based. A good deal of it survives, partly very well preserved by the Royal Society itself and much in other libraries and collections. My wife and I have so far edited well over 2,000 letters

---

* The Sydenham Lecture for 1970, given at Apothecaries' Hall, London, 4 November 1970.

[1] Kenneth Dewhurst, *Dr. Thomas Sydenham*, London, 1966, pp. 63–65, 85.
[2] 18 September 1665; A. Rupert Hall and Marie Boas Hall, *The Correspondence of Henry Oldenburg*, Madison, Milwaukee and London, University of Wisconsin Press, 1966–; hereafter cited as *Correspondence*), I, 512–13. Dewhurst, *Sydenham*, 32.
[3] *Correspondence*, III, 367; VI, 209, 286.
[4] *Correspondence*, IV, 80, 95.

Reprinted from *Medical History*, 15 (1971), pp. 111-125, by permission of the publisher. © 1971 the Trustees of the Wellcome Trust, London.

XIII

(by no means all of them complete letters, however) and we have still five more years to go with our work. The names of more than two hundred correspondents are known to us. Of these about eighty were resident in the British Isles, the remainder being widely scattered from India to Iceland and from the Bahamas to Stockholm. Not all in this latter and larger group were foreigners, however; a fair proportion were English travellers, merchants, diplomats and colonists resident overseas.

As a background to this large volume of correspondence I must this afternoon take the Royal Society's interest in medical matters largely for granted. At least a fifth of the Fellowship of the Royal Society was seriously concerned for the progress of medical knowledge at this time. I do not mean naturally that all of these were practising physicians or surgeons; only about a tenth of the Royal Society Fellows were also Fellows of the College of Physicians, for example. But taken in relation to the rather large inert mass of gentry, nobility, officials and lawyers, the medical element in the Royal Society was extremely strong; excepting the mathematicians it was the only coherent professional element. Apart from this common membership, including virtually all the most distinguished English medical men from Allen to Willis (with the exception of Sydenham), the Royal Society had almost no contact with the College of Physicians. It was not concerned with clinical medicine, Sydenham's preoccupation, or with the conduct and education of the profession. Its business was largely with anatomy, physiology, pathology and pharmacology. But of course these subjects are by no means clearly separable from 'experimental medicine' in a looser sense than Claude Bernard's, nor should they be taken as excluding a rather ghoulish taste for 'medical curiosities'. For even educated men of the seventeenth century the two-headed calf and the five-legged sheep had a fascination which the showman of today can—or could—only exploit in naive audiences.

Furthermore, the educated man shared with the common man a far more independent attitude to his body and its vagaries than is common in advanced societies at the present time. This point comes out frequently in the more personal letters of our *Correspondence*, and can be substantiated from many journals and other collections of private letters. For reasons which I think are well known the population of seventeenth-century England was far less healthy than that of England today; malaria of course was endemic, people suffered more or less regularly from fly- and water-borne diseases, they endured torments of toothache, rheumatism, bronchial infections, gout and the consequences of dietary rashness. They suffered too from the teachings of an ancient theory that the body is a machine in unstable equilibrium that requires constant tinkering to be kept in balance. Only the very greatest could employ daily professional attendance. Others dosed themselves. And for the poorest masses of the population there was no choice in the matter. While therefore a non-medical Fellow of the Royal Society, for example, would certainly have profound respect for the opinion of a distinguished physician like Sydenham or Edward Browne, he would also have almost daily recourse to a variety of household or traditional remedies and very probably be willing to experiment with a wide range of chemical preparations from antimony wine to calomel or a few drops of dilute sulphuric acid. Indeed I would guess—but no more—that there is almost an education- or class-distinction here in the later seventeenth century: the popular medicine of the upper classes tended to the inorganic, that of the lower classes to traditional organic sources of medicaments. But one can find enough of the latter even in the writings of Robert Boyle. Thus when we read that John Beale discovered how to improve his reading vision not by choosing empirically a pair of spectacles, but by adapting to his eyes a pair of paper cones which had the effect of greatly narrowing the field, this is a somewhat extraordinary and clumsy but still characteristic act of seventeenth-century self-medication.[5] Assuming that Beale suffered from severe astigmatism, for the correction of which contemporary spectacles were of no avail, one can understand the rationale of his strange expedient.

Between self-medication and the licensed practice of the College of Physicians there were many levels of medical practice. Many excellent provincial physicians who had proceeded to the Doctorate in Medicine were never members of the College, among them for example Henry Power, Malachi Thruston and Nathaniel Highmore. For those who had unqualified but perhaps real experience of medical practice the M.D. itself could be obtained by a perfunctory visit to a foreign university such as Leiden, or by procuring special dispensations. One who took the former course was the celebrated plant anatomist Nehemiah Grew, who figures a good deal in our *Correspondence* and who succeeded Henry Oldenburg in his Secretaryship; Grew matriculated at Leiden on 6 July 1671 and proceeded M.D. on 14 July with a thesis *De liquore nervosa* which he had obviously brought in his pocket. Another was Nathaniel Fairfax, of whom I shall say more in a moment, who had matriculated at Leiden on 21 June of the previous year and proceeded M.D. twelve days later. There were some eighteen Fellows of the Royal Society who practised medicine in or out of London without the licence of the College of Physicians. Some of these at least had no formal medical degree at all, John Locke, Shaftesbury's physician, being one of them. Another notorious unqualified practitioner was Henry Stubbe, ex-assistant to Bodley's Librarian, physician at Stratford-upon-Avon and Bath, virulent opponent of the Royal Society, and self-styled champion of the College of Physicians. Other men like Locke and of course Robert Boyle who had studied medicine profoundly used their knowledge to advise friends and relations without entering into normal practice. It was not unheard of, I believe, for beneficed clergymen to do the same and certainly in the early 1660s many ejected ministers took to medicine.

Nathaniel Fairfax (1637–90), many of whose letters we have printed in Vols. III–V of our *Correspondence*, was one of these. He was an M.A. of Corpus Christi College, Cambridge, and was therefore an educated man though one would hardly think so from his clumsy English and appalling Latin. He practised in Suffolk and was in some way a protégé of Dr. Thomas Browne. His freedom with anatomical truths that had been held since the time of Galen at least is illustrated by an amusing incident. Fairfax narrates the case of Goodwife Eliot of Mendlesham who passed by urine one of two caliver bullets which she had been induced by a neighbour to swallow for relief of her 'torment of the bowels'.[6] (This heroic measure or the exhibition

---

[5] See Beale's letters to Oldenburg in Vols. IV and V of the *Correspondence*, extracts from which were printed in the *Philosophical Transactions*.
[6] *Correspondence*, V, 47–49; Fairfax to Oldenburg, 18 September 1668. Dates throughout are in Old Style.

of a massive dose of mercury was a common last resort in cases of the iliac passion.[7])
Fairfax goes on:

> The main use that I would make of the instance (if it be worth mentioning) is to strengthen a
> suspicion that I have a long time had, of some other passage from the stomach to the bladder
> besides what anatomists have hitherto given account of. For that this bullet never came at the
> ureters through the veins, arteries, nerves or lymphducts (the only vessels that can be charged
> with it) is, I think, beyond dispute.[8]

And so forth; there is a good deal more. Now Oldenburg printed part of this letter
in the *Philosophical Transactions*,[9] as he usually did, for Fairfax is rich in 'curiosity'.
The curiosity of Fairfax's anatomical speculations did not escape the experienced
London anatomists, including Walter Charleton, and some months later Fairfax
wrote Oldenburg a humble exculpation for what was 'so hideously beyond dispute that
it was very unanatomical and a sorry weakness to hint it so'.[10] It is interesting that at
the same time he apologizes for mentioning his prescription of 'patent medicines'—
Lockyer's and Matthew's pills—'hereby giving occasion to strengthen the scandal
raised on the Society as too friendly to quacks and yourself [Oldenburg] as corresponding with a declared one'. Lest anyone should think that Fairfax was unique in
doubting one of the most famous conclusions of Greek experimental physiology, I
must add that exactly the same scepticism was shown by Pierre Daniel Huet, leading
light of the Scientific Society at Caen, later tutor to the Dauphin and Bishop of
Avranches.[11] Huet claimed that having ligatured the ureters of a dog the bladder
nevertheless filled with urine. In his reply Oldenburg was able to assure him that
when the ureters were effectively blocked in experiments by Dr. Edmund King no
urine entered the bladder:

> Our most learned physicians are convinced [he wrote] that there is no passage to the bladder
> except through the ureters in view of all the investigations which they say have been made
> with the greatest possible care to discover such a passage. To this they add that having thought
> it over carefully, they see no need of there being any other, considering the wonderfully rapid
> circulation of the blood and other fluids through the body and the swift fermentation and percola-
> tion of the same in the organs through which they pass.[12]

No doubt this last sentence goes to the root of the matter. It was the very swiftness
of the body's action in assimilating and distributing ingesta which made physiology
as it was taught by the sicentists of 1670 seem dubious to naive minds.

But to return briefly to Fairfax. Like many physicians he had a taste for natural
history—unsophisticated of course in him—and like almost everybody he was
excessively preoccupied with the poisonous attributes of spiders and toads. And this
gives me occasion to mention here that we have in the *Correspondence* besides
Fairfax's credulity the entirely rational and modern-sounding story of the attempt
by Tommaso Cornelio of Naples to discredit the extraordinary phenomena universally

---

[7] For the views of the physicians Allen, Clarke, Ent and Goddard on this use of mercury see
Correspondence, VI, 25–30; Oldenburg to Segni, 10 June 1669.
[8] *Correspondence*, V, 48; I have modernized spelling and punctuation in this and subsequent
quotations.
[9] See no. 40 (19 October 1668), 803–5.
[10] *Correspondence*, V, 505; Fairfax to Oldenburg, 30 April 1669.
[11] Ibid., VII, 206–9; Huet to Oldenburg, 20 October 1670.
[12] Ibid., VII, 394–97; Oldenburg to Huet, 16 January 1670/71.

English Medicine in the Royal Society's Correspondence: 1660-1677

linked with the tarantula of southern Italy. Like Redi's work on spontaneous generation, it is a minor episode in the progress of enlightenment. Unfortunately even such capable naturalists as Martin Lister were not quite able to dispel such traditional fables.[13] Fairfax also relates many medical histories, which give some notion of the byways of seventeenth-century medical practice; he describes a case of Siamese twins,[14] and another of hermaphroditism;[15] he tells of a strange case of attempted suicide by fasting, which was ended by the lady riding to Ipswich and eating buttered peas and a pint of strawberries ('which she told me made her sick'—not surprisingly);[16] and he reminds us of the rarity of the survival of multiple births under primitive medical conditions:

> Goodwife Rivers of Ipswich, a young woman, at her fifth conception last summer brought forth, with good labour, three living infants, two sons [and] one daughter, at a birth; all of which sucked of the mother and throve well for a week, but then fell into a wane and one after the other died within the month.

In another class, less learned but perhaps more skilled in their own way, were the provincial apothecaries and chemists. It would be interesting to try to form a picture of the magnitude and quality of their contribution to the health of seventeenth-century England. Here is a minute morsel. In a letter written from Durham a schoolmaster named Peter Nelson (who was acquainted with a man formerly in Boyle's service and also with John Webster, author of *Metallographia*) writes of the 'physicians' in that city:[17]

> first Dr. Wilson an ingenious man and a good scholar, for the most part a Methodist . . . a diligent peruser of your monthly [*Philosophical Transactions*]
> The next is Mr. Nicholson, a serious young man, and well educated, inclinable to chemistry but no great practitioner.
> Next I reckon one Mr. Selbie, who hath been as much beholding to fortune as education, but a civil man and well-spoken, has been born under a good thriving aspects and is fallen into a notable way of practise; he works sometimes in the fire and has a small laboratory in which he makes some of the medicines he uses.
> There is one Mr. Dancy, a man that is thought to have good skill and hath done divers handsome cures, but hath not had the luck to thrive and is not therefore so considerable as possibly he might have been.

Mutatis mutandis not unreminiscent of *Middlemarch*. Four practitioners of a sort in a remote and lightly populated region where humble coal-miners were already far more numerous than gentry and wealthy bourgeois does not seem an inadequate provision. Unfortunately I can say something more of only one of these four, since Dr. Wilson cannot be definitely identified as deserving the prefix, though local research might uncover more information. Robert Selbie himself also wrote to Oldenburg to give 'a general account of some more than ordinary success in my practice'. At the risk of over-quotation I must try to convey the flavour of this letter:

> The maladies I have observed (of those most feral and truculent) are your dropsies, convulsions and convulsive motions; as your emprostotonos and opistotonos which are much more terrible than a complete convulsion. A diabetes, and lately in twenty days' time a young man of a Scorbutic palsy. Consumptions as also the rickets. But especially an old gentlewoman of this town

[13] See *Correspondence*, Vols. VII and VIII, Index, s.v. 'Spiders'.
[14] Ibid., III, 491–97; Fairfax to Oldenburg, 28 September 1667.
[15] Ibid., V, 376–79; Fairfax to Oldenburg, 4 February 1668/69.
[16] Ibid., VI, 67–71; Fairfax to Oldenburg, 28 June 1669.
[17] Ibid., VII, 326–27; Nelson to Oldenburg, 15 December 1670.

past eighty years of a confirmed dropsy. And if I may further speak without ostentation equally successful in whatever distemper occurs with my fellow practitioners . . .As to medicines that I have used for this six or seven years are these principally, videlicet dissoluble magistry of coral prepared with your acetum philosophorum; your elixir proprietatis after several other menstruums made with spiritum vini subtilissimum et spiritum salis well incorporated by often drawing over; ens veneris in which sublimation I always receive a spirit first of good use; spiritum cornus cervi, antimonium diaphoreticum, salem antimonium. Also a pleasant tinctura antimonii imbodied with tartar which I use upon all occasion where vomits are required; volatile salt of tartar converted into a liquor with which I prepare several cathartics, with many others. My furnaces being for the most part constantly employed.[18]

This is rather Ben Jonson than George Eliot; not that Selbie's letter is at all abnormal by the standards of seventeenth-century pharmaceutical chemistry. It shows how far the teaching of Béguin, Van Helmont and Zwelfer had penetrated. No wonder one reads somewhere of a patient lying helplessly on the floor in an abandon of purgation both upwards and downwards.

Perhaps I ought to add a word on the general issue of Galenicals and chemicals, but really from our correspondence there is little to say. After the Plague the issue was no longer a live one. You will know that an attempt was made, after a good deal of controversy, to found a Society of Chemical Physicians in 1665, and that the attempt came to nothing. The failure was not significant. By now too much powerful influence—that of Boyle, Goddard and Willis, for example—was in favour of iatrochemistry for it to be other than respectable. There are a few trifling allusions to Stubbe's attempt to revive the debate by his attacks on George Thompson in 1670, which naturally were hardly grateful to the Royal Society, but in general chemical remedies seem to be accepted as perfectly normal, as they were by Peter Nelson, for example.

Other novel elements in therapy that appear are the practices of hot bathing and taking spa waters. The egregious Joseph Glanvill contributed a dull letter on the hot baths at Bath which was partly printed in the *Philosophical Transactions*.[19] There are a number of references to the book by Henricus ab Heer, *Spadacrene. Hoc est fons Spadanus accuratissime descriptus,* published in 1647, which seems to have set going the whole spa water movement; and about 1669–70 there was a special interest in the composition of spa waters and the chemical theory behind their effectiveness. Daniel Foote in a rather interesting letter tried to base such a theory on the acid alkali dualism of Otto Tachenius.[20] Robert Wittie (or Witty) and his writings naturally came up for discussion, much to the satisfaction of this defender of Scarborough Spa. Again, I need not enter into his controversy with William Simpson as Dr. Poynter has already done so.[21] Wittie wrote:[22] 'I must ever acknowledge my deep obligations to those noble gentlemen of the Royal Society for their candour and condescension to take notice of my weak endeavour, whom I wish I were able or worthy to serve in any thing.'

[18] Ibid., VII, 532–33; Selbie to Oldenburg, late March 1671.
[19] Ibid., VI, 47–51; Glanvill to Oldenburg, 16 June 1669; *Phil. Trans.*, no 49 (19 July 1669), 977–82.
[20] *Correspondence*, VI, 275–78; Foote to Oldenburg, 11 October 1669.
[21] F. N. L. Poynter, 'A Seventeenth Century Medical Controversy: Robert Witty versus William Simpson', in *Science, Medicine and History*, ed. E. Ashworth Underwood, London, Oxford University Press, 1953, II, 72–81.
[22] *Correspondence*, VII, 52; Wittie to Oldenburg, 4 July 1670. For his earlier letter see VI, 605–13. Wittie probably did not realise that the *Philosophical Transactions* were wholly controlled by Oldenburg.

English Medicine in the Royal Society's Correspondence: 1660-1677

I judge that in this case as in many others the Royal Society and the *Philosophical Transactions* (which in large measure was the contemporary published version of Oldenburg's correspondence), taken together, threw their weight on the side of innovation, properly controlled by analysis and experiment. The Royal Society rarely spoke for conservatism.

We see this in the episode I come to next. For it is time for me to move from the grass-roots of medical practice to the experimental and scientific aspects. The most dramatic of all these is, obviously, the story of transfusion. It has been told so many times that I may be brief. Its origins lie in new ideas about animal poisons, which in turn are not unrelated to the discovery of the swift circulation of the blood. By about 1660 many physicians and pharmacologists—but not all[23]—believed that when a snake (for example) bites, or an insect stings, a fluid poison is injected into the body which, carried by the blood, very quickly causes illness or death.[24] This, of course, contrasts with the alternative theory (seemingly supported by the puzzling contrast between the bites of healthy and 'mad' dogs) that poisoning was the result of the creature's rage, affecting its spirits. This new rationalist interpretation of poisoning immediately provokes a comparison with the action of ingested poisons. If, then, the same poisonous substances could be both ingested and injected—and experiments were performed to show that this was so, and that injected poisons acted more quickly—might it not also be true that health-giving rather than poisonous substances could work very well if injected into the body rather than ingested? Guided by this analogy Christopher Wren proposed and with Boyle carried out such injection experiments on animals not later than 1658. Independently, at a time when this early English initiative had lost impetus, the same notion occurred to Johann Daniel Major of Hamburg who published a little book about it in 1664 which he sent to the Royal Society.[25] And this in turn was followed quickly by the *Clysmatica nova* of Johann Sigismund Elsholtz.[26] We find Oldenburg assuring the continent that the English were first in the field by several years;[27] in fact, some further work on injection had been done in the interval by Timothy Clarke, a (to my mind) unlikeable royal physician, on which he had reported to the Royal Society on 16 September 1663.[28]

It was in the discussion of Clarke's paper that the idea of transfusing blood from an animal to another by means of a pipe was first mentioned. Not all those present could see the medical utility of the procedure, yet the naive logic is obvious enough: if the object of injected medicines was to purify the blood in a sick body, why not achieve the same object more directly by transferring good blood from a healthy animal? Naturally the idea that there might be crucial idiosyncratic differences between the blood of different individuals of the same species and still more between

[23] See *Correspondence*, Vol. VIII, Letters 1940 and 1944; Charas to Oldenburg, 28 and 30 March 1672; Vol. IX, Letter 2037, Platt to Oldenburg, 27 July 1672 and Letter 2038, Magalotti to Oldenburg, 28 July 1672.

[24] Redi held this view in print in 1664; compare Hooke and Merrett on the viper, 26 October and 2 November 1664, in Thomas Birch, *History of the Royal Society*, London, 1756, I, 479, 481, and the letters cited in the previous note. See also M. P. Earles, *Annals of Science*, 1963, **19**, 241 ff.

[25] J. D. Major, *Prodromus inventae a se chirurgiae infusoriae*, Leipzig, 1664; see *Correspondence*, II, 334-38.

[26] Ibid., II, 580.

[27] Ibid., II, 379-80; Oldenburg to Major, 11 March 1664/5.

[28] Ibid., II, 380, note 1; IV, 6, 363-4 and 368, note 8; Birch, *History*, I, 303. No copy of Clarke's paper survives, apparently; if it could be found it would be of extreme interest.

members of different species occurred to no one.[29] Long medical tradition emphasized distinctions between the temperaments or constitutions of individuals—essentially psychological or at least non-mechanistic characteristics—but not between their physiological mechanisms. As you know, even anatomists long found it difficult to accommodate the fact that the simple topographical anatomical structure of all individuals of the same species is not absolutely identical.

Here I must really cut short as we move on to very familiar ground, though our *Correspondence* adds many new details to the story. The English experiments were held up by technical difficulties, by Clarke's sluggishness, by the Plague of 1665, and by Wren's transfer of his allegiance from science to architecture. It is well known that the French surgeon Jean Denis first took the rash step of attempting to transfuse blood from a lamb into a human patient on 6 June 1667. Meanwhile the English had achieved seeming success in transfusion between animals of different species.[30] We have published in our *Correspondence of Henry Oldenburg* the whole of Denis' correspondence with the Royal Society concerning this event and its sequelae; and the inevitably many allusions to them in other letters. It is often supposed that the death of a patient after he had suffered a transfusion administered by Denis led to an official termination of such experiments in Paris. As Denis was at pains to point out to the Royal Society, he was at the subsequent inquiry found guiltless of causing the patient's death, and the official judgment was that transfusions should only be performed under the direction of the Medical Faculty of Paris.[31] In practice it must be admitted the result was the same. Nevertheless, Denis persisted in the defence of the transfusion operation for some time, and interest in injection and transfusion lingered on the continent for many years. There is a thorough and intelligent dissertation on the injection of fluids into the veins in a French version of Michael Ettmüller's *Nouvelle Pratique de Chirurgie* at least as late as 1691, and a fairly well-known engraving of a transfusion scene (lamb donor, human recipient) continued to appear at least until 1705.[32] After all, when a patient was pretty sick death at the surgeon's or physician's hands through the more extreme forms of treatment was no rarity; it was common enough among those cut for the stone. The Hippocratic saying, desperate cases justify desperate remedies, could be as well applied to transfusion as to lithotomy or the almost fabulous Caesarian section;[33] the rational way was to experiment carefully.[34] However, the English gave the business up, and we hear no more of it in our *Correspondence* after 1670 save as a subject of priority wrangles.[35] In reply to a rather full and sensible discussion of the general problems of injection therapy written him by the Venetian physician Francisco Travagino,[36] Oldenburg wrote (I translate):[37]

[29] For Denis' discussion of differences in blood between individuals and his conclusion that these are no more significant than the differences between the various sorts of food that enter the bloodstream, see his *Letter to Montmor* of 25 June 1667 [N.S.] and *Correspondence*, III, 480–83.
[30] See, besides *Correspondence*, III, *passim*, *Phil. Trans.*, no. 25 (6 May 1667).
[31] *Correspondence*, IV, 372–87; Denis' printed letter to Oldenburg of 5 May 1668.
[32] Reproduced in *Correspondence*, IV, Plate I.
[33] See ibid., VI, 362–63; Oldenburg to Rudbeck, 9 December 1669; VII, 95–96, Oldenburg to Rudbeck, 23 July 1670.
[34] See ibid., V, 480–83; Martel to Oldenburg, 11 April 1669.
[35] E.g. ibid., VII, 561, 564; Wallis to Oldenburg, 7 April 1671. Compare II, 484.
[36] Ibid., VI, 492–500; Travagino to Oldenburg, 13 February 1669/70.
[37] Ibid., VII, 557–9; Oldenburg to Travagino, 14 March 1669/70.

English Medicine in the Royal Society's Correspondence: 1660–1677

Your very learned and skilful remarks about the fluids or spirits to be mixed with the human blood by means of injection-surgery were very welcome to many of our Fellows, who allow as you do that the task of transferring injections of this sort with good success into the art of healing men is full of hazards. Meanwhile, if it shall prove possible to arrive at a more complete knowledge of both the human blood and the spagyrical fluids through all kinds of observations upon both, and through repeated experiments performed properly and faithfully by wise persons, then in my opinion it will be by no means necessary to despair of the outstanding usefulness of that kind of surgery.

So far as it goes this statement is unexceptionable. The Royal Society had clearly realized that, attractive as the potentialities of this new branch, or rather new branches, of medicine might be, they could not be developed safely until a great deal more basic science was known. Only we today can appreciate how much had, indeed, still to be learned.

One element is missing from this appreciation of the situation, however, on which I must dwell an instant: I mean, of course, that Oldenburg (and I presume his medical colleagues) does not also see any necessity for understanding the basic causes of diseases. Supposing the physician already in possession of a very complete knowledge of the constitution, functions and pathology of the blood, as also of the effects of a wide range of injected medicaments, does it not seem necessary that he should also appreciate the causes of pathological states in the blood or indeed elsewhere in the body? One might have thought that experiments with a totally new form of therapy would have stimulated fresh thoughts about the targets, so to speak, at which the new weapons were to be directed: apparently it did not. Either because the physicians of this time were quite unaware of the difference between treating the symptoms of the disease and treating its causes, or because they regarded the causes of disease as sufficiently well known (bad habits, a weak or unbalanced constitution, improper foods and so on) there seems to be in our *Correspondence* amid many case-histories, a multitude of pathological reports, and recurrent discussion of the value and preparation of a great number of drugs, little interest in the origins or communication of disease. Perhaps I may quote the only two that come readily to hand; a German physician, Michael Behm, writes in 1667:[38]

I have certainly observed that gout and arthritis are caused when the urinous corruption is not separated from the blood by the kidneys and by sweating but is circulated about the body with it, adhering to the colder ligaments around the joints; there it causes rather acute pain and even swellings by the accretion of salt, or because its viscosity occasion stiffness and calcification.

And he goes on to doubt the theory of de le Boë Sylvius that some diseases arise from the effervescence of the acid pancreatic juice with the bile in the duodenum.

In Behm's two examples—the one positive, the other negative—disease (or rather its symptoms) is assigned to physiological malfunction; the kidneys are disordered, or the alchemy of digestion has gone astray. Fair enough. To seventeenth-century medicine it was indeed obvious that the study of normal and pathological physiology is basic to a rational therapy. Did physicians then dismiss as hopeless or unnecessary any attempt to go a step farther back and ask why this patient's kidneys (but not all) ceased to do their work properly? I find this question puzzling.

[38] Ibid., III, 573, 575; Behm to Hevelius, 1 November 1667.

My second quotation is more straightforward. Jean Denis, the transfusion experimenter, replies to Oldenburg:[39]

> You wrote to me in your last that it would be of great convenience in putting transfusion to the test if certain diseases could be conveyed to animals artificially. Upon this I will impart a fact to you that came up here some time ago. A wet-nurse who was attacked by the smallpox communicated it to her child; when this was noticed the child was removed from her and suckled by a goat, which the child sucked at every day. This goat contracted the disease. For, some time afterwards, when milk from this goat was served to two different people who had been ordered to take goat's milk, both of them contracted the smallpox only by drinking its milk each morning.

I do not seek to account for this tale. But it is at least an observation on the *communication* of an identifiable disease from person to person in a rather precise way, which is rare, at least in the medicine of our *Correspondence*.

Reverting to the quotation from Oldenburg's letter to Travagino that I read a few moments ago, it is scarcely necessary to assert that all our evidence indicates a confident belief in the present and future progress of medicine. To hasten this progress was one of the great objects of the organized, co-operative investigation into Nature that Oldenburg tirelessly advocated in his letters of exhortation; it was *the* most obvious way in which this investigation was useful to mankind, more than the satisfaction of intellectual curiosity. As to the method of this investigation a significant geographical division between Oldenburg's correspondents (and the publications with which they were associated) may be noted. In England, France, Holland and Italy the scientific movement was, of course, composed of mathematicians and astronomers as well as physicians and iatrochemists; the two groups were members of the same societies and read the same journals. In eastern and northern Europe at this time the movement was almost entirely composed of physicians, as you may easily see by glancing at the *Miscellanea curiosa* or the *Acta Hafniensia*. (The *Acta Eruditorum* did not yet exist.) Moreover, while the scientific physicians of the west and south devoted their efforts to basic biological science—to comparative anatomy, physiology, microscopy, medical chemistry, embryology and so forth—those of the east and north were largely preoccupied with the rarities of clinical practice and pharmacology. The mysterious iatrochemistry of which J. J. Becher was the archpriest was much in vogue, while men like Malpighi, Redi, Bellini, De Graaf, Swammerdam, Croone, Willis, Lower or Lister were rare indeed beyond the Rhine.

I make this doubtless exaggerated and rash generalization simply to justify my contention that, although one may discover in the Royal Society's correspondence a farrago of medical curiosities and chemical wonder-drugs, such evidence of triviality or misguided enthusiasm among English medical practitioners of all levels of sophistication is not important when viewed in the context of the age and when set against the mass of learned publication devoted to basic science.

I suppose I could make some kind of a case for maintaining that natural history is the most basic of all sciences, even medical sciences, since it describes the inescapable environment of human life which (in seventeenth-century terms at least) not only occasions many ills but provides the cures for them. Certainly Oldenburg was fond of proclaiming that a true natural history is the *sine qua non* of sound

---

[39] However, the two creators of the science of plant anatomy, Malpighi and Grew, were both physicians by profession.

English Medicine in the Royal Society's Correspondence: 1660–1677

natural philosophy. It was routine in the Royal Society's correspondence to ask in this way about the medical experience of any region of the globe in relation to climate, topography and so on, seeking (somewhat ineffectually, it must be admitted) to assemble the elements of a medical geography. Moreover, although the age of geographical discovery was nearly two centuries old, the lure of the exotic was strong upon the philosophical physician. And that it should be so was not irrational. The botany and dietetic properties of non-European food-plants were almost unknown in England and elsewhere; even the potato and maize were still rare; cassava, yams, cocoa-palm, tropical nuts and fruits were hardly more than names and crabbed woodcuts; even the tea and coffee plants had not yet reached European herbaria. Though systematic botany was passing from the hands of physicians to the care of non-medical specialists like Ray and Tournefort[39] the belief was still general that a more thorough knowledge of plants, both European and exotic, would yield the discovery of many useful materials. This belief is very evident in the letters of Martin Lister, for example. With so many hitherto unvisited regions of the globe open to European commerce or settlement, there was a strong desire to learn more precise facts to substantiate the inadequate accounts of exotic drug plants that had already reached Europe. So, for instance, the Hamburg physician Martin Vogel continually urged Oldenburg to exploit English trade links with the East and with North America to obtain botanical specimens. If the boasted virtues of guaiacum had proved fraudulent, the physiological effects of Jesuits' bark, not to say tobacco, had proved perfectly real; and (as we now know all too well in some cases) so are those of the Indian *bhang* (marijuana), cocculus Indicus, or various species of *Hyoscyamus* and *Datura* about which we find Vogel inquisitive.[40]

As I hinted at the beginning, Oldenburg did establish frail lines of communication with Iceland, the Bahamas, New England (whence John Winthrop sent various parcels of natural curiosities and Indian craft to the Royal Society) and with British agents in the Near and Far East, but whatever geographical or ethnological fruits these secured him, they brought in little of medical interest. Nor were his pressing inquiries of the distinguished band of Oxford oriental scholars any more profitable. But the most extraordinary product of his efforts in this direction were the 'Inquiries for Brazil' which he concocted in August 1671.[41] The world-wide missionary activities of the Society of Jesus and notably the studious activities of Father Matteo Ricci in China were of course known in England, if not exactly well understood; Oldenburg long had it in mind to exploit these far-flung Jesuits as sources of scientific intelligence. Finally, through an English merchant in Lisbon named Thomas Hill, probably a younger brother of Abraham Hill, the Royal Society's treasurer, Oldenburg was promised communication with a learned and intellectually active Jesuit father at Bahia (that is, Salvador, then the capital city of Brazil). These inquiries were destined for his attention; if they had ever received adequate attention (which so far as we know at present they did not) they would have required the work of a lifetime. They are based on the books devoted to the natural history and medicine of the Indies published by Wilhelm Piso and Georg Marggraf in 1648, which are indeed of funda-

[40] See, for example, Vol. IX, Letter 2048, Vogel to Oldenburg, 13 August 1672.
[41] Vol. VIII, Letters 1747 (Hill to Oldenburg, 13 July 1671); 1780 and 1780a (Oldenburg to Hill, 19 August 1671).

mental historical importance to this day. Apart from many inquiries relating to ethnology and zoology (the skunk, the porcupine, the rhea, the humming-bird, the anaconda and all South American fishes were complete mysteries to Europeans) there is much of medical interest. Oldenburg naturally used the Indian plant names which he found in his sources. He inquires not only about food plants and dye-stuffs but about plants of the pilocarpus group (yielding pilocarpine), *Operculina macrocarpa,* a source of jalap, sarsaparilla, copaiba, *Pithecolobium avaremotemo* (the 'Brazilian astringent bark' of nineteenth-century pharmacy), nux vomica and other *Strychnos* species, ipecacuanha, and many more. While he can only think of the Indians of Brazil as savages, Oldenburg clearly believes that these remote primitives possess a potent herbal medicine and a mastery of poisons unknown to Europeans. Placed by God in a region of the world which was clothed by plants completely different from those of Europe and the Near East, plants possessed of different and perhaps more powerful virtues, they have learned how to convert these virtues to human use and misuse. There is more than a hint of the concept of the Noble Savage, with medical overtones:

> Is it true that the natives grow to puberty early and age slowly, and then without loss of hair or teeth? Do Brazilian mothers laugh at our way of dressing and bringing up children, which, they say, impedes the perspiration and causes much catarrh? Are no squinting, purblind, lame or hunchbacked persons found among them because infants are never swathed in linen or bound up in swaddling clothes, but are frequently washed with cold water? Are the Brazilians rarely affected with illhealth? Do the more thoughtful among them attribute their good health and longevity to these causes, namely, that they have strength from birth, and are exposed to the excellent calmness and constancy of the air and winds, as also, that they hardly know what care is, what is heaviness of heart or bodily delights; that they always enjoy the same dress and diet, and those of the simplest? . . . Do the natives mostly employ as their usual healthful drink the very clear water of their rivers and springs, which even when drunk copiously cause no wind nor pains in the belly or abdomen, and far from weakening the stomach fortifies it remarkably?

I confess to complete ignorance of the long and complex story of the introduction of exotics into European medicine, nor could I say whether primitive simplicities have influenced medical thinking; but Oldenburg's tremendous epistle is at least worth noting in the former context.

I have only a few minutes left in which to refer briefly to anatomy, physiology and embryology as they appear in the Royal Society's correspondence. As I remarked before, we now deal with correspondence between or concerning the authors of well-known books. You will not be surprised to learn that we are publishing the full correspondence of Malpighi and Oldenburg, or Oldenburg and Regnier De Graaf. In the late 1660s and 1670s virtually all scientific communication between England and the continent passed through Oldenburg's hands and was known to the Royal Society. I need hardly remind you that Swammerdam dedicated a part of his work on the human uterus to the Society, or that (under Oldenburg's management) the Society published both of Malpighi's embryological studies. In some cases our correspondence adds further details about these various exchanges and corrects established errors. In others it contains opinions about the significance of the work of such English and continental medical scientists.

I will venture on two general comments. Firstly, these men did not nearly so often as one might idealistically imagine visualize their scientific work as related to their

English Medicine in the Royal Society's Correspondence: 1660–1677

medical practice. All too often—and we can judge how human this is—the daily business of the physician appeared a mere drudgery, necessary to support life and family, which merely impeded the urgent task of scientific research. Vogel, Malpighi, Thomas Bartholin, Rudbeck, all voice this complaint. Swammerdam actually gave up medicine altogether in order to pursue microscopy. English physicians do not so complain—whether because the already greater wealth of English upper-class society in effect gave them greater leisure, or for other reasons, I do not know.

Secondly, one discovers a virulence of national pride. In the Germans it took the form of emulation of France, Britain and Italy and envy of the rich patronage which they knew the Italians and French and believed the English to have enjoyed. The French enjoyed a calm sense of superiority in philosophy and civilization over the rest of Europe; though they admired candidly English experimental achievements in medical science they were not above strident (and sometimes doubtful) claims of priority. The English were extremely vociferous in asserting their own discoveries. 'Philpatris comme tous les Anglois', Huygens once remarked. Swammerdam never wrote more welcome words than when he addressed the Royal Society in sending his presentation copy of *Miraculum naturae*:[42]

> I am not unaware how fate has brought it about that, just as Christendom owes no slight advancement of its religion to the English people, so in these recent very difficult times there was discovered among them the method of setting aside the empty disputations of the Schoolmen and of placing the useful arts and sciences on a solid basis; And as this is not the least part of Britain's glory, so it is the reason why no one dares or ought to dare, in matters of natural philosophy, to resort to any other tribunal than the Royal Society.

We shall fail to understand scientific communication in the seventeenth century if we fail to take the operation of this intense feeling of nationalism into account. If on the one hand, true native roots might be ascribed to a seemingly foreign innovation, it might flourish. Thus, although a German like Ettmüller will fairly allow Wren's priority in injection therapy, he and other Germans found the *effective* origin of this innovation (at least for German medicine) in Daniel Major's *Prodromus;* they took it up—rather in theory than in practice—as a German technique. Similarly Denis was able to (so to speak) naturalize transfusion in France by going from experiments on animals to the bold step, before which others had hesitated, of experiments on man, which the English could then only tamely imitate. It was of little use for Timothy Clarke to write a verbose statement of the English priority in injection and transfusion, and complain of the way in which the wily foreigner grasped for himself discoveries in anatomy and medicine first made by Englishmen.[43] The positive nationalism of the foreigners had led to further advances (if such, for the sake of argument, they may be termed). On the other hand *passive* nationalism, hugging a little bit of trivial priority to one's national pride so as to exclude a foreign investigation, could produce nothing but obscurantism. This seems to have happened—but the matter would be worth fuller investigation—to De Graaf's work on mammalian reproduction so far as England is concerned. De Graaf, whose conduct so far as I

---

[42] See *Phil. Trans.*, no. 84 (17 June 1672), 4098, and *Correspondence*, IX, Letter 1996, Oldenburg to Swammerdam, 13 June 1672.
[43] See *Correspondence*, IV, 350–69 reprinting and translating from *Phil. Trans.*, no. 35 (18 May 1668), 672–82, and many related references.

can judge was honest, patient and modest, was passionately anxious to have his researches properly esteemed in England, with its galaxy of medical talent and Royal Society. He was met with the charges—from Timothy Clarke chiefly but at first from other English physicians also—that in so far as his discoveries had not been previously known in England they were false, and in so far as they were true they had been anticipated. Clarke brought in a battery of names—Vesalius, Riolan, Tilman Trutwin, Glisson, Wharton—in an attempt to convict De Graaf of ignorance of the previous anatomical literature;[44] but the point at issue, of course, is not whether anyone before De Graaf had described the anatomy of the testis with approximate accuracy, but whether De Graaf had improved significantly on these earlier descriptions and their interpretation. I take it—but I am no anatomist—that the modern view is that he had.

To some extent De Graaf's originality was vindicated in English eyes during the subsequent wrangle, especially after De Graaf had sent to London the testis of a dormouse prepared in his own special way. But because of this wrangle his second and more important work on the female reproductive organs and the mammalian ovum (as he saw it) seems to have had a cool reception in England. Hence also, perhaps, the exaggerated fervour of Swammerdam's letter that I quoted just now. Confidence in their own national priority appears to have convinced English physicians and anatomists that they had nothing to learn from foreigners in the theory and anatomy of reproduction. The similar and far more serious case involving Newton is well known in the history of mathematics.

With Malpighi there was no such sad history. Technically, in the study of plant anatomy Nehemiah Grew had priority over him; William Croone could claim if not priority at least independent observation of the chick in the unincubated stage of the egg.[45] But the situations were different from those of De Graaf. Oldenburg, as our *Correspondence* shows, handled them with great tact, whereas previously he had submitted to Clarke's authority. Moreover, though Malpighi was by no means of a placid phlegmatic temper, he was not minded to make priority an issue; he was confident in the originality and importance of his observations. It was soon evident that while Malpighi's and Grew's study of plant tisssues did not produce violent conflict, they were in many ways different—in fact Malpighi's is much superior. Hence Oldenburg after much soft-pedalling when he finally despatched a copy of Grew's *Anatomy of Vegetables Begun* to Malpighi at Bologna could predict that

> you may assure yourself [by examining the book itself] that you have developed this investigation most worthily by another method and also extended your observations further.

Hence, he goes on, the Royal Society was most anxious to have from Malpighi his drawings elucidating the text, so that Malpighi's essay could be properly printed in

[44] See *Correspondence*, V, 268–72, Clarke to Oldenburg, 20 December 1668 and the many subsequent exchanges between De Graaf and Clarke via Oldenburg. Also Malpighi's letter of 10 November and notes (Vol. VII, 243–45).
[45] Croone's paper *De formatione pulli in ovo* was mentioned by himself on 29 February 1671/72 and read on 28 March; it is printed in Birch's *History of the Royal Society*, III, 30–40. Malpighi's first embryological essay had been read on 22 February: see *Correspondence*, Vol. VIII, Letter 1879, Malpighi to Oldenburg 22 January 1671/72, and subsequent correspondence, also Birch, *History*, III, 16.

London.[46] As for Croone, as an embryologist he was not in the same league with Malpighi. This was at once pointed out with great fairness by John Wilkins when Croone stated his case:[47]

> The Bishop of Chester desired that, notwithstanding this, Signor Malpighi might have the honour of this discovery, since Dr. Croone had never brought into the Society an account or a figure of this discovery, as Signor Malpighi had now sent to them an accurate description of this discovery, accompanied with very neat and laborious schemes.

We now know that Croone's observation was quite false, based on an accidental conformation of the vitelline membrane within the egg.[48]

If I have dwelt at some length on the investigation of reproduction and embryology it is because this investigation figures largely in the correspondence of the late 1660s and early 1670s with which we have been concerned in the last few years. I cannot also consider the scraps of information concerning the study of respiration, of muscle-nerve action, of histology, and of the brain, since time does not permit. It would be interesting too to review the attitude of English physicians to the iatromechanical theory developed by Descartes, Bellini, and G. A. Borelli. Then, at a more directly medical level, there is the promising episode of the attempt to find a really effective styptic, but this we have not come to yet in our work. I need hardly add that such questions of medical history cannot be studied in our *Correspondence* alone, but must be followed in the publications of the men concerned, Birch's *History*, and the *Philosophical Transactions*, as well as in much other correspondence which is not our immediate concern. Nor have I touched on the history of the relationship of English medicine—or sometimes the frustration of an attempted relation—with such distinguished foreign investigators as Rudbeck and Thomas Bartholin, Steno,[49] Pecquet, the *Academia curiosorum* of Leipzig, and so forth.

In conclusion, may I say how grateful I am for this opportunity to convey to medical historians something of the interest for them which may lie in the fruits of the labours which have engaged my devoted wife and myself for over a decade; it has been sometimes an arduous task, and therefore one hopes a profitable one. The wheels of scholarship grind slowly, and it is only after a long lapse of time that one begins to perceive that the bread one has cast upon the waters is nourishing the ducks. In this lecture I have of set purpose touched on a multitude of facets of medical history to catch your attention, and omitted a great deal of agonising detail. Let me with my last words ask your indulgence; in our edition of the *Correspondence of Henry Oldenburg* we have doubtless omitted much, and made many errors with respect to bibliography, medicine, physiology, anatomy, zoology, and botany. In committing at least two million words to paper in eleven years not every one can be beyond reproach.

[46] See *Correspondence*, IX, Letter 1969, Oldenburg to Malpighi, 26 April 1672.
[47] Birch, *History*, III, 17.
[48] F. J. Cole, *Early Theories of Sexual Generation,* Oxford, 1930, p. 47; Joseph Needham, *A History of Embryology,* Cambridge, 1934, p. 146. Malpighi's relations with the English physicians are also considered in his biography of Malpighi by Howard B. Adelmann, *Marcello Malpighi and the Evolution of Embryology,* Ithaca, Cornell University Press, 1966, Vol. I.
[49] See Dr. F. N. L. Poynter's recent papers, 'Nicolaus Steno and the Royal Society of London', *Analecta Medico-Historica,* 1968, 3, 273–80; and 'Italian Doctors and the Royal Society' in *Communicazione presentata al XXI Cong. Int. di Storia della Medicina, 1968,* Rome, 1969, 325–33.

XIV

MEDICINE AND THE ROYAL SOCIETY

The most obvious statement to make concerning the place of medicine and medical men in the seventeenth-century history of the Royal Society is a flatly negative one: as an institution the Royal Society had no direct concern with the practice of medicine or the development of the medical profession in England, with one or two experimental exceptions. The Royal Society never trod upon the delicate ground of diagnosis and prognosis, it interfered not at all with the established methods of medical education, and it kept itself quite aloof from the professional concerns of the College of Physicians, though of course many men were Fellows of both the Society and the College.

The Royal Society and the College of Physicians

The best-known story concerning the relations between these two bodies has been told so often that I hardly need do more than allude to it. In 1668 a difficult and venial scholar named Henry Stubbe intervened in a current literary quarrel by publishing a violent attack on the Royal Society, which he renewed and extended in subsequent years.[1] The force of Stubbe's attacks had nothing to do with medicine, and though among his many instances of the impertinence and ignorance he attributed to the

[1] Sir George Clark, *A History of the Royal College of Physicians of London* (Oxford: Clarendon Press, 1964), Vol. I, pp. 311–312, and especially Rosemary H. Syfret in *Notes Rec. Roy. Soc.*, 1950, 7:254 ff.; 8:24 ff. There are many details about Stubbe's and others' criticisms of the Royal Society in A. Rupert Hall and Marie Boas Hall, *The Correspondence of Henry Oldenburg*, Vol. IV:*1667–1668* (Madison: University of Wisconsin Press, 1968) and subsequently.

XIV

Fellows of the Royal Society he included their perversion of traditional medical wisdom, he was careful not to appear as an official
champion of the physicians. It is clear that Stubbe was motivated
by personal spleen (if I may be forgiven this antique medical
metaphor), for he had tried and failed to worm his way into the
Society by writing a long natural-historical account of his voyage
to Jamaica, which had actually been printed in the *Philosophical
Transactions of the Royal Society*. But it is also known that he
had been particularly encouraged by Baldwin Hamey the Younger,
an eminent member of the College of Physicians, who felt that
the Royal Society trod too "close upon the heels of the Aescula-
pians . . . whereas all matters within ye sphere of Medicine,
Anatomy and Surgery most properly should belong to ye Royal
College of Physicians, whilst ye larger purlieus of Natural Philosophy and Mathematics, exclusive of ye former, might have found
ye Societists Exployment and Enquiry enough." And according
to the writer of these words—Hamey's nephew—because Stubbe's
efforts failed, this actually happened: the Royal Society flourished
without control, and "left the College no transactions to make
'em famous in print, but what are indigene and private within
their own walls."

Hamey probably and Stubbe certainly linked the Royal Society with the "empiricks and quacksalvers" and "cheating mounte-
banks" who not only endangered the health of the nation by their
illicit practice, but threatened to overthrow the whole profession
of medicine. That this was not a widespread view is not merely
obvious from the activity of many physicians within the Royal
Society, as we shall see, but is specifically asserted in the complaint
of Christopher Merret, equally distinguished in the Society and
the College, that Stubbe was secretly befriending the apothecaries
against both, Merret holding that the practice of medicine by
apothecaries was the worst possible evil. However, there is absolutely no reason to imagine that the encouragement Hamey gave
to Stubbe was other than purely private; according to Sir George
Clark there are no references of any sort to the Royal Society in
the College's record, and support for Stubbe among medical men
was virtually nonexistent, though he won the approval of a number of conservative academics. The amusing thing is that Stubbe

was himself a quack; he conducted a practice of medicine at Stratford-on-Avon and Bath, being quite innocent of any medical qualification, degree, or license.

After traversing all this vapor, let us consider a few solid facts. When a philosophical assembly was first formed in London toward the end of 1660 it was imagined that its relations with the College of Physicians would be close and cooperative; indeed, it was suggested that the College might be a convenient place for the assembly and that as a return courtesy any Fellows of the College who so desired might be admitted to the assembly on the same easy terms as noblemen, provided that they would agree to undertake the "particular works or tasks that should be allotted to them." [2] The writer of the records of these early days—the assembly's "register," William Croone—was, by the way, himself a physician. However, no such provision was repeated in the Society's First Charter. Rather this marks a move toward independence of the College, for the infant Royal Society was granted the right to obtain the bodies of executed criminals for dissection on the same terms as the College and the Barber-Surgeons. One committee of the Royal Society under the chairmanship of Christopher Merret did, however, meet for a time in the College of Physicians.

If the Royal Society was to develop its own interest in physiology and other aspects of medical science independently of the College of Physicians, as we shall see later, an examination of its membership suggests far more interpenetration than formal records disclose. Of the 425 Fellows elected to the Royal Society in its early years—and I conclude these, as Thomas Birch did, in 1687, the year of publication of Newton's *Principia* (though that great event is hardly relevant to our present concerns)—at least a fifth had some sort of connection with medicine. I should refer Robert Boyle to this category, for example, though obviously he was not a professional physician; his erstwhile assistant Robert Hooke, awarded an M.D. from Oxford; and Christopher Wren for his early interest in transfusion. When one recalls that some-

2. Clark, *History*, Vol. I, p. 311; Thomas Birch, *History of the Royal Society* (London, 1756; reprinted with an Introduction by A. Rupert Hall, New York and London: Johnson Reprint, 1968), Vol. I, p. 5.

thing like a third of all the Fellows elected were of the nobility and gentry, playing little part in the Society's business, this proportion of medical men is a high one. Compared with about eighty medical men in a very general sense, there were some seventy drawn from the professions (church, law, armed forces, civil service) and about the same number of those whom we might call scientists—teachers of mathematics like Newton himself, astronomers, and so on. Thus the medical men actually formed the strongest group in the Royal Society, and their personal distinction exceeded even their numerical strength. Some dozen or so were distinguished foreigners, Marcello Malpighi and Regnier de Graaf among others. Some score or more of the British "medical" Fellows were *not* in addition Fellows of the College of Physicians—among them Robert Boyle, Robert Hooke, John Locke, Malachi Thruston, Henry Power, William Aglionby, Sir Alexander Frazier, and James Molines—but nearly fifty were Fellows of both institutions. I have not computed the total membership of the College of Physicians during these years, but at a guess this fifty would represent a fifth or more of the College, and again in terms of professional distinction far more than that proportion. It would be tedious to recite a long list of names, but those of Edward Browne, Walter Charleton, Sir George Ent, Francis Glisson, Nehemiah Grew, Martin Lister, Walter Needham, Sir Charles Scarburgh, Hans Sloane, and Thomas Willis stand out conspicuously.

Given the recognition of this simple statistical fact, and a very little sampling of the pages of Birch's *History*—which will show that these men and their publications did figure regularly in the business of the Royal Society—any further search for possible tension or lack of understanding between the two bodies becomes superfluous. One might well ask why the scientific and indeed strictly medical activities of these seventeenth-century physicians is historically associated with the Royal Society rather than with the College of Physicians of which they were equally distinguished members. There are at least two possible answers to this question. At a superficial level we have at our disposal surviving evidence of the Royal Society's activities in minutes and correspondence showing that it did actually display interest in these physicians' researches. We have no such evidence relating to the College. Consider the case of Edward Browne, eldest son of Dr. Thomas Browne. We know how much interest the mineralogical and technological information he collected on his wide-ranging continental travels excited in the Royal Society. What do we know of him in relation to the College of Physicians? That he was distinguished by receiving its high offices as Censor, Treasurer, and President, and that he conditionally left half his estate to the College in his will. (Edward Browne had become a very successful and fashionable physician.) But there is no evidence that the College valued his early scientific work, and perhaps there is no reason why it should have done so.

Further, at a rather deeper level one finds a real distinction between the two bodies. Although it would have been perfectly proper and logical for the College to encourage research and education in such subjects as human anatomy, physiology, and pharmacology, it did not in fact do so. While the regular series of anatomy lectures was apparently continued until about 1690 (with a reduction in their annual number from six to three in 1678), they seem to have been delivered without any distinction except by George Ent in the 1660s. Of the years between 1666 and 1679 Sir George Clark remarks in his history of the College of Physicians, "In matters relating to medical thought and science little activity can be traced."[3] One might go further and say that it is hard to see what intellectual, social, or administrative function was served by the existence of the College during these dark years. It was far otherwise with the Royal Society. Not only did the Society attend critically at its meetings to reports of new studies in the medical sciences, not only were some of its leading figures (notably Robert Boyle and Robert Hooke) themselves active in this direction, but it encouraged the work of the university physicians, especially those of the Oxford school led by Thomas Willis.

Openness of Research and Dissemination of Results

Moreover, it was the Royal Society, not the College of Physicians, which cultivated relations with medical men overseas and sought information about exotic remedies whose prospective vir-

3. Clark, *History*, Vol. I, p. 345.

426

tues excited a number of the more learned physicians. Henry Oldenburg, the Royal Society's highly active Secretary, maintained a correspondence with Sylvius at Leiden, with Thomas Bartholin and Olaus Rudbeck, with the physicians of the Academia Curiosorum in Germany, as well as with de Graaf, Malpighi, Swammerdam, and Leeuwenhoek. He wrote to Iceland, India, the Near East, the Bahamas, Brazil, and New England in search of medical experience and new remedies.[4] In short, whereas the College of Physicians seems to have been wholly parochial in its outlook, the Royal Society took a worldwide and inclusive view of the development of the medical sciences, both directly and indirectly laying before English physicians the fruits of its Fellows' experiments, of its inquiries into all quarters of the globe, and of the researches and case histories reported over all Europe.

The Royal Society believed in the openness of research, and none believed more than its Secretary. In his correspondence Oldenburg commonly spoke of reporting something at the *consessus publicos* of the Royal Society, an expression we have translated as "ordinary meetings" (strictly the meetings were not public, though any properly accredited person, including many foreigners, might be introduced as a visitor). For the first time that I know of in the history of any society minutes were kept of communications and discussions, while written papers were, with the author's consent, preserved in the archives for consultation by any Fellow. Moreover, Oldenburg was in general liberal in communicating to Fellows and indeed to non-Fellows matters of scientific interest that occurred in meetings or were imparted to him in correspondence, unless the speaker or writer specifically requested reticence (for example, because he wished to pursue a subject further before making it public). Perhaps Oldenburg was sometimes too liberal in communication, but that is a matter which I have no wish to argue now; I will only say that having read some three thousand letters in Oldenburg's correspondence, I feel convinced that practically everyone understood the situation and, realizing that communication was Oldenburg's and the Society's function, approved of his discretion in separating private affairs from matters of gen-

4. I dealt with the foreign medical relations of the Royal Society in "English Medicine in the Royal Society's Correspondence: 1660–1677," *Med. Hist.*, 1971, *15*: 111–125.

eral intellectual interest and trusted him to keep quiet when asked. I need hardly add that such a working distinction between private and scientific communications had long been observed.

This openness of discussion—the dissemination of work in progress as well as formal, finished treatises—was new and provocative in all branches of science, but particularly in the medical sciences where the tradition of secrecy with respect to methods, ideas, and investigations was at least as old as the Hippocratic oath. What medical men were actually doing and thinking from week to week had never before been recorded, much less spread abroad. If it were not for an historical accident, for example, we should know nothing of Harvey's *Prelectiones,* and Harvey would be known to us only as the author of two printed books—the second, one might say, snatched from him. But work in the medical sciences from 1660 onward was known and discussed all over Europe long before the relevant books appeared, if indeed such books ever appeared at all.

Again, the role of Henry Oldenburg as the first and perhaps the greatest of scientific journalists is one on which I need not dwell. The example of the *Philosophical Transactions* was followed by the *Journal des Sçavans,*[5] by the *Giornale de'Letterati,* by the *Mémoires* and *Conférences* of Jean Denis, by the *Miscellenea Curiosa,* by the *Acta Hafniensis,* and the *Acta Eruditorum.*[6] The only point I would make here is that the *Philosophical Transactions* contained a good deal of matter relating to medical case histories, iatrochemistry, physiology, and other branches of medical science—all this brought into the open for the benefit of medical men everywhere and the public at large. I might add obviously that books of similar tenor were of course regularly epitomized, sometimes at considerable length. The old closed world of medical investigation was suddenly thrown open, even if medical practice remained (for the most part) as secret as ever.

Alignment with Iatrochemistry

Before I proceed further, I should explain the various positions available to a thoughtful physician of the late seventeenth century,

5. Of course the *Journal des Sçavans* was a few months older than the *Philosophical Transactions.* But the comment is fair.
6. All these journals found some space for medical and biological topics; the *Miscellenea Curiosa* and *Acta Hafniensis* were exclusively medical.

428

as I understand them. First, he might adopt a conventional scholastic, or if you will, Galenical line, as Guy Patin in Paris did; to what extent this was still orthodox medical education in the England of Charles II I hardly know, but Locke for example read only modern authors from Vesalius through Sydenham (*Methodus curandi febres*) with a heavy emphasis on the iatrochemical and physiological authors of the seventeenth century, and one may imagine that many other young men followed this pattern.[7] Secondly, he might be a nihilist, as some French physicians of the early nineteenth century were, denying merit to all theories and attempted explanations in medicine, relying instead on simple empiricism molded by experience. Sydenham was, I take it, a nihilist. I do not know whether nihilism is admirable in itself— medical historians have often seemed to praise it in Sydenham— but it is surely obvious that it can only flourish when theoretical medicine is wholly inadequate. Thirdly, the physician might place all his hope in the new theory of still and furnace with its gamut of remedies, which were certainly capable of producing dramatic effects on the patient, though the relation of these effects to the disease was mercifully and irretrievably buried in the enormous verbiage of J. B. van Helmont and his predecessors. If, however, the physician felt that some knowledge of physiology and pathology must precede the cure of disease, he could see the advancement of his profession as depending on either the line of anatomical inquiry glorified by Vesalius and Harvey or on other forms of experiment and chemical investigations into the working of the human body.

Now, having set all that out, I can ask the question: where did the Royal Society stand in relation to all these possibilities? It had certainly no leanings toward either conservatism or nihilism: Boyle's advocacy of suspension of belief in any particular dogma was not equivalent to the view that no theory is tenable by a rational man.[8] The Society's position was in general the optimistic

7. Kenneth Dewhurst, *John Locke (1632–1704). Physician and Philosopher* (London: Wellcome Historical Medical Library, 1963), Ch. I.
8. See *The Usefulness of Experimental Natural Philosophy* (Oxford, 1663), Pt. II, Sec. I, "Of its Usefulness to Physic" (an early work); the *Memoirs for the Natural History of Humane Blood* (London, 1684), and in general Marie Boas, *Robert Boyle and Seventeenth Century Chemistry* (New York: Cambridge University Press, 1958).

one that systematic knowledge will be attainable at the end; again, one finds Oldenburg in his correspondence frequently reiterating the inductivist view that an accurate natural history is essential as the foundation for a true natural philosophy. The stories of anatomical abnormalities and strange medical case histories that the Royal Society collected into its budget of curiosities were, of course, pieces in this mosaic of natural history, justified by Bacon's notion that the philosopher might learn more from Nature's by-ways than from her plain and ordinary course.

Many Fellows of the Society were dazzled by the bright hope of iatrochemistry. Not least among them was the chief of British scientific physicians, Thomas Willis; but though Willis had been a Fellow since November 1663 and was in practice in London from 1666 onward, he was never active in the Society's affairs, perhaps because (as Anthony Wood relates) "he became so noted and so infinitely resorted to in his practice, that never any physician before went before him, or got more money yearly than he." [9] Something of the aloof character of such a great medical man may be illustrated by the following anecdote. In June 1671 the German physician P. J. Sachs, founding father of the Academia Curiosorum (the "Leopoldina"), wrote to Oldenburg requesting the cooperation of English medical men in his enterprise:

Italians, Danes, Bohemians, Hungarians, Swiss and other foreign medical men share some practical observations in medicine with us, and perhaps the same generosity may be expected from the very excellent English physicians, among whom in this century medicine seems to claim a certain pre-eminence . . . since they can improve the whole shape of medicine remarkably by communicating some observations appertaining to both their most skillful practice and their very ingenious investigation of the causes of disease, and not infrequently their highly perceptive dissections of the bodies of those carried off by disease.[10]

After this general and flattering invitation, Sachs particularly requested information "under pledge of secrecy" and with the promise of a like arcanum in return, concerning Willis' preparation

9. William Munk, *Roll of the Royal College of Physicians* (London, 1861), Vol. I, p. 323.
10. Hall and Hall, *Correspondence of Oldenburg*, Vol. VIII: *1671–1672* (1971), p. 108.

"that without being a corrosive solvent so disposes iron, crabs'-eyes, coral, etc., that they transfer their effectiveness to any fluid and are dissolved in it." [11] In hs reply Oldenburg's account of Willis' attitude speaks rather of the great physician than of the scientist; it is not proper, he wrote, to interrupt such a one inopportunely in the business of his practice, and Willis answered that he had had good reasons for not imparting his secret method in the past and did not propose to do so now.[12]

It is perhaps curious that while Willis the physician enjoyed fame as a philosopher, Boyle the philosopher had a tremendous reputation as a physician. While he probably limited his practice to family and close friends, there is no doubt that his advice was very much sought after.[13] In his earlier years, when more obviously under the influence of the iatrochemical school than he was to be later, Boyle showed a rather nonchalant inclination to criticize unspecified physicians for excessive narrowness or caution. So he censured them for being "wont to reject, if not to deride, the use of such specificks, as seem to work after a secret and unknown manner, and not by visibly evacuating peccant humours (or by other supposedly manifest qualities) . . ." a difficulty which "the naturalists may do much towards the removal of." [14] In contrast with this assignment to the naturalist of the role of teacher and to the physician that of pupil, in later years Boyle found it easier to tolerate the distinct experience of the physician, acknowledging

11. *Ibid.*, p. 110. The allusion is to Thomas Willis, "Of Fermentation," Ch. IX, trans. S. Pordage, *Remaining Medical Works of that Famous and Renowned Physician Dr. Thomas Willis . . .* (London, 1681), p. 35.

12. Hall and Hall, *Correspondence of Oldenburg*, Vol. VIII, pp. 417, 418.

13. Since Boyle's writings are full of such stories as the following, his views were widely spread: "And I remember, that a gentleman of great note coming to bid me farewel because of a long and troublesome journey he was taking to mineral waters, which he intended to drink for many weeks, to ease him of a very painful sharpness of urine: I . . . desired him, before he went, to make use of this powder [of gum arabic] once, or (if there should be need) twice a day; which when he had done, it so relieved him, that he thought himself quite cured, and forebore his intended journey, not only that year, but the next" (Thomas Birch, *Works of the Honourable Robert Boyle*, London, 1772, Vol. V, p. 112). But Boyle also observed (1684): "Having resided for many years last past [i.e., since 1668] in a place so well furnished with physicians as London is, I was careful to decline the occasions of entrenching upon their profession" (*ibid.*, Vol. IV, p. 637). For an outrageous attempt to get free advice from "the venerable Esquire Boyle" see Hall and Hall, *Correspondence of Oldenburg*, Vol. VII:*1670–1671* (1970), p. 533.

14. *The Usefulness of Experimental Natural Philosophy*, Pt. II, Essay V; Birch, *Boyle*, Vol. II, p. 170. See also Boas, *Boyle*, pp. 61–62.

that the "medicinal art and this science [physics, or natural philos-
ophy] will be conversant about the same subject, though in differing ways and with differing scopes," the naturalist "directing his
speculations to the discovery of truth" and the physician "his theory to the recovery of health." [15] The tract on *The Reconcileable-
ness of Specific Medicines to the Corpuscular Philosophy* (1685)
was in fact written to prove that the physician's inductive conviction of the reality of specifics was deductively justifiable from the
tenets of the corpuscular philosophy, not the other way round.

Naturally Boyle disapproved of polypharmacy. Whether a remedy be natural—like gum arabic—or prepared by chemical processes, it was, he believed, best taken simple and in sufficient quantities to have some possibility of being effective. But a preference
for iatrochemistry by no means excluded an appreciation of the
older and often nastier remedies. In the *Usefulness of Experimental Natural Philosophy* Boyle assures us in a few pages of the
virtues of the juices of horse dung and of ivy berries, and particularly of the medical efficacy of woodlice in treatment of the eyes
and bodily sores.[16] Whether with good reason or not, these household remedies seem to have been less vulnerable than the "trea-
cle," or syrup, of the five roots of traditional polypharmacy.

Having such powerful patronage as that of Boyle and Willis
from the late 1650s onward, iatrochemistry was never seriously in
jeopardy in England from that time, nor was the Royal Society's
evident alignment with this modern school of medical practice a
cause of disparagement in the eyes of reasonable men. Hence I
need no more than refer in passing here to the brief history of
the Society of Chemical Physicians, or to Stubbe's charge that it
was the Royal Society that had promoted "the Anti-Colledge of
Pseudo-Chymists, encouraging Odowde and his ignorant Adher-
ents in opposition to the Physicians." [17] It is at least certain that

15. Birch, *Boyle*, Vol. V, p. 74.
16. *Ibid.*, Vol. II, pp. 130–131. Richard Lower (June 24, 1664) concludes a long
medical letter with a note on an ophthalmic cure wrought with woodlice: "And this
I must acknowledge I received from the reading of your book concerning millipedes,"
and on a similar cure of a strangury effected by Willis, also with woodlice (Birch,
Boyle, Vol. VI, p. 474).
17. Henry Thomas, "The Society of Chymical Physitians," in E. A. Underwood,
ed., *Science, Medicine and History* (London: Oxford University Press, 1953), Vol.
II, p. 64.

Jonathan Goddard, one of the pioneers of the Royal Society, was also active among the Chemical Physicians and that several of the noble Fellows, like the earls of Albemarle and Anglesey, were also among the patrons of the new society.

The examination of this polarization, however, would only be of significance if the issues dividing the Galenists from the Helmontians had remained important. They did not—not so much because of the Plague of 1665 as because of the entry of younger men into the medical profession. The use of chemical remedies seems to have been regarded on all sides as perfectly normal, or even praiseworthy, in the 1660s and 1670s. So we find the practioners of Durham classified by a local schoolmaster, Peter Nelson, as "a scholar but for the most part a methodist" (a Galenist, I assume) and "a serious young man, and well educated, inclinable to chemistry but no great practitioner," while the last and most thriving "works sometimes in the fire and has a small laboratory."[18] I do not pretend to explain the wave of medical fashion that swept across England and all Europe, but I think we must regard the victory of iatrochemistry as something that happened quite independently of the Royal Society, though the Society set the seal on a victory that would have been won in any case.

Physiochemical Theories

Much the same may be said of physiochemistry—that is, the chemical interpretation of physiological phenomena—for this again goes back to Oxford and the circle of Willis (may I be forgiven the pun?). The *Two Diatribes*, in which appeared Willis' own first important contribution to the subject, *De fermentatione*, were published in 1659, when the Royal Society was in embryo. Willis asserted,

It is so certain that the Bodies of Animals, consist of the aforesaid [fermentative] principles, that it wants no proof. For they so plentifully swell up, with Spirit, Salt and Sulphur, that their Particles are obvious to the Sense: Wherefore they are moved with a more swift motion, and more excellent senses of Life, and Functions of Heat, in the Subjects, in which they are implanted, are inlarged.[19]

18. Hall and Hall, *Correspondence of Oldenburg*, Vol. VII, p. 326.
19. Willis, *Remaining Medical Works*, Ch. V, p. 13.

Further, "We are not only born and nourished by the means of Ferments, but we also die: Every Disease acts its Tragedies by the strength of some Ferment." Hence we must "endeavor the Cure of Diseases [also] by the help of Fermentation: For to the preserving or recovering of the Health of man, the business of a Physician and a Vintner, is almost the same . . . ," the physician's business being to temper the fermentative processes of the body, in the blood especially, when they become too fierce, and to encourage them when they become too cold and feeble.[20]

In *De fermentatione* Willis did not venture into the favorite topic of neuromuscular action, but took it up in his tract *De motu musculari* in 1670. He now followed in the steps of William Croone, though Croone's theory and Willis' are by no means identical. Whether Croone, who was certainly (as Gresham Professor of Rhetoric) associated with these earliest stages of the Royal Society of which he served for a time as "register," was in any way associated with Willis, I do not know. Croone was a Cambridge man; he received his M.D. by royal mandate, and nothing seems to be known of his medical education. However, an association with Willis is rendered likely not only by the joint publication of their works (from which it seems probable that the anonymous *De ratione motus musculorum*—London, 1664, Amsterdam, 1666 and 1667—may have been attributed to Willis himself) but by a remark made by Croone in writing to his Amsterdam printer about the second edition of his work: "lest your edition seem to the ordinary reader to be a faulty one of Willis (also a Fellow of the same [Royal] Society) you may if you like add this letter in place of a preface." As Willis' own theory of neuromuscular action was not to be published for another three years, it is not easy to see why the ordinary reader should be so confused, except by the appearance of *De ratione* with *Cerebri anatome*.[21] However,

20. *Ibid.,* p. 16.
21. Leonard G. Wilson, "William Croone's Theory of Muscular Contraction," *Notes Rec. Roy. Soc.,* 1961, *16*:177, n.38. Croone's authorship of *De ratione motus musculorum* was not acknowledged when it was published with Willis' *Cerebri anatome* at London in 1664 and at Amsterdam in 1666. It was unknown to Huygens and possibly to Oldenburg himself in 1669, though Croone's name had appeared on the title page of the second Amsterdam edition of 1667 (Hall and Hall, *Correspondence of Oldenburg,* Vol. V:*1668–1669* (1968), pp. 435, 451). According to Wilson, Borelli too did not know of Croone's authorship.

XIV

434

Richard Lower's criticism of the theory that muscles swell in contracting and of "explosion" as the cause of swelling seems to be plainly directed against Willis as well as against Croone and may indicate that Willis' particular hypothesis was well known before its appearance in print.[22]

I must refrain from pursuing these interesting details, which do however confirm my view that the Royal Society had little share in shaping either Willis' or Croone's physiological theories, and that Willis' "Oxford" circle—among whom we may include Boyle, Hooke, Lower, and Mayow, besides Croone—was independently applying chemical principles to physiological processes. Of course the work of this group was increasingly assimilated by the Royal Society itself, especially after first Hooke, then Willis, and Boyle established themselves in London. There is no question but that this kind of science (whatever the personal disagreements among individuals as to the validity of particular hypotheses) was extremely acceptable to the Royal Society and was widely regarded as yet another successful manifestation of the mechanical philosophy.

Scientific Investigation and Experiments with Animals

Perhaps in view of the constraints of space I may be forgiven for not dwelling on other aspects of "scientific medicine" with which the Royal Society was much concerned. The most interesting of these was the study of respiration, by which Harvey's discovery of the circulation of the blood was taken an important stage further: the connection between the movement of the blood and the inspiration of air became apparent, and the motion of the lungs, so long a matter of controversy, was at last explained in a plausible hypothesis. Again, all this work began in Oxford, among Willis' associates, and the most frequent speakers on this topic at Royal Society meetings in London were Boyle and Hooke.[23]

22. Richard Lower, *Tractatus de corde* (London, 1669), pp. 76–78. Lower did not mention either Croone or Willis as author of the theory of muscle he there confuted. K. J. Franklin's English version of the text unfortunately quite destroys the allusion to Willis, which is clear in the Latin.

23. The story was reviewed not very many years ago by Douglas McKie under the title "Fire and the Flamma Vitalis" in Underwood, ed., *Science, Medicine and History*, Vol. I, pp. 469–488. McKie omitted the work of Lower (*Tractatus de corde*, 1669), perhaps because it was purely physiological; indeed, this work seems never to have been esteemed as highly as it should have been and would repay further study. But see Everett Mendelsohn, *Heat and Life* (Cambridge, Mass.: Harvard University Press, 1964).

Among the problems raised in this long discussion was that of fetal circulation and respiration, on which Hooke made some experiments, but this problem (not unnaturally) defied solution at a time when the notion of a gaseous interchange between maternal and fetal blood could not be formulated. It is interesting to note that as yet the distinction between respiration in air and in water was itself imperfectly understood, and hence the idea of whelps "breathing" in the amniotic fluid did not seem wholly incongruous. However, this notion was rejected by Hooke as a result of the experiment he reported to the Royal Society on December 19, 1667, which led him to conclude that "the foetus in the womb has its blood ventilated [by the help of the dam] . . . and methinks also that it may seem very manifestly to prove the continual and necessary communication of the blood of the dam with that of the foetus, and of the immediate dependance of the one upon the other." [24]

The study of human anatomy had no outstanding exponents in England at this time other than Willis in his work on the brain, and it figures little in the proceedings of the Royal Society except for the reporting of strange anomalies. Only one topic aroused lively interest, and that was Regnier de Graaf's investigation of mammalian reproduction. From the correspondence between de Graaf and Oldenburg it is clear that the first suspicion of de Graaf was aroused by the belief that he was claiming for himself discoveries first made by such Englishmen as Wharton and Joyliffe.[25] Moreover, the Royal Society anatomists disputed the contention of de Graaf and van Horne that the mammalian testis consisted of very long threads and so could not properly be considered a gland. Timothy Clarke, Thomas Allen, Edmund King, and Oldenburg himself combined in an experimental examination of this structure, and a negative result was reported on November 12, 1668.[26] A series of acrimonious exchanges followed, calling many points of male anatomy in dispute; this finally died down (it can hardly be

24. See Urban Hjärne to Oldenburg, Nov. 26, 1671, in Hall and Hall, *Correspondence of Oldenburg*, Vol. III, pp. 373–382; the writer finds it natural to suppose that the lungs can absorb some air from water in cases of drowning. For Hooke, see Birch, *History*, Vol. II, p. 233.

25. Hall and Hall, *Correspondence of Oldenburg*, Vol. IV, pp. 358–359, 366–367, Timothy Clarke to Oldenburg, Apr./May 1668.

26. Birch, *History*, Vol. II, p. 321.

436

said to have been resolved) when de Graaf's technique and opinions were better understood and he himself had dispatched to London the prepared testis of a dormouse.[27] Allen and King at least seem to have admitted in the end that de Graaf's account of the testis was not unjustified, though most of the other points were never satisfactorily cleared up.

It must be said that the role of the Fellows of the Royal Society in this dispute was not an admirable one: they displayed impatience, patriotic complacency, and an excessive fondness for *verba* in place of *acta*. The whole business illustrates the truth that an appeal to "observation" is by no means sufficient to answer all questions, even in a descriptive science. It is true that de Graaf received a hearing and that his views were put to what seemed to be adequate tests; perhaps in the end he received rough justice, but the consequence was that when both de Graaf and Jan Swammerdam took up female reproductive anatomy, the latter—who proceeded with extreme tact—won a far better hearing from the Royal Society.[28]

This topic calls to mind one of the series of vivisection experiments made by the Royal Society which had a potentially immediate medical application but at that time could only produce a suggestion that no surgeon would be tempted to put into practice. Lithotomy (cutting for a stone in the bladder) was not exceptionally rare but, as readers of Pepys will recall, an operation of the greatest hazard; stone in the kidneys was a no less painful and more dangerous condition. Could surgery provide a remedy? John Wilkins thought it might, and he suggested that the experiment be made on a dog of wounding a kidney, to see if it would heal. The physicians present at this early meeting of the Royal Society —it was on August 5, 1663—protested (no doubt on behalf of the possible human patient) that surgery on the kidney was too deep for safety and loss of blood too dangerous;[29] nevertheless, Walter

27. Hall and Hall, *Correspondence of Oldenburg*, Vol. VI:*1669–1670* (1969), pp. 118, 122; Birch, *History*, Vol. II, p. 397. The preparation was illustrated in *Phil. Trans.*, Oct. 17, 1669 (No. 52).

28. The letters bearing on this will appear in subsequent volumes of our *Correspondence of Oldenburg* (in press).

29. These physicians, besides Charleton, were Francis Glisson, Jonathan Goddard, and William Quatremain.

Charleton was requested to try the experiment at the next meeting (which so far as I know he never did). More than five years later the Bishop of Chester had better luck with Malachi Thruston at Exeter, whom he persuaded to wound a dog in one kidney; it recovered, but to little purpose, for it was slaughtered to discover what progress had been made, at which point the kidney was found to be perfectly healed. After this, in the autumn of 1668, Edmund King performed an even more thorough and brutal piece of surgery on another dog, which was nevertheless found to have healed its internal wound almost completely.[30]

Many other experiments of the same sort were tried on animals, the human extrapolation being obvious enough. At one opening of Birch's *History* one can read of transfusion experiments, another on the thoracic duct, a third on extirpation of the spleen, and a fourth on skin grafting.[31] The trouble with this approach to medical research was the obvious one that the more recklessly one experiments on animals, the bigger the gap to cross before attempting the same experiment on man. And whereas in physic the dosage can be adjusted, in surgery it is a case of all or nothing. And in fact it was nothing.

Blood Transfusion and Injection Therapy

Only in the one instance of blood transfusion was a flamboyant medical experiment transferred from animals to men; or rather I should say *two* instances, for the injection therapy, or as it was then called, *chirurgia infusoria,* was also attempted on men, though the evidence for this is not strong. The two procedures should be distinguished, of course (though seventeenth-century medical men did much to confuse them): injection therapy was the subject of several dissertations; transfusion of blood had a short and scandalous history.

Again, in this story the Fellows of the Royal Society seem to have been more intent on defending their priority than on publishing any sensible account of their researches. Let me briefly remind you of the principal English documents connected with injection therapy. It is first mentioned by Boyle in a postscript to

30. Birch, *History,* Vol. I, p. 292; Vol. II, pp. 317, 338.
31. *Ibid.,* Vol. II, p. 173; May 2 and 9, 1667.

Essay 2, Part 2, of the *Usefulness of Natural Philosophy,* written perhaps in the spring of 1663.[32] Boyle there wrote of the experiments as "much noised," as practiced on dogs and an "inferior domestick," and as originating in a conversation (date unspecified) between Wilkins, Wren, and himself. He gives Wren the credit for first contriving an experimental injection into a dog. Next, at a meeting of the Royal Society on September 16 Timothy Clarke read a paper, which has not survived, on his own injections of several fluids into dogs. There was some skepticism among those present as to the the value of the procedure and some fear that it would "cause strange symptoms in the body."[33] Clarke was supposed to prepare a proper report on his investigations; as he never did so, he had only himself to blame if they subsequently received little credit. In fact two Germans, Johann Daniel Major and Johann Sigismund Elscholtz, were the first to describe the "new surgery" in print.[34] Their dissertations stimulated the note on English priority (based on the two documents just mentioned) which Oldenburg inserted in the *Philosophical Transactions* for December 1665, in which Clarke is highly praised, though Wren's priority is again asserted vigorously.[35] Oldenburg dated Wren's first attempt no later than 1659.

About two and a half years later Clarke wrote Oldenburg a long account of the whole affair, in which he took its beginning back to "the end of the year 1656, or thereabouts," and his own association with injection to the following year, using "waters, various kinds of beer, milk and whey, broths, wines and alcohol."[36] Finally, in *De corde* (1669) Lower spoke of his own involvement with these same injection experiments, "injecting into the veins of living animals various opiate and emetic solutions, and many medicinal fluids of that sort."[37] Unfortunately, like all the other English, Lower thought it out of place "to describe the individual results and outcomes of these experiments," so apart from Boyle's

32. See Birch, *Boyle,* Vol. II, pp. 88–89. This book was presented to the Royal Society on June 24, 1663.
33. Birch, *History,* Vol. I, p. 303.
34. J. D. Major, *Prodromus inventae a se chirurgiae infusoriae* (Leipzig, 1664); J. S. Elscholtz, *Clysmatica nova* (Hamburg, 1665).
35. *Phil. Trans.,* Dec. 4, 1665 (No. 7): 128–130.
36. Hall and Hall, *Correspondence of Oldenburg,* Vol. IV, pp. 355–357, 364–366.
37. Lower, *Tractatus de corde,* pp. 171–173.

brief and early notice, they are lost to us. Only one solid point emerges: once again experimental medicine, as discussed by and demonstrated before the Royal Society, is firmly linked with the Oxford school and the pupils of Willis.

To Lower, of course, injection was merely a preliminary to transfusion, and so I will describe the origins of the latter procedure briefly before attempting any analysis. Apart from Clarke's casual claim to have thought of blood as a possible substance to be injected, with the admission that he failed to effect such a transfusion, the earliest allusion to transfusion is at the meeting of the Royal Society, already noted, on September 16, 1663, when the proponents of this operation are left unnamed. There seems no reason to doubt Lower's own statement that he first achieved transfusion between two dogs, without ill effect, in February 1665. Lower reasoned that the healthy blood of a donor would mix more naturally with the defective blood of a recipient than any medicament could do, or it might replace it altogether.[38] Lower further expressed his firm conviction that where a human patient had lost much blood by venesection or hemorrhage, animal blood could take its place "safely and advantageously," and that in the case of arthritic patients and lunatics "perhaps as much benefit is to be expected from the infusion of fresh blood as from withdrawal of the old."[39] When he wrote these words Lower had effected his first and only transfusion with a human recipient. In this he had been preceded by Jean Denis at Paris, who had taken the same rash step on June 6, 1667.

I should perhaps make it clear that quite apart from the subsequent drama of the death of a patient under Denis' care, not all the animal transfusions were successful. On October 29, 1668—a bit late in the day, it is true—John Wilkins reported how a dog into which a large quantity of sheep's blood was transfused "fell into a great disorder and agony and died . . . his heart was found full of coagulated blood, and the stomach black and bloody, and all his veins exceedingly distended."[40]

I do not mean to trace the history of transfusion, so often dis-

38. *Ibid.*, p. 173.
39. *Ibid.*, pp. 191–192.
40. Birch, *History*, Vol. II, p. 316.

440

cussed, any further.[41] It is clear enough that, as with injection therapy, the Fellows of the Royal Society had failed to push the investigation to a conclusion; and so they were scooped by Denis, to his own cost. One may speculate that in consequence Englishmen asserted their priorities the more loudly. It is certainly original to think of injecting drugs directly into the blood and to extend this procedure to transfusion. Much skill was exercised in perfecting the techniques required. But to develop, in the first place through experiments on animals, a systematic body of knowledge by employing such techniques would have required more patience and a more systematic plan of experiment than these early experimenters demonstrated. There is a hint of the planning of a series of experiments in the words of both Clarke and Lower, but we do not know how far they implemented any such plan. One actually reads of little more than crude "live or die" trials; so that, quite apart from those fallacies which would in any case have condemned the double enterprise to failure, I think we may more historically criticize these experimentalist physicians on purely methodological grounds. If one compares this piece of experimentation with Newton's contemporaneous prism experiments on light, the former seems jejune, blundering, uncritical. No doubt Newton's was the easier task, and the time for Claude Bernard was not yet. The whole art of medical experiment had to be refined before it could give sensible answers to questions.

Undoubtedly, then, the English experimenters of the Royal Society were right not to rush into practice upon human subjects: Lower's confidence, his assurance that "the subject [of transfusion was] worthy of recommendation to the care of all doctors, and to the whole world, whenever an opportunity occurs of trying it" was ill founded, on his own minute experience as well as our much greater knowledge. But by not pressing on, they weakened the logic of their own enterprise; if experimental science produced results that promised well for the alleviation of man's suffering, why hang back. If human lives were to be improved by the prog-

41. We have added some details from the letters of Denis and elsewhere in our *Correspondence of Oldenburg*, and I have reviewed the question from a different point of view in *Med. Hist.*, 1971, *15*:117–119.

ress of science, then men had to put their trust in science. There was indeed a dilemma here. If experimental science genuinely makes additions to human knowledge, then there comes a point at which one must trust this knowledge in practice—by building a bridge on new scientific principles, by leaping into the air in a flying machine, by putting a new medical remedy to work on a human patient. If society never gets to the point of so applying its knowledge, then that knowledge, however excellent in explanation, can never claim to be useful. But if at any point the scientist does not yet possess this confidence in his knowledge, he must either know how to pursue his investigation further, extending and confirming his knowledge until it does win his complete trust, or admit defeat and stay content with the old ways. I think that as regards injection therapy and transfusion the Royal Society was essentially in this position in 1666. It could not plan a program of further research, yet human trials seemed perilous. After a brief euphoria contributed by Denis, on the morrow of his catastrophe no one had much stomach for either further trials or further experiments. They still did not know how to plan the research for further advantage. And who can blame them for that?

If we consider the pathological speculation underlying these two procedures, there is of course a logical distinction: injection therapy could be supposed to embrace the view of blood as a vehicle for distributing suitable medicaments about the body, the seat of disease being elsewhere, while transfusion therapy would presuppose that the seat of the disease is in the blood itself. In fact the distinction is more apparent than real, the blood being regarded as the most sensitive index to bodily health. As Boyle put it, "the mass of the blood being vitiated, or . . . disorderly moved, is the seat of divers and the cause of most diseases, whose cure consequently depends mainly on the rectifying of the blood"; [42] hence the swift transfer of interest from the indirect therapy of injection to the more direct one of transfusion, once the technique for effecting transfusion had been perfected. Boyle in the same *Memoirs for the Natural History of Humane Blood* (1684) again underlines the importance of Willis' teaching, for he notes "how ordinary it is, especially since the learned Dr. Willis's writings

42. Birch, *Boyle*, Vol. IV, p. 595.

came to be applauded, to look upon fevers as inordinate fermentations of the blood."[43]

Attitudes toward the Study of Blood

Generally the Fellows of the Royal Society paid little attention to the aetiology of disease, though whether their lack of interest was due to an implicit belief that this was sufficiently well understood (in the humoral theory and so forth), or whether it was due rather to an empiricist belief that it was more important to cure the manifestations of diseases than to search further for their causes, I do not know. Perhaps the mechanical philosophy itself—by suggesting a comparison between the organism and a machine which breaks down under excessive strain or long service—may have tended to distract attention from the search for more suitable pathological causes. Certainly nothing had happened in seventeenth-century medical science to detract from the importance attached to the blood; rather the reverse—so that besides its roles in nutrition and respiration its supreme significance as the source of heat and life and the preserver of the body's healthful stability was accepted without question. Everything that could be learned about the blood, its circulation in the body, and about the various humors entering into its composition was of high interest.

As Adelmann has already noted, Malpighi's *De polypo cordis* (Bologna, 1666) contains many new observations on the blood; indeed, this little work contains the first definite description of the red corpuscles, which Malpighi called "atoms."[44] This essay, together with the *Tetras anatomicarum epistolarum,* to which it was appended, aroused so much interest in London that, having been reviewed at some length by Oldenburg in the *Philosophical Transactions,* it was reprinted by the Royal Society's printer, doubtless at Oldenburg's suggestion.[45] (Curiously enough there is no reference to this first appearance of Malpighi's work in London

43. *Ibid.,* Vol. IV, p. 619. Compare Malpighi's views of 1665 on the relation of the blood to fever as summarized in Howard B. Adelmann, *Marcello Malpighi and the Evolution of Embryology* (Ithaca, N.Y.: Cornell University Press, 1966), Vol. I, p. 273.
44. Adelmann, *Malpighi,* Vol. I, pp. 266–267.
45. *Phil. Trans.,* Feb. 15, 1668/1669 (No. 44): 888–891. Martin's edition—mentioned here—followed later in 1669. I have not referred here to Malpighi's other histological investigations, despite their evident importance for medical science, because they were not pursued further by other investigators in the Royal Society.

in the surviving Royal Society correspondence with Malpighi.) [46]
The discovery of the blood corpuscles—so favorite a demonstration
of subsequent microscopists—seems not to have aroused much attention, nor did Malpighi himself prize it. It is natural enough to
find no mention of the corpuscles in Lower's book (1669) but more
suprising that they are mentioned neither by Walter Needham
(1675) nor by Boyle in his little treatise, for both of them were
particularly concerned with the composition and properties of
blood. Moreover, at the time Needham and Boyle wrote, Leeuwenhoek's first paper on the microscopic study of blood had already
appeared in the *Philosophical Transactions;* this is properly regarded as the real foundation of knowledge of the red blood
corpuscles.[47]

Leeuwenhoek's direct correspondence with Oldenburg began
in August 1673, and his first extant reference to the blood corpuscles (apparently in complete ignorance of any earlier description) is in a letter to the elder Constantijn Huygens of April 5,
1674.[48] He told Oldenburg about them in a letter written two
days later; [49] from this letter and subsequent communications the
first *Philosophical Transactions* notice was compiled.[50] At this
stage he saw the corpuscles as spherical, for indeed he was inclined
to see globules everywhere, possibly as a result of aberrations in
his first microscopes and improper illumination. He was not surprised (he wrote later) that his discovery of "particles" provoked
incredulity in both France and England, but he repeated his assurance that he could observe them as plainly as one could see grains
of sand with the naked eye.[51] It was only in a letter to Oldenburg
of October 5, 1677, that Leeuwenhoek first reported seeing "thin
rods" in the blood of an eel, an appearance he attributed to mechanical deformation of the globules, while in 1682 he observed

46. One letter from Oldenburg is in fact missing at this point; see Hall and Hall,
Correspondence of Oldenburg, Vol. VI.
47. *Phil. Trans.,* Sept. 21, 1674 (No. 106): 121–124.
48. *Alle de Brieven van Antoni van Leeuwenhoek,* Vol. I (Amsterdam: Swets &
Zeitlinger, 1939), p. 67.
49. *Ibid.,* p. 75.
50. *Ibid.,* pp. 93–105, 121–125.
51. *Ibid.,* pp. 211–213. It is just worth noting that Leeuwenhoek's insistence on his
nonobservation of pores or pipes in sections of nerve (e.g., the optic nerve of a cow)
was similarly doubted, for obvious theoretical reasons. Among his critics was, apparently, Willis (*ibid.,* pp. 217–219).

444

for the first time "flat, oval particles" (lenticular corpuscles) in the blood of fishes.[52] Even two years later, when he recorded similar particles in the blood of "various birds," he still maintained that his earlier description of globules held good for men, oxen, sheep, and rabbits, while the lenticular shape was to be seen only in "fishes, birds and animals that live in water." [53]

Looking at the third and fourth volumes of Birch's *History* one can be in no doubt of the Society's intense interest in Leeuwenhoek's extraordinary discoveries. As early as February 1677 Oldenburg was instructed to seek more detail of Leeuwenhoek's method of observation in order to attempt confirmation of his results, while in April Nehemiah Grew was exhorted to look for the animalculae that Leeuwenhoek had reported in pond water.[54] After Oldenburg's death (early in September 1677) Hooke, his successor as Secretary, was urged to maintain the correspondence with the Dutch microscopist and was encouraged to prepare equipment of his own with which to emulate Leeuwenhoek's work. In the next few months Hooke submitted numerous reports to meetings on his progress, and at last (on January 24, 1678) showed by demonstration how "the blood consists of two substances, the one containing liquor undetermined and undistinguishable as to its parts, flowing about and incompassing the other, which consists of an infinite number of exceedingly small parts, which were plainly perceived to be globular." [55] One might therefore date full recognition of Leeuwenhoek's discovery from this report, though as we have just seen the microscopic appearances were still incorrectly interpreted.

However, the microscope did not provide the only novel means of studying the blood, while the old distinction between the serous and the fibrous or clotted parts of stale blood could be seen as

52. *Ibid.*, Vol. II, pp. 243–245; Vol. III, pp. 405–407. The former letter was published by Hooke in *Lectures and Collections* (London, 1678) omitting the passage on blood corpuscles but including another from a letter of Jan. 14, 1678, in which Leeuwenhoek (temporarily) supposed all globular blood corpuscles capable of being extended into rods. The latter letter was included in Hooke's *Phil. Coll.*, Feb. 5, 1681/1682 (No. 5), the relative passage being on p. 158.

53. *Alle de Brieven*, Vol. IV, pp. 241–243. Published in *Phil. Trans.*, Nov. 20, 1684 (No. 165): 790 (i.e., 780)–789.

54. Birch, *History*, Vol. III, pp. 333–334, 338.

55. *Ibid.*, pp. 346, 349, 374–375, 379–380.

parallel to that between the clear serum and red corpuscular matter of fresh blood seen through the microscope. Now the tendency of blood to thicken or clot could be regarded as a possible pathological manifestation if it should occur unnaturally in the vessels and not in a healthy way so as to close a wound exposed to the air, and this tendency could be investigated chemically. How old the attribution of disease to a pathological clotting may be, I do not know, but one can certainly find this notion in Malpighi's *De polypo cordis.* For English readers Oldenburg provided a clear précis of Malpighi's theory in the *Philosophical Transactions;* the "polyps" found in the right ventricle of the heart arise from the fact that

the returning mass of the blood is now, by the long continued nutrition of the parts and by transpiration, depauperated of the spirituous and finer particles, such as the sulphureous and the red; and whilst it is freshly compounded with the chyle and other liquors, yet different from the nature of blood, the white and ragged parts thereof, being precipitated by the contiguity of those unlike parts, are in the large folds of the heart's right ventricle or auricle, by their ruggedness and little chinks entangled; whence ... they grow to a greater bulk; as happens in the generation of the stone in the pelvis of the kidneys, or in the concretion of tartar in water conduits.[56]

The formation of these polyps could be accelerated by poisons, fevers, corruptions of the air, plague, and infectious distempers, "where it happens, that such steams or juices are, by the corrupt ferments of the viscera, mixed with the blood, which disturb its texture." Malpighi then showed how one could experimentally thus "disturb" the normal constitution of the blood, for by addition of sulfuric acid or alum, blood *in vitro* was coagulated, while conversely saltpeter, alcohol, salt, and sulfur tended to maintain its fluidity. Therefore, he supposed, the coagulants produced by disease might be of the character of the former agents, while the latter suggested themselves as remedies against it.[57]

56. *Phil. Trans.,* Feb. 15, 1668/1669 (No. 44): 891.
57. Compare also the long passage translated by Adelmann, *Malpighi,* Vol. I, pp. 312–313. Malpighi seems here both to adopt the action of a "balsam of life" passing from the air into the blood, popularized by various writers after Michael Sendivogius (H. Guerlac, "John Mayow and the Aerial Nitre," *Actes du VIIe Congrès International d'Histoire des Sciences,* Jerusalem, 1953, pp. 332–349) and to anticipate Boyle's idea of a "spirit of blood."

446

Malpighi's chemical investigation of the blood was obviously rudimentary, though he had a clear idea of the material complexity of the humor itself. A more elaborate attempt upon the chemistry of blood was made by Needham in 1675.[58] The serum and the clotted part of bullocks' blood, whose red color he attributed to the action of heat in the digestive process, possessed, he wrote, a similar composition of a phlegm, a spirit, a volatile salt, an oil, a fixed salt, and an earth. However, in a quantity of human blood serum, taken presumably from sick persons, he found neither volatile salt nor oil, a discovery leading him to the suggestive comment (not followed up because the experiments were discontinued), "I question, if the persons had been examined, from which the serum was taken, whether their diseases might not give some light to the reason of this difference." [59]

Having thus distinguished and chemically analyzed the two components of the blood, Needham devoted the remainder of his essay to a demonstration that it is the serum which is functionally superior, deriving from the nutriment in the digestive process, and hence containing "a great variety of parts . . . these parts being the products of the nourishment eaten, and the materials out of which all the parts of the body and all the excrements of the second and third concoction are made; this serum is the proveditor-general of the body, the instrument of all the concoctions in it." It becomes then "the matter out of which all the parts of the body, and all the juices of it, whether noble or ignoble, do receive their origin," [60] for (Needham continued) it is the serous and not the red part of the blood that passes into the nerves and the reproductive organs. The red, fibrous part of the blood seems to have no easily determinable function, unless it is that which maintains heat in the body.

Boyle's Treatise on Human Blood

Obviously the chemical investigations at the opening of Needham's essay have little enough to do with the generalizations that follow about the function of the blood, and these are rather justi-

58. Birch, *History*, Vol. III, pp. 233–241.
59. *Ibid.*, p. 235.
60. *Ibid.*, pp. 237, 239.

fied by Needham's appealing to anatomical considerations. Nevertheless, they do to some extent anticipate the better-known tract of Boyle on human blood, which I have already quoted. I confess I cannot share John Fulton's enthusiasm for the *Memoirs for the Natural History of Humane Blood* (1684). Fulton described this work as "the most important of Boyle's medical writings," marking "the beginning of physiological chemistry." Boyle inaugurated, wrote Fulton, "the method of study which, in recent years, has become universally adopted, and had those who followed him in the eighteenth century listened more attentively to the note which he sounded, the Liebigs and Poggendorffs might have come a century earlier." [61] Apart from the question of the claims of "physiological chemistry" to a much earlier origin in the writings of Paracelsus and van Helmont, not to mention those of the Oxford school (including Boyle himself in earlier years) so often mentioned here, the suggestion that the chemical advances of the late eighteenth and early nineteenth century could have been bypassed seems a very doubtful historical judgment; and more fundamentally one may ask whether Boyle's chemical investigation of blood is, though longer, much more profound than that of Needham.

After a particularly tedious example of his customary apologies for lost papers and inconclusive experiments Boyle began these *Memoirs* by setting out the plan for a systematic treatise, which included sections on the artificial coagulation or liquefaction of blood (a subject whose study was begun by Malpighi, as we have seen), enlarged on the well-known "natural analysis'" of blood into a serous and a fibrous part, and then passed into discussion of the components obtained by chemical analysis—that is, a spirit, a volatile salt displaying crystalline forms, a phlegm, two kinds of oils, a fixed salt, and a *caput mortuum*—the same components, of course, as described by Needham. Next, in lieu of the planned treatise, Boyle describes a series of experiments relative to its various sections, some (like those on specific gravity) relating to the whole blood, some to its fibrous part, and some to its serum. Like Needham, Boyle gave qualitative analyses of these com-

61. John F. Fulton, *Bibliography of the Honourable Robert Boyle* (2d ed., Oxford: Clarendon Press, 1961), p. 99.

ponents. He related reactions between, for example, the salt of blood and spirit of niter; or its oil and oil of vitriol; and described the effect of adding alcohol to the serum or pouring the serum on copper filings. If any one can make sense of these experiments he must be much cleverer than I, or Robert Boyle.

Finally, much the longest part of the treatise is devoted to the spirit of human blood, in which Boyle found great medical virtues whether used externally or internally.[62] I shall not linger over his reasons for attributing medical value to a fluid so natural, penetrating, and alkalizate, nor over his case histories of its use. The strange thing is that this spirit of human blood is ammonia. Boyle knew this but was reluctant to admit the fact. He assures us, rather elaborately, that it is a volatile akali. Is it then, he asks, identical with such other volatile alkalis as spirits of urine and hartshorn, all more or less pure solutions of ammonia? This is a question he finds difficult of decision: because first, a very highly refined and purified volatile spirit differs in its greater simplicity from one less refined; second (even if these spirits should really be identical in their highest state of purification), as ordinarily prepared they seem to be distinguishable by the nose and otherwise.[63] Now as propositions both these statements may be factually correct, but Boyle can hardly have examined the logic of this passage deeply. For he is telling us that on the one hand ordinary chemistry presents us with three (or more) different spirits, while on the other hand a superior chemistry may be capable of reducing them all to one. So that whether or not the spirit of human blood is identical with the other volatile alkalis is not a question determined by chemical analysis but by what kind of chemist performs the analysis.

Now this does not help at all to convince one that chemical analysis can elucidate the nature of physiological substances like blood; it simply invites us to choose which kind of chemical analysis we believe to be possible. Or to put it another way, it seems to me that Boyle could not say definitely "blood, like urine and other organic materials, yields x" (for x we would write ammonia), because when put to the test, he is not at all sure what

62. Birch, *Boyle*, Vol. IV, pp. 637–645 (misnumbered 745).
63. *Ibid.*, pp. 622–623.

the statement "this sample of x is (or is not) identical with that sample of x" means in terms of an operational definition of chemical character. This is, of course, a fundamental chemical deficiency.[64] It is a matter of chemistry, not physiology, to know whether this substance obtained from urine is identical with that obtained from blood. Boyle's chemistry was not able to answer such a question or—in order to conceive how it might be answered —to visualize a distinction between a "substance" and its accompanying "contaminants," for example. Therefore there could really be no chemical understanding of the composition of physiological substances until such basic chemical matters had been decided.

Boyle's final reason for deciding that spirit of blood is not identical with spirit of urine is of interest, and not only because it reverses the more familiar course of argument: instead of arguing from chemical facts to a physiological interpretation, Boyle argued from medical facts (or what he accepted as such) to a chemical interpretation. If "we credit the famous Helmont," he wrote, "there is a considerable difference between the spirit of human blood and that of human urine, since he somewhere expressly notes (though I remember not the place, nor have his book at hand) that the spirit of human blood cures epilepsies, which is a thing the spirit of urine will not do." [65] This again gives occasion to remark on an aspect of the experimental approach to medicine which was far from clear to Boyle and most of his contemporaries; because they were unsure on this issue, their efforts to promote experimental medicine were rendered the less effective and deci-

64. It can hardly be necessary to say that this criticism arises from the logic of Boyle's own arguments and not from superior knowledge. One might add that Boyle did not know whether or not gold and silver may be proved identical at some level of chemical skill: while the *ordinary* analysis of the assay master amply showed them as distinguishable metals, Boyle's concepts of matter did not permit him to be sure that through some higher chemical art some part of silver might not be turned into gold or the existence of some constituent common to both metals be demonstrated.

65. *Ibid.*, p. 623. *Joannis Baptistae van Helmont opera* (Leiden, 1667), p. 122, par. 16: "Cruor quoque per distillationem, dat hunc spiritum salsum, plane volatilem, ac nequicquam a spiritu urinae distinguibilem. In eo tamen, essentiali ambos proprietate, differre consideravi, quod spiritus salis cruoris, curet epilepsiam, etiam adultorum non item spiritus urinae." See also the English translation by J[ohn]. C[handler]., *Oriatrike, or Physick Refined* (London, 1662), p. 195, par. 16. Of course this is further evidence that the analysis of blood by fire was begun neither by Boyle nor by Needham.

450

sive. The experimentalist must believe *a priori* that his results, obtained through the most exact procedures available to him, are *within the limits of their relevance* fully valid; if he seeks to amend or modify them by the results of clinical experience, he runs counter to the logic of his own scientific method. For in so doing the experimentalist is, in fact, preferring the less certain, less constant, and less comprehensible clinical knowledge to the knowledge gained by experiment, which is more certain, more constant, and more understandable; for if experimental knowledge does not possess within its limits these decisive virtues, then there is no merit in the experimental approach to medicine at all.

Boyle's weakness was that he did not have the courage of his experimental convictions. As a chemist he perceived the identity of spirit of blood and spirit of urine, recognizing that the difference in their origin was irrelevant. But as a physician he preferred to rely on the authority of van Helmont and others. And as a writer he was prone to give way to the opinions of those whom he should have been instructing and to concede as possibly true what his own experimental logic showed to be false because he thought his readers might insist upon it. Presumably the *Memoirs* on human blood were to be read by physicians like Needham and by Helmontian physiologists, and so Boyle did not care to insist—in the face of their supposed *clinical* experience—on the plain lessons of his own chemical experimentation.[66]

Conclusions

To conclude, then, we have seen that medical science from human and comparative anatomy through physiology to experimental medicine itself came within the wide-ranging activities of the early Fellows of the Royal Society. I have pointed out from time to time how important was the role of Thomas Willis and his Oxford associates—not, by the way, because I think this bears on

66. Malpighi, a physician and not a chemical philosopher, since he was one of those interested in Boyle's strong claims for the utility of the spirit of blood, wrote to London for further details of its clinical application; in reply, Boyle declared that investigation of its medical uses was outside his province, which was pure philosophy. This reply makes his appeal to van Helmont's supposed clinical experience—and indeed all the latter part of the *Memoirs*—the more illogical. Malpighi also wished to know Boyle's opinion of the function of the red corpuscles (*atometti*). His request received no answer (Adelmann, *Malpighi*, Vol. I, pp. 497–499).

the question of the organizational history of the Royal Society, but rather to the contrary because an examination of Willis' influence permits the historian to make a proper distinction between the intellectual activities of the Fellows and the history of the Society as an institution.

Finally, I have tried to raise some questions about the experimental approach to medical science as it was developed in the second half of the seventeenth century. It is my view, which I have tried to illustrate here, that an historical interpretation which sees this development as basically similar to the experimental investigation of physiology and medicine taken up a century or more later is radically mistaken. At least such an interpretation is mistaken if it tries to convince us that Willis, Lower, Boyle, Mayow, Hooke, and others were simply "precursors" and that their work could, but for various unfortunate accidents, have led rapidly to the majestic results obtained in the nineteenth century. On the contrary, if the experimental physiology and medicine attempted by the early Fellows of the Royal Society led nowhere, if the discussion of vital heat or of muscular activity later turned into other paths, it was not because of historical accident or willful blindness among the successors of these early Fellows, but rather because their experimental approach was inherently inadequate and could not be otherwise. It was never shown, for example, that chemical reactions supposed to occur *in vivo* did actually occur, and seldom that they could occur *in vitro*. One cannot logically speak of "reductionism" in this context, for the chemical effects to which vital phenomena might be reduced were not known to occur; they were postulated hypothetically. And it is not merely the case that chemical *knowledge* was too frail to bear the burden of reductionist physiology, but rather that basic *concepts* in terms of which alone any debate about the identity of chemical and physiological processes makes sense did not exist.

Furthermore, the position of experimental medicine in relation to clinical medicine was very imperfectly understood, so that (as in the case of injection therapy) it was almost impossible for it to fulfill a creative function. To oversimplify, one might say that seventeenth-century experimental medicine was no more than seventeenth-century clinical medicine applied to animals; and one must recollect how experimental such philosophical physicians as van Helmont, Boyle, and Willis were in their treatment of human patients also. Its fundamental intellectual allegiance was, I suggest, to clinical medicine (and how chaotic its principles were at this time I need not emphasize), not to the basic sciences; hence it was in principle impossible that experiment in medicine should serve as a vehicle for the introduction of more exact knowledge and better-substantiated theories. In other words, the structure of the thinking of the seventeenth-century theorists was not identical with that of their nineteenth-century successors; even if further extended, it therefore could not have produced the same advances in medical science.

XV

THE FIRST HUMAN BLOOD TRANSFUSION: PRIORITY DISPUTES

A. RUPERT HALL AND MARIE BOAS HALL*

IT IS notoriously difficult to deal clearly and accurately with the history of controversy over priority, and the dispute over priority in blood transfusion recently explored by Dr. A. D. Farr in this journal is no exception.[1] The story has often been told from various points of view:[2] and we ourselves printed many documents connected with the controversy in our edition of *The correspondence of Henry Oldenburg*.[3] The aspect with which Dr. Farr is specifically concerned is the history of *human* blood transfusion, and he has found it difficult to understand why many of the English felt touchy about this, since the French physician Jean Denis (d. 1704) was clearly the first to practise it.[4] In particular he has been puzzled by the reaction of Henry Oldenburg, secretary of the Royal Society, to the publication of an English translation of Denis' printed letter (addressed to Hebert de Montmor and dated 25 June 1667 (N.S.)) in what purported to be no. 27 of the *Philosophical Transactions*. To clarify this matter, and to set the record right in a number of points, it is necessary to consider first, the career of Henry Oldenburg; second, the history and status of the *Philosophical Transactions*; and finally the pre-history of human blood transfusion and the feelings of nationalism aroused by it.

The article on Oldenburg in the *Dictionary of national biography* (publ. 1894–5) was by Herbert Rix, assistant secretary of the Royal Society from 1885 to 1896, and is exceptionally full, with references to both manuscript material and secondary sources. *Pace* Dr. Farr, much more information about Oldenburg's private life is available than is indicated there; further, Rix was far from historically accurate in his understanding of the workings of the Royal Society in the seventeenth century, nor was he

---

* A. Rupert Hall, M.A., Ph.D., Litt.D., F.B.A., and Marie Boas Hall, A.B., M.A., Ph.D., 14 Ball Lane, Tackley, Oxford OX5 3AG.

[1] A. D. Farr, 'The first human blood transfusion', *Med. Hist.*, 1980, **24**: 143–162.

[2] Dr. Farr relies particularly upon Geoffrey Keynes (editor), *Blood transfusion*, Bristol, Wright, 1949. The *Isis cumulative bibliography*, vol. 3, London, Mansell, 1976, lists some fifteen items for the period before 1965. See also, e.g. M. Nicolson, *Pepys' Diary and the New Science*, Charlottesville, University Press of Virginia, 1965, ch. II, pp. 55–99. In fact Keynes in the first, historical chapter of the book he edited printed facsimiles of pages from *Phil. Trans.* no. 20 (with the earliest published English work, 1666) and false no. 27, which he says Oldenburg "suppressed". Far more evidence is given by Keynes for English work than Dr. Farr notes.

[3] Especially in vols. III and IV, Madison, Milwaukee, and London, University of Wisconsin Press, 1966 and 1967, respectively.

[4] Ironically Denis was later for some years physician to Charles II.

Reprinted from *Medical History*, 24 (1980), pp. 461–465, by permission of the publisher. © 1980 the Trustees of the Wellcome Trust, London.

inclined to question contemporary or near-contemporary sources. Thus the date of Oldenburg's birth in Bremen is given as 1615?, correcting the earlier, also incorrect date of 1626; as we have tried to show,[5] 1618 or 1619 is the most probable date, based upon the known facts of his education and his declaration at his first marriage in 1663 that he was "aged about 43 yeares". He is known to have graduated from the *Gymnasium Illustre* of Bremen in 1639 with the degree of Master of Theology, and Rix knew nothing of his career after that until his appointment as agent for the Senate of Bremen to Cromwell in 1653 which brought him to England, except for the rumour, current in 1653, that he had lived in England from 1640–48 and was variously described as in favour with the Parliamentarians and banished as a Royalist. What Oldenburg did in 1640 is not clear, but in 1641 he went from Bremen to Holland with highly commendatory letters of introduction to the learned G. J. Vossius in Leiden.[6] He subsequently wrote to Vossius with a young man's natural discovery that one needs a job to live; in his case he resolved to combine business with self-education and look for a post as tutor to some young man about to make a European tour. That he succeeded, and that at least some of these young men were English, is clear from surviving drafts of letters to many of them,[7] and it was presumably from them that he learned the near-perfect English that was to astonish Milton. Possibly he did indeed come to England in these years, but certainly he never came into conflict with authority. When he came on his diplomatic mission he was warmly welcomed by men close to Cromwell, to some of whom, like Milton, he may have had introductions from scholars abroad. His mission completed, he found himself with a wide circle of acquaintances, among them at least two members of the Boyle family: Katherine, Lady Ranelagh, a friend of Milton's, and her young brother Robert Boyle, not yet known as a natural philosopher. In 1656 he became tutor to Lady Ranelagh's son (he was never a tutor to Henry O'Brien as has been claimed), and spent the next four years on the Continent, whence he sent back a stream of letters, to Boyle and others, showing an increasing involvement with the "new science" of the time. He returned at the Restoration, was quickly elected into the newly formed, still unofficial Royal Society, and, helped greatly by Boyle's patronage, became a Secretary under its first and second Charters. He received irregular remuneration for some years, and an annual salary (of forty pounds a year) from 1668.

In October 1663 he married one Dorothy West "aged about 40 yeares and a mayden of her owne Disposing," who died at the beginning of February 1664/5. She had some money of her own (400 pounds), administered by two baronets (so she was of a good family), and had possibly but not certainly some connexion with the wife of John Dury. Dury, a Protestant divine who spent most of his life abroad and whom Oldenburg may have known either there or in England, had married in Holland in

    [5] 'Some hitherto unknown facts about the private career of Henry Oldenburg', *Notes Rec. R. Soc. Lond.*, 1963, **18**: 94–103.
    [6] See *The correspondence of Henry Oldenburg*, vol. I, Madison and Milwaukee, University of Wisconsin, 1965, Letter 1, pp. 3–7; and vol. XIII, London, Mansell [in press], Letter 1a. Oldenburg's surviving correspondence for this period, including that to the Senate of Bremen, is published in *Correspondence*, vol. I.
    [7] Ibid., *passim*; vol. XIII, Letter *1 bis*; vol. XI (London, Mansell, 1977), p. 407.

1645 Dorothy King (d. 1664), widow of Arthur Moore, son of Viscount Moore of Drogheda, and hence connected with many of the great Anglo-Irish families of the time.[8] At Mrs. Dury's death her daughter, Dora Katherina (1654–77) was entrusted to the Oldenburgs as a ward. It is not clear what happened to the child at the first Mrs. Oldenburg's death, but in August 1668, when she was little more than fourteen years old, she was married to Henry Oldenburg. She it was who brought to Oldenburg the "estate in the marshes of Kent" which he was known to possess – two farms now in Bexley Parish, although Oldenburg always referred to his country retreat as in Crayford, and went there for a late summer holiday in most years. In fact he (and almost immediately his wife) died there in early September 1677, leaving two small children, Rupert and Sophia; their subsequent history is obscure, but Rupert died in pathetic circumstances in 1724, having sold his patrimony to secure a commission in the army in 1699.[9]

From 1662 to his death in 1677 Oldenburg was the active and industrious "working secretary" of the Royal Society. But he received little remuneration, and although he earned some money as editor and translator for Boyle, he needed more. One source of income came from work he did for Joseph Williamson, then assistant to the Secretary of State, Lord Arlington, later (1674) Secretary himself. In the second charter the Royal Society was given the right to correspond freely with foreign lands, and by 1666 many incoming letters were delivered to the Office of the Secretary of State, where postage was (almost certainly) paid. This was done by the device of having correspondents address their letters simply to "Mr. Grubendol, London", a device which at first puzzled contemporaries as much as it has done historians. Such letters were then sent on. usually unopened, and in return Oldenburg sent political news to Williamson.[10] He also utilized this news for various newsletters edited by Williamson, and often served as translator. In 1676 through Williamson he received a warrant to license books, presumably a potentially remunerative post, but he soon relinquished it because he found it led to accusations of disloyalty.[11] The foundation of the *Philosophical Transactions* was, as Dr. Farr notes, also intended to be remunerative, but he has confused the issue by assuming that *Phil. Trans. R. Soc.* came into being with the first issue. The reality is very different indeed.[12] The *Philosophical Transactions: Giving Some Accompt of the Present Undertakings, Studies, and Labours of the Ingenious in many Considerable Parts of the World*, clearly indicates by its title that it was *not* intended to be the "transactions" of the Royal Society alone, but those of the whole world of learning. True, that world persisted in believing that the *Transactions* had a very close connexion with the Society, a point on which Oldenburg was sensitive. Thus at the end of no. 12 (7 May 1666) he explicitly denied that these were "publish't by the Royal Society"; as was correct, for he was the sole initiator and editor ("publisher" in seventeenth-century terms – or "author" as Oldenburg

---

[8] See A. Rupert Hall and Marie Boas Hall, 'Further notes on Henry Oldenburg', *Notes Rec. R. Soc. Lond.*, 1968, **23**: 33–42, which discusses the family connexions of Mrs. Dury.

[9] See ibid., pp. 34, 40–41.

[10] *Correspondence*, vol. III, pp. 127–128. Oldenburg corresponded with Williamson steadily from 1666.

[11] *Correspondence*, vol. XII, Letters 2890 and 2890a.

[12] This is clearly brought out in E. N. da C. Andrade, 'The birth and early days of the *Philosophical Transactions*', *Notes Rec. R. Soc. Lond.*, 1963, **20**: 9–27.

denominated himself in the presentation inscription to the Royal Society of Volume I). Oldenburg's ownership of *Phil. Trans.* appears clearly in the Minutes of the Council (as published in Birch's *History of the Royal Society*) as also in the only independent account of its origin, namely a letter from Sir Robert Moray to Christiaan Huygens.[13] Perhaps it might be added that as late as 1683 when *Phil. Trans.* appeared after a lapse of four years the editor, Robert Plot, felt it necessary to begin by emphasizing that "the Writing of these *Transactions*, is not to be looked upon as the Business of the Royal Society . . .".[14] And indeed *Phil. Trans.* did not become the official journal of the Society until 1752 (vol. 47).

In the light of these facts, Dr. Farr's account of the spurious no. 27 is inevitably inaccurate. No one had a right to publish *Phil. Trans.* without Oldenburg's permission. It was no business of Wilkins, whom Dr. Farr believes responsible (he gives no reason, but it was certainly Wilkins – probably acting in his capacity as sole secretary in Oldenburg's absence – who produced the account at the Society's meetings. It seems in fact likely that the original printed letter had been sent to Oldenburg, and even possible that he had made the translation which was published). It was equally no business of the stationer, upon whom Oldenburg laid the blame.[15] Presumably the stationer tried to distribute it to all those normally subscribing (not, as Dr. Farr surmises, the Fellows of the Society automatically, for it went to anyone who subscribed, and the "free list" was in Oldenburg's private beneficence. There was a print-run of about 500). At any rate the obscure Suffolk physician Nathaniel Fairfax, never F.R.S. although for some years an ardent correspondent of Oldenburg's, had a copy, as he told Oldenburg in September.[16]

It seems possible that Oldenburg had intended to print the letter with editorial comment; certainly he would never have printed it without. But there can be no question of his "suppressing" the false no. 27 in the conventional sense; rather he disowned it, when he came to make up his own next issue.

But why the fuss, which Dr. Farr finds necessary to ascribe to Oldenburg's personal adherence to the English cause? First, as Oldenburg's letter to Boyle already referred to (note 15) makes plain, his reaction was not merely personal; he had been blamed by many and especially by Lower for printing Denis' claim to have instituted transfusion ten years earlier, in defiance of English claims for priority. What justification was there for this feeling, unless it merely resulted from English xenophobia? That certainly did exist, and accounts, for example, for such claims as that put forward by Dr. Timothy Clarke in 1668, that the idea of blood transfusion had first been suggested by Francis Potter (F.R.S. 1663) about 1639.[17] What is more certain, as Clarke went on to say, is that in 1656 Christopher Wren and others had begun experiments on the injection of "various liquors into the mass of the blood of living animals"

---

[13] *Oeuvres complètes de Christiaan Huygens*, vol. V, The Hague, Nijhoff, 1893, p. 232.

[14] No. 143 of January 1682/3.

[15] See for example Oldenburg's letter to Boyle of 24 September 1667, *Correspondence*, vol. III, pp. 480, 483, note 2.

[16] Letter of 28 September 1667, *Correspondence*, vol. III, p. 497.

[17] *Correspondence*, vol. IV, pp. 364–365; John Aubrey in his *Brief life* of Potter said the same. For another claim for English priority see John Wallis to Oldenburg, *Correspondence*, vol. III, p. 373.

as Clarke put it, and Clarke claimed to have tried various experiments himself in 1657. The first *transfusion* of blood from one animal to another (dogs) was performed by Richard Lower in late 1665 and an account read at the Royal Society some six months later.[18] Lower published an account of his experiments to justify his claim to priority since "a certain *Denis*, seeks in a recently published letter to deprive me of priority in the discovery of this experiment, and to claim it for himself." There can be no question but that Lower had performed the transfusion experiments in dogs when he claimed to have done so, nor that this preceded French work. When on 21 March 1666/7 Oldenburg at a meeting of the Society produced an account "which he had received from Paris, of the success, which the curious had met with there" it was because there was a lively discussion going on the same subject; and there were endless further discussions in subsequent months.

There can therefore be no doubt that in animal transfusion the English were right to claim priority. Equally there is no doubt that the French were the first rashly to venture on human transfusion, which the English did not attempt until late November 1667. (To the credit of the physician in charge of Bedlam he "scrupled" to expose his patients to the ordeal when it was earlier proposed to him.)[20] Oldenburg was clearly right to set the record straight in his *Philosophical Transactions*, nor could he have faced his colleagues and fellow members of the Royal Society had he failed to repudiate responsibility for the false *Phil. Trans.* no. 27. But neither he nor the Society was otherwise hostile to Denis, with whom Oldenburg was to correspond freely in 1668.[21] At that time Oldenburg in fact published translations of two of Denis' printed letters on the subject (one addressed to himself) without arousing any further reactions from the Society: these are straightforward accounts of events – including the death of one of the human recipients of animal blood, and the judgment of the Paris court thereon — and involve no questions of priority.

The English at this period were, to say the least, touchy about questions of national prestige and priority in scientific discovery, and in many cases it is difficult to arrive at the truth. But here it seems that their claims were clearly justified, and Lower quite rightly resentful of Denis' claim. No wonder that Oldenburg repudiated the false *Phil. Trans.*, as he probably would have done even if no question of national prestige had been involved, since the *Philosophical Transactions* were his private enterprise, and his standards as editor high.

[18] Meeting of 20 June 1666. See T. Birch, *History of the Royal Society of London*, 4 vols., London, 1756, vol. II. There is a more detailed account in the letter from Boyle to Lower printed (pp. 177–179) in the latter's *Tractatus de corde*, London, 1669, reprinted, with English translation, in vol. IX of R. T. Gunther (editor), *Early science in Oxford*, Oxford, [The Author], 1932, and London, Dawson, 1968. Lower's account of his work on transfusion is in ch. 4.

[19] Birch, op. cit., note 18 above, vol. II, p. 161.

[20] Minutes of the Royal Society's meeting of 31 October 1667.

[21] There is no evidence for Dr. Farr's surmise that the two were in correspondence in 1667. It is not known precisely who in Paris received *Phil. Trans.* at this time, other than Henri Justel, the editor of the *Journal des Sçavans*, and Christiaan Huygens, but several copies were in circulation.

INDEX

Academia curiosorum: XIV 426, 429
Académie Royale des Sciences: II 334, 339;
    III 134; IV 98; IX 22; XI 37
Accademia del Cimento: X 126–7
Adrian, Lord: VII 651
aeolipile: V 19, 22
Agricola, G: I 18; II 336; IV 98
Aikin, Arthur: VII 648
airgun: IX 9
Albert, Prince Consort: VII 648
Algarotti, F.: III 144
Allen, Thomas: XIV 435
Amontons, G.: III 142, 144
anatomy: XIII 123–4; XIV 435–7
Anderson, R.: IX 21; X 121–3, 128, 133
Anon., *De rebus bellicis*: VIII 2
Archimedes: I 11, 12
Arkwright, Richard: III 147; VII 645
Arsenal (Venice): II 340; XI 35

Bacon, Francis: I 5, 6, 8, 9, 10, 16; VII 642;
    X 131; XIV 429
Bacon, Roger: VIII 2
Baker, Henry: VII 641–2
ballistics: II 333–4, 339; III 136; IX 15–23;
    X 111–36
Barker's mill: V 23
Barrow, Isaac: I 6
Bashforth, Francis: IX 4
Bate, John: V 17; X 113
Beale, John: XIII 113
Beaufoy, Mark: XI 43–4
Behm, Michael: XIII 119
Belidor, B. F. de: III 146; IX 22
Bell, Patrick: VII 643
Benedict XII, Pope: VIII 1
Bergmann, Tobern: III 145
Bernoulli, Johann: XI 37, 39
Berthollet, C. L.: III 137–8, 145
Birch, Thomas: XII 8
Biringuccio, V.: I 18; IV 98; IX 9
Black, Joseph: III 137, 138–40
Blackheath: X 123–4
Blake, William: XII 13
Blondel, Francois: II 340; IX 21–2; X 133
blood: XIV 442–50; XV 461–5

Borel, Pierre: XII 9–10
Borelli, G. B.: X 126
Bossut, Charles: XI 43
botany: XIII 120–22
Bouguer, Pierre: XI 39–40, 44–5
Boulton, Mathew: VI 200, 202; VII 644
Bourne, William: IX 15
Boyle, Robert: I 20; II 337; X 126; XII 6–9,
    30; XIII 113; XIV 424, 425, 430–31,
    434, 437–8, 441, 446–50; XV 462–4
Bramah, Joseph: VII 644
Branca, Giovanni: V 19, 23
Bremen: XV 462
Bridgewater, Duke of: VII 644
Brindley, James: VII 644
Brouncker, Lord: X 123–5, 130, 134; XII 5
Browne, Edward: XIII 112; XIV 424–5
Browne, Dr Thomas: XIII 113
Brunel, I. K.: XI 34
Brunel, M. I.: VII 644

Campani, Giuseppe: XII 12
Campano di Novaro: IX 14
Carnot, Lazare: IV 99, 144
Carnot, Sadi: III 142, 144, 146
Carron Iron Co.: VI 204
Cartwright, Edmund: VII 655n
Cassini, J.-D.: X 133
Castelli, Benedetto: II 336
Cavalieri, B.: IX 21, 22; X 114
Cellini, B.: I 18
cells: XII 24–7
Chadwick, Edwin: VI 196–7
Channel tunnel: VII 653–4
Chapman, Frederick af: XI 34, 42–3
Charles the Fair: VIII 1
Charleton, Walter: X 127; XIII 114; XIV 437
chemical industry: III 137–41, 144–5, 148;
    VII 645–6
Chichele, Sir Thomas: X 123
child labour: VI 204–5
chimney sweeps: VII 646, 647
City and Guilds of London Institute: VII 650
Clark, Sir G. N.: XIV 422, 425
Clarke, Timothy: XIII 117–8, 123; XIV
    435, 438–9; XV 464–5

Printed and bound by CPI Group (UK) Ltd, Croydon, CR0 4YY

17/10/2024

01775690-0004